Proceedings of the Sixth Asian Logic Conference

Proceedings of the Sixth Asian Logic Conference

Beijing, China
20–24 May 1996

Editors

C T Chong
National University of Singapore

Q Feng
National University of Singapore

D Ding
Nanjing University

Q Huang
Academia Sinica

M Yasugi
Kyoto Sangyo University

World Scientific
Singapore • New Jersey • London • Hong Kong

SINGAPORE UNIVERSITY PRESS
NATIONAL UNIVERSITY OF SINGAPORE

Published by

World Scientific Publishing Co. Pte. Ltd.

P O Box 128, Farrer Road, Singapore 912805

USA office: Suite 1B, 1060 Main Street, River Edge, NJ 07661

UK office: 57 Shelton Street, Covent Garden, London WC2H 9HE

and

Singapore University Press (Pte) Ltd.
Yusof Ishak House
National University of Singapore
10 Kent Ridge Cresent
Singapore 119260

British Library Cataloguing-in-Publication Data
A catalogue record for this book is available from the British Library.

PROCEEDINGS OF THE SIXTH ASIAN LOGIC CONFERENCE

ISBN 981-02-3432-5

This book is printed on acid-free paper.

Printed in Singapore by Uto-Print

Foreword

The Sixth Asian Logic Conference was held from 20 to 24 of May, 1996 in Beijing, China. There were one hundred and fifty participants from six countries.

This volume collects written versions of invited as well as contributed talks presented at the conference. An article by M. Yasugi on the Asian Logic Conference, first appeared in Japanese, is included as it provides a glimpse into the history and development of the series of conferences.

We wish to thank all those who help referee the submitted papers, working within a tight deadline. On behalf of the organizers we also wish to thank the Institute of Software, Academia Sinica, China and the Lee Foundation of Singapore for their generous financial support.

Editors 31 October 1997

LIST OF INVITED SPEAKERS

K. Ambos–Spies	Genericity and Randomness in Complexity Theory
S. Buss	Algebraic Proof Systems
S. B. Cooper	Oracles, Enumerations, and Relative Computability
D. Ding	On d–r.e. Degrees
Q. Feng	Principles of Stationary Reflection
K. Hauser	Sets of Reals, Infinite Games and Core Model Theory
A. Kechris	Borel Equivalence Relations and Classifications of Countable Models
S. Lempp	Decidability and Undecidability in the Enumerable Turing Degrees
W. Li	A formalization of Inductive Inference and its Rationality
H. Ono	Recent results on Substructural Logic
A. Pillay	Differential Fields and Model Theory
G. E. Sacks	Infinite Orbits and Vaught's Conjecture
M. Yasugi	A System of Classical logic and natural Reasoning

ORGANIZING COMMITTEE

DING DECHENG	Nanjing University, Nanjing
GAO HENGSHAN	Graduate School, Academia Sinca, Beijing
HUANG QIEYUAN	Institute of Software, Academia Sinica, Beijing
LI WEI	Beijing Univ. Aeronautics and Astronautics, Beijing
LI XIANG	Guizhou University, Guiyang
LU YIZHONG	Nanjing Aeronautics and Astronautics Univ., Nanjing
MUO SHAOKUI	Nanjing University, Nanjing
SHEN ENSHAO	Shanghai Jiaotong University, Shanghai
SHEN FUXING	Beijing Normal University, Beijing
WANG SHIQIANG	Beijing Normal University, Beijing
YANG DONGPING	Institute of Software, Academia Sinica, Beijing

ADVISORY COMMITTEE

CHONG CHI TAT	National University of Singapore, Singapore
CROSSLEY, J. N.	Monash University, Australia
TUGUÉ, TOSIYUKI	Konan Woen's Junior College, Japan
YASUGI, MARIKO	Kyoto Sangyo University, Japan

Table of Contents

Seminormal Fine Measures on $\mathcal{P}_\kappa\lambda$

Yoshihiro Abe *

Department of Mathematics, Kanagawa University,
Kanagawa-ku, Yokohama 221, Japan

Abstract. We first show that κ is λ-supercompact if seminormal fine measures on $\mathcal{P}_\kappa\lambda$ exist. Second, assuming a supercompact cardinal, we prove the existence of non-normal ideals whose seminormal closures are normal. We also discuss prepicitousness and saturation of the minimal κ-normal ideal.

Mathematical Subject Classification: 03E05, 03E55

Keywords: Fine measure; Seminormality; κ-normality; nowhere prepicitous; saturation

1. Introduction

The $\mathcal{P}_\kappa\lambda$ context of closed unbounded sets was investigated in Jech [7], and the structure of normal ideals and filters on $\mathcal{P}_\kappa\lambda$ have been studied in various aspects. Some weakening of normality was also presented. Seminormal ideals were considered in Johnson [9], [10] in relation to the partition property. Their basic structure is dealt with in [2] and ideals with partial normality are treated. We discuss here large cardinal features of seminormal ideals. But our interest is not in $\mathcal{P}_\kappa\lambda$-combinatorics but in more familiar large cardinal properties such as fine measures, and saturated and prepicitous ideals.

In 3, we project a seminormal fine measure to a normal measure, showing the former is in fact a supercompactness measure. We also mention seminormal fine measures in the Rudin-Keisler orderings.

In 4, assuming κ is supercompact and λ is regular, we build a nonnormal ideal I on $\mathcal{P}_\kappa\lambda$ such that every seminormal ideal extending I is normal, answering a question in [2].

In the last section we argue saturation of the minimal κ-normal ideal on $\mathcal{P}_\kappa\lambda$ and its prepicitousness. Such an ideal was also studied in [2]. Results

*This research was partially supported by Grant-in-Aid for Scientific Research (No.06640178), Ministry of Education, Science and Culture of Japan.

in [13] are used.

2. Definition and preliminaries

Throughout this paper, $\kappa < \lambda$ are uncountable cardinals and κ is regular. For such a pair (κ, λ), $\mathcal{P}_\kappa\lambda = \{x \subset \lambda : |x| < \kappa\}$.

F is a *filter* on $\mathcal{P}_\kappa\lambda$ if F is a collection of subsets of $\mathcal{P}_\kappa\lambda$ such that

(1) $\emptyset \notin F$ and $\mathcal{P}_\kappa\lambda \in F$.

(2) If $X \subset Y \subset \mathcal{P}_\kappa\lambda$ and $X \in F$, then $Y \in F$.

(3) If $X \in F$ and $Y \in F$, then $X \cap Y \in F$.

I is an *ideal* on $\mathcal{P}_\kappa\lambda$ if I is a collection of subsets of $\mathcal{P}_\kappa\lambda$ such that

(1) $\emptyset \in I$ and $\mathcal{P}_\kappa\lambda \notin I$.

(2) If $X \subset Y \subset \mathcal{P}_\kappa\lambda$ and $Y \in I$, then $X \in I$.

(3) If $X \in I$ and $Y \in I$, then $X \cup Y \in I$.

A filter F on $\mathcal{P}_\kappa\lambda$ and an ideal I on $\mathcal{P}_\kappa\lambda$ are *dual* to each other if the following holds:

$$\text{for every } X \subset \mathcal{P}_\kappa\lambda, \quad X \in F \text{ iff } \mathcal{P}_\kappa\lambda - X \in I\ .$$

The dual filter (ideal) of I (F) will be denoted by I^* (F^*) and each member of I^* (F) is called I (F)-*measure one.* Let $I^+ = \mathcal{P}\mathcal{P}_\kappa\lambda - I = \{X : X \notin I\}$. $X \in I^+$ is called I-*positive.* Also let $F^+ = \mathcal{P}\mathcal{P}_\kappa\lambda - F^* = \{X : \mathcal{P}_\kappa\lambda - X \notin F\}$. $X \in F^+$ is called F-*positive.*

For $X \in I^+$, $I|X = \{Y \subset \mathcal{P}_\kappa\lambda : Y \cap X \in I\}$ is an ideal $\supset I$.

A filter F on $\mathcal{P}_\kappa\lambda$ is κ-*complete* if it is closed under intersection of less than κ many members. F is *fine* if for each $\alpha < \lambda$, $\{x \in \mathcal{P}_\kappa\lambda : \alpha \in x\} \in F$. For the sake of convenience, throughout this paper, by 'filter', we mean a 'fine κ-complete filter'.

Most properties defined for filters are translated into properties for ideals using duality.

We often say a *measure* for an ultrafilter. That is, a filter F is a measure iff for all $X \in \mathcal{P}_\kappa\lambda$, either $X \in F$ or $\mathcal{P}_\kappa\lambda - X \in F$. The dual ideal of a measure is called a *prime ideal.*

A function $f : \mathcal{P}_\kappa\lambda \to \lambda+1$ is *the least unbounded function* for a measure U if the following hold.

(1) $\{x \in \mathcal{P}_\kappa\lambda : f(x) \geq \alpha\} \in U$ for all $\alpha < \lambda$.

(2) If $\{x \in \mathcal{P}_\kappa\lambda : g(x) < f(x)\} \in U$, then $\{x \in \mathcal{P}_\kappa\lambda : g(x) < \gamma\} \in U$ for some $\gamma < \lambda$.

Note that the range of the least unbounded function is included in λ if $cf(\lambda) \geq \kappa$.

A filter F on $\mathcal{P}_\kappa\lambda$ is *weakly normal* if for any regressive function $f : X \to \lambda$ with $X \in F^+$ there is $\gamma < \lambda$ such that $\{x \in X : f(x) < \gamma\} \in F^+$.

Thus, a fine measure is weakly normal iff its least unbounded function assigns x to $\sup(x)$.

For any $x \in \mathcal{P}_\kappa\lambda$, $\hat{x} = \{y \in \mathcal{P}_\kappa\lambda : x \subset y\}$, and $X \subset \mathcal{P}_\kappa\lambda$ is *unbounded* iff $X \cap \hat{x} \neq \emptyset$ for all $x \in \mathcal{P}_\kappa\lambda$.

Let $I_{\kappa\lambda} = \{X \subset \mathcal{P}_\kappa\lambda : X$ is not unbounded $\}$. $I_{\kappa\lambda}$ is the minimal ideal on $\mathcal{P}_\kappa\lambda$. Hence $FSF_{\kappa\lambda} = I^*_{\kappa\lambda}$ is the minimal filter on $\mathcal{P}_\kappa\lambda$, and $\hat{x} \in FSF_{\kappa\lambda}$ for any $x \in \mathcal{P}_\kappa\lambda$.

$X \subset \mathcal{P}_\kappa\lambda$ is *closed* iff $\bigcup D \in X$ for any increasing $\subset$ −sequence $D \subset X$ such that $|D| < \kappa$. $C \subset \mathcal{P}_\kappa\lambda$ is said to be a *cub* iff it is closed and unbounded. X is *stationary* iff $X \bigcap C \neq \emptyset$ for every cub C.

Set $NS_{\kappa\lambda} = \{X \subset \mathcal{P}_\kappa\lambda : X$ is not stationary$\}$. So $NS_{\kappa\lambda}$ is the dual ideal of the cub filter on $\mathcal{P}_\kappa\lambda$, $CF_{\kappa\lambda}$, generated by cub sets.

$X \subset \mathcal{P}_\kappa\lambda$ is *strongly closed* iff $\bigcup D \in X$ for any $D \subset X$ with $|D| < \kappa$ and we call $C \subset \mathcal{P}_\kappa\lambda$ a *strong cub* iff it is strongly closed and unbounded.

$SCF_{\kappa\lambda}$ is the filter generated by strong cub sets, and its dual ideal is $SNS_{\kappa\lambda} = \{X \subset \mathcal{P}_\kappa\lambda : X \cap C = \emptyset$ for some strong cub set $C \subset \mathcal{P}_\kappa\lambda\}$.

Such notions were generalized to $\mathcal{P}_\kappa\lambda$ by Jech [7]. Further study was made in Menas [14] and Carr [3] where the following was proved:

$$I_{\kappa\lambda} \subsetneq SNS_{\kappa\lambda} \subsetneq NS_{\kappa\lambda}$$

Definition. For $\delta \leq \lambda$, the *diagonal union* and the *diagonal intersection* of $\{X_\alpha : \alpha < \delta\}$ are defined respectively by:

$$\nabla_{\alpha<\delta} X_\alpha = \{x \in \mathcal{P}_\kappa\lambda : x \in X_\alpha \text{ for some } \alpha \in x\}$$
$$\Delta_{\alpha<\delta} X_\alpha = \{x \in \mathcal{P}_\kappa\lambda : x \in X_\alpha \text{ for any } \alpha \in x\}$$

A filter F on $\mathcal{P}_\kappa\lambda$ is *δ-normal* iff $\Delta_{\alpha<\delta} X_\alpha \in F$ for any $\{X_\alpha : \alpha < \delta\} \subset F$. So, F is *normal* iff it is *λ-normal.*

An ideal I on $\mathcal{P}_\kappa\lambda$ is δ-normal iff $\nabla_{\alpha<\delta} X_\alpha \in I$ for any $\{X_\alpha : \alpha < \delta\} \subset I$. Note that I is δ-normal iff I^* is δ-normal.

$\nabla^\delta I = \{X \subset \mathcal{P}_\kappa\lambda : X$ is a diagonal union of $\{X_\alpha : \alpha < \delta\} \subset I\}$ and $\Delta^\delta F = \{X \subset \mathcal{P}_\kappa\lambda : X$ is a diagonal intersection of $\{X_\alpha : \alpha < \delta\} \subset F\}$

For $\delta \leq \lambda$ and $f : \delta \to \mathcal{P}_\kappa\lambda$ ($f : \delta \times \delta \to \mathcal{P}_\kappa\lambda$), let $C_f = \{x \in \mathcal{P}_\kappa\lambda : \bigcup f''x \subset x\}(\{x \in \mathcal{P}_\kappa\lambda : \bigcup f''(x \times x) \subset x\})$.

The following are easy generalizations of Menas' results in [14] and are proved in [2].

Lemma 1.1. (1) *$\Delta^\kappa FSF_{\kappa\lambda}$ is the minimal κ-normal filter on $\mathcal{P}_\kappa\lambda$.*

(2) *If $\kappa^+ \leq \delta \leq \lambda$, then $\Delta^\delta FSF_{\kappa\lambda}$ is not κ-normal and $\Delta^\delta\Delta^\delta FSF_{\kappa\lambda}$ is the minimal δ-normal filter on $\mathcal{P}_\kappa\lambda$.*

(3) *For every $\delta < \lambda$, $FSF_{\kappa\lambda} \subseteq \Delta^\delta FSF_{\kappa\lambda} \subseteq SCF_{\kappa\lambda}$ and $SCF_{\kappa\lambda} \not\subset \Delta^\delta\Delta^\delta FSF_{\kappa\lambda} \not\subset SCF_{\kappa\lambda}$.*

(4) *For every $X \in \Delta^\delta FSF_{\kappa\lambda}$, there is $f : \delta \to \mathcal{P}_\kappa\lambda$ such that $C_f \subset X$. And, for each $X \in \Delta^\delta\Delta^\delta FSF_{\kappa\lambda}$, there is $f : \delta\times\delta \to \mathcal{P}_\kappa\lambda$ such that $C_f \subset X$.*

$\nabla^\delta I_{\kappa\lambda}$ is the dual ideal of $\Delta^\delta FSF_{\kappa\lambda}$. We say $X \subset \mathcal{P}_\kappa\lambda$ is *δ-stationary* if $X \in (\nabla^\delta\nabla^\delta I_{\kappa\lambda})^+$. By Lemma 1.1-(1), X is κ-stationary iff $X \in (\nabla^\kappa I_{\kappa\lambda})^+$.

Definition. A filter F on $\mathcal{P}_\kappa\lambda$ is *seminormal* if it is closed under diagonal intersection of less than λ many members.

So, F is seminormal iff it is δ-normal for any $\delta < \lambda$ iff any regressive function $f : X \to \delta < \lambda$ with $X \in F^+$ is constant on some $Y \in \mathcal{P}(X) \cap F^+$.

Definition. The *seminormal closure* of I is the minimal seminormal ideal extending I if exists.

For an ideal I consider a poset $< P_I, \leq >$ where $P_I = I^+$ and $X \leq Y$ iff $X - Y \in I$. If G is a V-generic filter on P_I, then we can form an ultrapower of V in $V[G]$, $Ult(V,G) = \{[f] : f \in V \text{ is a function with domain } \mathcal{P}_\kappa\lambda\}$.

We say I is *prepicitous* if $Ult(V,G)$ is well-founded. I is *nowhere prepicitous* if $I|X$ is not prepicitous for any $X \in I^+$.

$X, Y \subset \mathcal{P}_\kappa\lambda$ is *almost disjoint* with respect to I if $X \cap Y \in I$. I is *η-saturated* if there is no pairwise almost disjoint family of η many I-positive sets.

κ is *λ-(super)compact* if there is a fine (normal) measure on $\mathcal{P}_\kappa\lambda$. κ is *strongly compact* (*supercompact*) if it is λ-(super)compact for all $\lambda \geq \kappa$. κ is *measurable* if it is κ-compact, equivalently κ-supercompact. κ is *huge* if there is a normal measure on $\mathcal{P}_{\kappa^+}\lambda$, which is equivalent to the existence of an elementary embedding $j : V \to M$ such that M is closed under $j(\kappa)$ sequences. κ is *almost huge* if there is an elementary embedding $j : V \to M$ such that M is closed under $< j(\kappa)$ sequences.

We have tried to use standard notations and our set theory is ZFC. Any matters not mentioned above would be referred to in Jech [8].

3. Seminormal fine measures

In this section we shall show the existence of a seminormal fine measure is equivalent to the hypothesis of supercompactness.

We state an easy lemma without proof.

Lemma 3.1. (1) *If $cf(\lambda) < \kappa$, then every filter on $\mathcal{P}_\kappa\lambda$ is weakly normal.*

(2) *If F is seminormal and weakly normal, then it is normal.*

Theorem 3.2. *If there is a seminormal fine measure on $\mathcal{P}_\kappa\lambda$, then there is a normal measure on $\mathcal{P}_\kappa\lambda$.*

Proof. Let U be a seminormal fine measure on $\mathcal{P}_\kappa\lambda$. If $cf(\lambda) < \kappa$, then U is normal by Lemma 3.1. Assume $cf(\lambda) \geq \kappa$ and f is the least unbounded function for U and define F by

$$X \in F \quad \text{iff} \quad X \subset \mathcal{P}_\kappa\lambda \quad \text{and } \{x \in \mathcal{P}_\kappa\lambda : x \cap f(x) \in X\} \in U.$$

Since f is the least unbounded function for U, F is weakly normal. Moreover, seminormality of U is inherited by F. Hence F is normal and κ is λ-supercompact. □

If we try to characterize such a κ as λ-compact, α-supercompact for all $\alpha < \lambda$, yet not λ-supercompact by single measure, a natural one will be a non-normal seminormal fine measure on $\mathcal{P}_\kappa\lambda$. Theorem 3.2 shows it is impossible.

Corollary 3.3. (1) *Suppose that κ is α-supercompact for all $\alpha < \lambda$, yet not λ-supercompact. Then there is no seminormal fine measure on $\mathcal{P}_\kappa\lambda$.*

(2) *Suppose that κ is almost huge, but not huge. Then there is no seminormal fine measure on $\mathcal{P}_{\kappa^+}\lambda$ for any λ.*

In [2] we proved that, for λ singular, any seminormal filter extending $SCF_{\kappa\lambda}$ is normal. The next proposition is a counterexample for regular λ.

Proposition 3.4. *If κ is λ-supercompact and λ is measurable, we have a seminormal, non-normal measure on $\mathcal{P}_\kappa\lambda$ extending $CF_{\kappa\lambda}$.*

Proof. Let W be a λ-complete non-normal measure on λ containing $CF_\lambda \cup \{\alpha < \lambda : \alpha \text{ is singular }\}$ and U_α a normal measure on $\mathcal{P}_\kappa\alpha$ for each $\alpha < \lambda$. We define U by

$$X \in U \quad \text{iff} \quad X \subset \mathcal{P}_\kappa\lambda \quad \text{and} \quad \{\alpha < \lambda : X \cap \mathcal{P}_\kappa\alpha \in U_\alpha\} \in W.$$

Let $f : \mathcal{P}_\kappa\lambda \to \delta < \lambda$ be regressive. Since U_α is normal, we have a $\gamma_\alpha < \delta$ for each $\delta < \alpha < \lambda$ such that $f^{-1}(\{\gamma_\alpha\}) \cap \mathcal{P}_\kappa\alpha \in U_\alpha$. Since W is λ

complete and $\delta < \lambda$, there is a γ such that $\{\alpha < \lambda : \gamma_\alpha = \gamma\} \in W$. Then $f^{-1}(\{\gamma\}) \in U$ and U is seminormal.

If α is singular, then $\{x \in \mathcal{P}_\kappa\alpha : \overline{x}$ is singular $\} \in U_\alpha$. Hence $\{x \in \mathcal{P}_\kappa\lambda : \overline{x}$ is singular $\} \in U$. This shows that U is not normal.

Finally, assume $C \subset \mathcal{P}_\kappa\lambda$ is a cub. Since $\{\alpha < \lambda : C \cap \mathcal{P}_\kappa\alpha \in CF_{\kappa\alpha}\} \in CF_\lambda \subset W$ and $CF_{\kappa\alpha} \subset U_\alpha$ for every α, we have $C \in U$. □

In Menas [14], we can find a counterpart of the above stated as Magidor and Solovay's result.

Proposition 3.5. *If there is a seminormal, non-normal measure on $\mathcal{P}_\kappa\lambda$ assigning measure one to the cub subsets of $\mathcal{P}_\kappa\lambda$, then there is a measurable cardinal δ with $\kappa < \delta \leq \lambda$.*

It was also shown there that for any $\{C_\alpha : \alpha < \kappa\} \subset SCF_{\kappa\kappa^+}$ there exists a non-normal fine measure U on $\mathcal{P}_\kappa\kappa^+$ such that $[< x \cap \kappa | x \in \mathcal{P}_\kappa\lambda >]_U = \kappa$ and $C_\alpha \in U$ for all $\alpha < \kappa$. ($[f]_U$ means the transitive collapse of the equivalence class of a function f in the usual ultrapower $V^{\mathcal{P}_\kappa\lambda}/U$.)

We can not have such a fine measure for some $\{C_\alpha : \alpha < \kappa^+\} \subset SCF_{\kappa\kappa^+}$ unless κ is κ^+-supercompact.

Theorem 3.6. *There is a family of κ^+ many strong cub subsets of $\mathcal{P}_\kappa\kappa^+$ $\mathcal{A}$ such that κ is κ^+-supercompact if there is a fine measure U on $\mathcal{P}_\kappa\kappa^+$ with $U \supset \mathcal{A}$ and $[< x \cap \kappa | x \in \mathcal{P}_\kappa\lambda >]_U = \kappa$.*

Proof. For each $\alpha < \kappa^+$, let $h_\alpha : \alpha \to \kappa$ be injective and $C_\alpha = C_{h_\alpha}$. Let $\mathcal{A} = \{C_\alpha : \alpha < \kappa^+\}$. Suppose that U is a fine measure on $\mathcal{P}_\kappa\kappa^+$ such that $U \supset \mathcal{A}$ and $[< x \cap \kappa | x \in \mathcal{P}_\kappa\lambda >]_U = \kappa$. We only have to show U is seminormal by Theorem 3.2.

For any $\beta < \kappa^+$ and regressive function $f : \mathcal{P}_\kappa\kappa^+ \to \beta$, define $g : \mathcal{P}_\kappa\kappa^+ \to \kappa$ by

$$g(x) = h_\beta(f(x)) \quad \text{for all } x \in \mathcal{P}_\kappa\kappa^+.$$

Since $C_\beta \in U$, $\{x \in \mathcal{P}_\kappa\kappa^+ : g(x) \in x \cap \kappa\} \in U$. Hence there is a $\gamma < \kappa$ with $g^{-1}(\{\gamma\}) \in U$ because $[< x \cap \kappa | x \in \mathcal{P}_\kappa\lambda >]_U = \kappa$. So, $f^{-1}(\{h_\beta^{-1}(\gamma)\} \in U$.
□

Definition. For a fine measure U on $\mathcal{P}_\kappa\lambda$ and $f : \mathcal{P}_\kappa\lambda \to \mathcal{P}_\kappa\lambda$, let $f_*(U) = \{X \subset \mathcal{P}_\kappa\lambda : f^{-1}(X) \in U\}$. We say U is *projected* to W if $W = f_*(U)$ for some $f : \mathcal{P}_\kappa\lambda \to \mathcal{P}_\kappa\lambda$. U is *minimal* if f is one to one on some $X \in U$ whenever $f_*(U)$ is a fine measure.

Theorem 3.2 says that every seminormal measure on $\mathcal{P}_\kappa\lambda$ can be projected to some normal measure on $\mathcal{P}_\kappa\lambda$.

For a given normal measure U on $\mathcal{P}_\kappa\lambda$, we easily find a non-normal seminormal measure to which U is projected provided that $cf(\lambda) \geq \kappa$.

Proposition 3.7. *If $cf(\lambda) \geq \kappa$, then every normal measure on $\mathcal{P}_\kappa\lambda$ is isomorphic to a non-normal, seminormal measure which does not extend $SCF_{\kappa\lambda}$.*

Proof. Let U be a normal measure on $\mathcal{P}_\kappa\lambda$ and $f(x) = x \cup \{sup(x)\}$ for each $x \in \mathcal{P}_\kappa\lambda$.

Since U is normal, there is an $X \in U$ such that f is injective on X. So, $U \cong f_*(U)$ and $f_*(U)$ is clearly seminormal by definition. Also $SCF_{\kappa\lambda} \not\subset f_*(U)$ since $\{x \in \mathcal{P}_\kappa\lambda : sup(x) \in x\} \in f_*(U)$. □

We know that every normal measure is minimal if λ is regular or $cf(\lambda) < \kappa$. For the proof see [14].

Suppose that λ is regular. $f_*(U)$ in the previous proposition is thus minimal. It was stated in [1] that any non-weakly normal fine measure extending $SCF_{\kappa\lambda}$ is not minimal. So we have:

Proposition 3.8. *If λ is regular, then any non-normal, seminormal measure extending $SCF_{\kappa\lambda}$ is not minimal.*

4. Seminormal closures and supercompact cardinals

In [2] it was asked whether seminormal closures of non-normal ideals can be normal for regular λ. We partially answer the question assuming supercompactness. $S_m(I)$ denotes the seminormal closure of I, the minimal seminormal ideal containing I.

Theorem 4.1. *If κ is supercompact, then for any $\lambda > \kappa$ there is a non-normal ideal I such that $S_m(I)$ is normal.*

The reader should note that any seminormal ideal $\supset SNS_{\kappa\lambda}$ is normal if λ is singular. For the proof, see [2]. Before proving the theorem, we make some definitions and prove some lemmas.

Definition. An ideal I is *η-subnormal* if there is an η-normal ideal $J \supset I$.

A function f is *I-small* if $f^{-1}(\{x\}) \in I$ for any x.

The following is a $\mathcal{P}_\kappa\lambda$-version of well-known fact for ideals on κ.

Lemma 4.2. *I is not η-subnormal if there is an I-small regressive function $f : X \to \eta$ for some $X \in I^*$.*

Lemma 4.3. *If $\delta < \lambda$ and I is δ^+-normal, and not δ^+-saturated, then there is a $J \supset I$ which is not δ^+-subnormal.*

Proof. By our hypothesis there is a disjoint partition of $\mathcal{P}_\kappa\lambda$ $\{X_\alpha : \alpha < \delta^+\} \subset I^+$. J is defined by

$$X \in J \quad \text{iff} \quad X \subset \mathcal{P}_\kappa\lambda \quad \text{and} \quad |\{\alpha < \delta^+ : X \cap X_\alpha \in I^+\}| < \delta^+$$

It is easy to see J is an ideal extending I. To see that J is not δ^+-subnormal, let $X = \bigcup\left\{X_\alpha \cap \widehat{\{\alpha\}} : \alpha < \delta^+\right\}$ and $f(x) = \alpha$ for each $x \in X_\alpha \cap \widehat{\{\alpha\}}$. Then $f : X \to \delta^+$ is regressive, and $X \in J^*$ since $(\mathcal{P}_\kappa\lambda - X) \cap X_\alpha \cap \widehat{\{\alpha\}} = \emptyset$ for every $\alpha < \delta^+$. Since each $X_\alpha \in J$, f is J-small. Hence J is not δ^+-subnormal. $\square$

By the same argument we have:

Corollary 4.4. (1) *If I is not κ-saturated, then there is a $J \supset I$ which is not κ-subnormal.*

(2) *If κ is strongly compact and I is not κ-saturated, then I can be extended to a non-normal prime ideal.*

Definition. Two ideals I and J are *coherent* if there is an ideal $K \supset I \cup J$.

Lemma 4.5. *Let $\{I_\alpha : \alpha < \eta\}$ be a family of ideals on $\mathcal{P}_\kappa\lambda$, $\nu \geq sup\{sat(I_\alpha) : \alpha < \eta\}$, and $I = \bigcap\{I_\alpha : \alpha < \eta\}$.*

(1) *I is an ideal. I is normal if each I_α is normal.*

(2) *If $cf(\nu) > \eta$, then I is ν-saturated. So, I is ν^+-saturated if $\eta \leq \nu$.*

(3) *If $\eta \leq \kappa$ and I_α and I_β are not coherent whenever $\alpha \neq \beta$, then I is not η-saturated.*

(4) *If $\eta \leq \lambda^+$, each I_α is normal, and I_α and I_β are not coherent whenever $\alpha \neq \beta$, then I is not η-saturated.*

Proof. (1) is trivial.

To prove (2), let $\eta < cf(\nu)$ and $\mathcal{A}$ be an almost disjoint family for I. For each $\alpha < \eta$, $\mathcal{A}_\alpha = \mathcal{A} \cap I_\alpha^+$ is an almost disjoint family for I_α. Hence $|\mathcal{A}_\alpha| < \nu$. Now $|\mathcal{A}| = |\bigcup\{\mathcal{A}_\alpha : \alpha < \eta\}|$ and the conclusion is clear.

(3) Suppose to the contrary that I is η-saturated. Since I is κ-saturated and $I \subset I_\alpha$, we can find an $X_\alpha \in I^+$ with $I_\alpha = I|X_\alpha$ for each $\alpha < \eta$. If $X_\alpha \cap X_\beta \in I^+$, then $I_\alpha \cup I_\beta \subset I|(X_\alpha \cap X_\beta)$, contradicting our assumption. So $\{X_\alpha : \alpha < \eta\}$ is an almost disjoint family for I. Hence I is not η-saturated.

(4) is proved similarly. □

Corollary 4.6. (1) *If $\eta \le \kappa$, I_α is η-saturated for every $\alpha < \eta$, I_α and I_β are not coherent whenever $\alpha \ne \beta$, then $\bigcap\{I_\alpha : \alpha < \eta\}$ is an η^+-saturated ideal which is not η-saturated.*

(2) *If κ is $2^{\lambda^{<\kappa}}$-supercompact, then, for any $\eta \le \lambda^+$, there are $2^{2^{\lambda^{<\kappa}}}$ many normal η^+-saturated, non-η-saturated ideals on $\mathcal{P}_\kappa\lambda$.*

Question. *Suppose that $\kappa < \eta \le \lambda^+$ and each I_α is not subnormal. Can it be proved that $\bigcap\{I_\alpha : \alpha < \eta\}$ is not η-saturated ?*

We quote a result in [2].

Lemma 4.7. *Let λ be regular. A seminormal ideal I on $\mathcal{P}_\kappa\lambda$ is λ-saturated iff every seminormal ideal extending I is of the form $I|A$ for some $A \in I^+$.*

Proof of Theorem 4.1. Let κ be supercompact and λ a regular cardinal $> \kappa$. Choose a family of κ many normal prime ideals on $\mathcal{P}_\kappa\lambda$, $\{I_\alpha : \alpha < \kappa\}$ (we can find them by a theorem in [12]) and set $J = \bigcap\{I_\alpha : \alpha < \kappa\}$.

J is a normal λ-saturated non-κ-saturated ideal on $\mathcal{P}_\kappa\lambda$ by Lemma 4.5 and Corollary 4.6. Using Lemma 4.3, we can find a non-κ-subnormal ideal $K \supset J$. Since κ is strongly compact, there is a non-normal prime ideal $L \supset K$ by Corollary 4.3. Let $I = I_0 \cap L$.

We show I is a desired ideal. Since I is 3-saturated, $L = I|A$ for some $A \in I^+$. Hence I is not normal. $S_m(I)$ is proper since $S_m(I) \subset I_0$. Finally $S_m(I)$ is normal by Lemma 4.7 since $S_m(I) \supset I \supset J$ and J is λ-saturated and normal. □

The author could not answer the following question.

Question. *Are large cardinals necessary to have such an ideal I in Theorem 4.1 ?*

5. Prepicitousness and saturation of κ-normal ideals

We discuss here saturation and prepicitousness of $\nabla I_{\kappa\lambda}$, the minimal κ-normal ideal on $\mathcal{P}_\kappa\lambda$. Our tool is Lemma 5.1 due to Matsubara and Shioya [13].

Definition. For any ideal I and $\mathcal{F} \subset I$, we say $\mathcal{F}$ *generates* I if for every $X \in I$ there exists a $Y \in \mathcal{F}$ such that $X \subset Y$.

Lemma 5.1. *If I is an ideal such that $min\{|X| : X \in I^+\} = min\{|\mathcal{F}| : \mathcal{F}$ generates $I\}$, then I is nowhere prepicitous.*

For normal ideals, we know the following:

Remark. *$NS_{\kappa\lambda}$ is not λ-saturated, nowhere κ-saturated. If J is λ^+-saturated and $J \subset I$ is normal, then $I = J|A$ for some $A \in J^+$.*

The following are κ-normal versions of this.

Proposition 5.2. *Let I be a κ-normal ideal on $\mathcal{P}_\kappa\lambda$.*

(1) *If there is a κ^+-saturated ideal $J \subset I$, then $I = J|A$ for some $A \in J^+$ hence I is also κ^+-saturated.*

(2) *$\nabla^\kappa I_{\kappa\lambda}$ is not λ-saturated and nowhere κ-saturated.*

Proof. (1) is an adjustment of the above remark to κ-normal ideals.

(2) Since $NS_{\kappa\lambda}$ is not λ-saturated, neither is $\nabla I_{\kappa\lambda}$. Moreover, almost the same argument in Solovay's proof in [15] showing every stationary subset of κ splits into κ many disjoint stationary sets also tells us $\nabla I_{\kappa\lambda}$ is nowhere κ-saturated. □

So a natural question is whether $\nabla^\kappa I_{\kappa\lambda}$ is nowhere λ-saturated. By Gitik's result in [6] it is consistent that $NS_{\kappa\lambda}|S$ is κ^+-saturated for some $S \in NS^+_{\kappa\lambda}$ relative to a supercompact cardinal. The author does not know if Gitik's construction is applicable to $\nabla^\kappa I_{\kappa\lambda}$. We only show some cardinal power condition is necessary for $\nabla^\kappa I_{\kappa\lambda}$ to be somewhere κ^+-saturated. Note that every κ-normal κ^+-saturated ideal is prepicitous by Lemma 10 in Foreman [5].

For each $f : \kappa \to \mathcal{P}_\kappa\lambda$, let $X_f = \mathcal{P}_\kappa\lambda - C_f = \{x \in \mathcal{P}_\kappa\lambda : f(\alpha) \not\subset x$ for some $\alpha \in x\}$. Then $\{X_f : f : \kappa \to \mathcal{P}_\kappa\lambda\}$ generates $\nabla^\kappa I_{\kappa\lambda}$ and has cardinality λ^κ. If $2^{<\kappa} \leq \lambda$, then every unbounded subset of $\mathcal{P}_\kappa\lambda$ has cardinality $\lambda^{<\kappa}$. Now the following is clear by lemma 5.1.

Theorem 5.3. *If $2^{<\kappa} \leq \lambda$ and $\lambda^{<\kappa} = \lambda^\kappa$, then $\nabla^\kappa I_{\kappa\lambda}$ is nowhere prepicitous. Hence every κ-stationary set splits into κ^+ many disjoint κ-stationary sets.*

So if $G.C.H.$ holds and $cf(\lambda) \neq \kappa$, then $\nabla^\kappa I_{\kappa\lambda}$ is nowhere prepicitous and every κ-stationary subset of $\mathcal{P}_\kappa\lambda$ is a disjoint union of κ^+ many κ-stationary sets.

Corollary 5.4. (1) *If* $2^\kappa = \kappa^+$, *then* $\nabla^\kappa I_{\kappa\kappa^+}$ *is nowhere prepicitous.*

(2) *If there is a* κ*-stationary subset of* $\mathcal{P}_\kappa\kappa^+$ *which does not split into* κ^+ *many disjoint* κ*-stationary sets, then* $2^\kappa > \kappa^+$.

By adding κ^{++} subsets of κ to a model of $V = L$, one can see the converse of Theorem 5.3 does not hold.

Theorem 5.5. (1) *Let* V *be a model of G.C.H. and* $cf^V(\lambda) \neq \kappa$. *Then* $\nabla^\kappa I_{\kappa\lambda}$ *is nowhere prepicitous in any* κ*-c.c. extension of* V.

(2) *Assume* $cf(\lambda) < \kappa$ *and G.C.H. holds in* V. *Then* $NS_{\kappa\lambda}$ *is nowhere prepicitous in any* κ*-c.c. extension of* V.

Proof. Assume that *G.C.H.* holds in V, the ground model, and $P \in V$ is κ-c.c..

Let $f \in V[G]$ be such that $f : \kappa \to (\mathcal{P}_\kappa\lambda)^{V[G]}$. Since P is κ-c.c., there is an $\tilde{f} \in V$ such that $\tilde{f} : \kappa \to (\mathcal{P}_\kappa\lambda)^V$ and $f(\alpha) \subset \tilde{f}(\alpha)$ for all $\alpha < \kappa$. Thus we have

$$V[G] \models \text{"}\nabla I_{\kappa\lambda} \text{ is } (\lambda^\kappa)^V \text{ generated".}$$

If $cf^V(\lambda) > \kappa$, then $(\lambda^\kappa)^V = \lambda$ by G.C.H.. Since every unbounded subset of $\mathcal{P}_\kappa\lambda$ has cardinality $\geq \lambda$, $V[G] \models$"$\nabla I_{\kappa\lambda}$ is nowhere prepicitous" by Lemma 5.1.

If $cf^V(\lambda) < \kappa$, then $(\lambda^\kappa)^V = \lambda^+$. By Proposition 2.2 in [4], $cf(min\{|X| : X \in I^+_{\kappa\lambda}\}) \geq \kappa$. Hence every κ-stationary set has cardinality $\geq \lambda^+$. By Lemma 5.1, $\nabla I_{\kappa\lambda}$ is nowhere prepicitous in $V[G]$.

(2) By the same argument one can see, in any κ-c.c. generic extension of V, $NS_{\kappa\lambda}$ is $(2^\lambda)^V$-generated and nowhere prepicitous. □

The same argument as Kakuda's proof for normal ideals on κ in [11] gives the following:

Theorem 5.6. *Let* P *be a* κ*-c.c. poset.*

(1) *For any* κ*-normal ideal* I *on* $\mathcal{P}_\kappa\lambda$, I *is prepicitous iff* $\vdash_P$ *"the ideal generated by* I *is prepicitous".*

(2) $\nabla^\kappa I_{\kappa\lambda}$ *is prepicitous iff* $\vdash_P$ "$\nabla^\kappa I_{\kappa\lambda}$ *is prepicitous".*

References

[1] Y. Abe, *Weakly normal filters and the closed unbounded filter on $\mathcal{P}_\kappa\lambda$,* Proc. Amer. Math. Soc. 104 (1988), 1226-1234.

[2] Y. Abe, *A hierarchy of filters smaller than $CF_{\kappa\lambda}$,* Arch. Math. Logic, to appear.

[3] D. M. Carr, *The minimal normal filters on $\mathcal{P}_\kappa\lambda$,* Proc. Amer. Math. Soc. 86 (1982), 316-320.

[4] H. D. Donder and P. Matet, *Two cardinal versions of diamond,* Isr. J. Math. 83 (1993), 1-43.

[5] M. Foreman, *Potent axioms,* Trans. Amer. Math. Soc. 294 (1986), 1-28.

[6] M. Gitik, *Nonsplitting subset of $\mathcal{P}_\kappa\kappa^+$,* J. Symbolic Logic 50 (1985), 881-894.

[7] T. Jech, *Some combinatorial problems concerning uncountable cardinals,* Ann. Math. Logic 5 (1973), 165-198.

[8] T. Jech, *Set theory,* Academic press, New York (1978).

[9] C. A. Johnson, *Seminormal λ-generated ideals,* J. Symb. Logic 53 (1988), 92-102.

[10] C. A. Johnson, *Some partition relations for ideals on $\mathcal{P}_\kappa\lambda$,* Acta Math. Hung. 56 (1990), 269-282.

[11] Y. Kakuda, *On a condition for certain extensions which preserve prepicitous ideals,* J. Symbolic Logic 46 (1981), 296-300.

[12] M. Magidor, *There are many normal ultrafilters corresponding to a supercompact cardinal,* Isr. J. Math. 10 (1971), 186-192.

[13] Y. Matsubara and M. Shioya, *Nowhere prepicitousness of some ideals,* preprint.

[14] T. K. Menas, *On strong compactness and supercompactness,* Ann. Math. Logic 7 (1974), 327-359.

[15] R. M. Solovay, *Real-valued measurable cardinals,* In, Axiomatic Set Theory, Proc. Symp. Pure Math. 13, I (D. Scott, ed.), Amer. Math. Soc., Providence, Rohde Island (1974),397-428.

Effective Baire Category Concepts

Klaus Ambos-Spies and Jan Reimann
Mathematisches Institut
Universität Heidelberg
D-69120 Heidelberg, Germany

Abstract

Mehlhorn [Me73] introduced an effective Baire category concept designed for measuring the size of classes of computable sets. This concept is based on effective extension functions. By considering partial extension functions, we introduce a stronger concept. Similar resource-bounded concepts have been previously introduced by Ambos-Spies et al. [AFH88] and Ambos-Spies [Am96]. By defining a new variant of the Banach-Mazur game, we give a game theoretical characterization of our category concept.

1 Introduction

The classical Baire category concept gives a classification of the subclasses of the Cantor space 2^ω (and other topological spaces) according to their size. The small classes in this sense are the *meager* classes. All finite classes are meager, any subclass of a meager class is meager, and the countable union of meager classes is meager again. The consistency of the system follows from Baire's theorem which implies that the complement of a meager class is not meager.

Since the class **REC** of the recursive (computable) sets is countable, hence meager, Baire category is too coarse for classifying classes of computable sets. Mehlhorn [Me73] has shown, however, that the category concept can be effectivized to obtain a classification schema for classes of recursive sets, which has similar properties if we replace arbitrary countable unions by uniformly computable unions (in an appropriate sense). Mehlhorn's concept is based on the characterization of Baire category in terms of total extension functions and it admits only recursive extension functions.

The extension function characterization of Baire category shows the close relations between this concept and the finite extension method, a fundamental diagonalization technique in computability theory (see e.g. Odifreddi

[Od89], Chapter V.3). Correspondingly, Mehlhorn's concept can be viewed as a formalization of the effective finite extension method.

It turned out, however, that the effectivization of certain finite extension arguments requires slightly more complex diagonalization arguments, namely slow diagonalizations which are also called effective wait and see arguments. The typical feature of these arguments is that, in contrast to a standard finite extension argument, the requirements are not met in order. In computational complexity theory these more flexible diagonalizations become necessary for studying almost everywhere complexity and the related bi-immunity concepts (see [Am96] for details).

An effective Baire category concept capturing slow diagonalizations can be obtained by admitting partial extension functions. For resource-bounded diagonalizations, such concepts have been introduced by Ambos-Spies, Fleischhack and Huwig [AFH88] and Ambos-Spies [Am96]. In particular, the latter contains a detailed discussion of the role of the above diagonalization concepts in complexity theory and their relation to Baire category. There, the interest focused on complexity classes, however, whence the general recursive case was not discussed.

Here we introduce the corresponding effective category concept, called extended effective (e.e.) category. We prove some basic facts for this new concept. In particular we show that it properly extends Mehlhorn's concept. Moreover, by introducing a new variant of the Banach-Mazur game, we give a game theoretical characterization of e.e. category.

The outline of the paper is as follows. In Section 2 we review Mehlhorn's effective category concept. There we give a new characterization of the effectively meager classes in terms of the arithmetical hierarchy of classes which easily implies many of Mehlhorn's results on effective meagerness.

In Section 3 we introduce our new e.e. category concept. By showing that **REC** is not e.e. meager, we show that this concept induces a consistent category notion for classes of computable sets. Moreover, we show that e.e. category is a proper refinement of effective category.

Section 4 is devoted to the game theoretical characterization of Baire category in terms of Banach-Mazur games. We introduce a new variant of this game, in which the second player is allowed to cut back certain moves of the first player, and we characterize e.e. meagerness in terms of determinateness of the effective variant of this extended Banach-Mazur game.

Finally, in Section 5 we shortly comment on resource-bounded category.

We conclude this section by introducing some notation.

Let ω be the set of natural numbers, let $2^{<\omega}$ be the set of (finite) binary strings, and let 2^ω be the set of infinite binary sequences (the Cantor space). Lower case letters $\ldots, v, w, x, y, z$ from the end of the alphabet will denote strings while the other letters denote numbers with the exception of f, g, h and l which are reserved for functions. For a function $f : \omega \times 2^{<\omega} \to 2^{<\omega}$,

f_n will denote the n-th branch of f: $f_n(x) = f(n, x)$. A subset of ω is called a *set* while sets of sets are called *classes*. Capital letters denote sets, boldface capital letters denote classes.

For a string x, $x(m)$ denotes the $(m+1)$th bit in x, i.e., $x = x(0)\dots x(n-1)$, where $n = |x|$ is the length of x. λ is the empty string and $x\hat{\ }y$ denotes the concatenation of x and y. The initial segment of x of length n is denoted by $x\restriction n = x(0)\dots x(n-1)$. If y extends x then we write $x \sqsubseteq y$ and $x \sqsubset y$ expresses that the extension is proper.

The notation for strings is canonically extended to infinite sequences. Moreover, we identify a set A with its characteristic sequence, i.e., $n \in A$ iff $A(n) = 1$ and $n \notin A$ iff $A(n) = 0$. Hence a class is a subset of the Cantor space. The class of all sets extending a string x is denoted by $\mathbf{B}_x = \{A : x \sqsubset A\}$.

We assume familiarity with the basic concepts of recursion theory. From complexity theory we will only consider the complexity classes **P** and **NP** of deterministic respectively nondeterministic polynomial-time. Here $A \in \mathbf{P}$ ($A \in \mathbf{NP}$) if there is a (non)deterministic Turing machine M computing A, where the running time of M on input n is bounded by $f(|n|)$, where $|n|$ is the length of the binary representation of n and f is some polynomial.

2 Mehlhorn's Effective Category Concept

The classical Baire category concept for the Cantor space 2^ω can be defined in terms of extension functions. An *extension function* f is a function $f : 2^{<\omega} \to 2^{<\omega}$ such that $x \sqsubseteq f(x)$ for all strings x. A subclass $\mathbf{C}$ of 2^ω is *nowhere dense* via f if $\mathbf{B}_{f(x)} \cap \mathbf{C} = \emptyset$ for all strings x. Then a class $\mathbf{C}$ is *meager* if it is contained in the countable union of nowhere dense classes, and the complement of a meager class is *comeager*. Intuitively, meager classes are small and, correspondingly, comeager classes are large. This intuition is justified by the following facts.

(2.1) Every finite class is meager.

(2.2) Every subclass of a meager class is meager.

(2.3) The countable union of meager classes is meager.

(2.4) 2^ω is not meager.

By (2.1) and (2.3) every countable class is meager whence Baire category is too coarse for classifying the size of classes of recursive sets. Mehlhorn [Me73], however, introduced an effectivization of the category concept which provides such a classification. By considering only computable extension functions, Mehlhorn defines effectively nowhere dense classes and he lets the effectively meager classes be the countable unions of *uniformly* effectively nowhere dense classes.

Definition 2.1 *(Mehlhorn [Me73])*

(a) *An* effective extension function f *is a total recursive function* $f: 2^{<\omega} \to 2^{<\omega}$ *such that* $x \sqsubseteq f(x)$ *for all strings* x. *A class* $\mathbf{C}$ *is* effectively nowhere dense via f *if* f *is an effective extension function and* $\mathbf{B}_{f(x)} \cap \mathbf{C} = \emptyset$ *for all strings* x, *and* $\mathbf{C}$ *is* effectively nowhere dense *if* $\mathbf{C}$ *is effectively nowhere dense via some function* f.

(b) *An* effective system of extension functions *is a recursive function* $f: \omega \times 2^{<\omega} \to 2^{<\omega}$ *such that* f_e *is an effective extension function for all* $e \geq 0$. *A class* $\mathbf{C}$ *is* effectively meager via f *if* f *is an effective system of extension functions and* $\mathbf{C}$ *is the union of classes* $\mathbf{C}_e$, $e \geq 0$, *where* $\mathbf{C}_e$ *is effectively nowhere dense via* f_e. $\mathbf{C}$ *is* effectively meager *if* $\mathbf{C}$ *is effectively meager via some function* f, *and* $\mathbf{C}$ *is* effectively comeager *if the complement* $\overline{\mathbf{C}}$ *of* $\mathbf{C}$ *is effectively meager.*

Mehlhorn's category concept refines the classical concept in the sense that every effectively nowhere dense (meager, comeager) class is nowhere dense (meager, comeager), but the converse fails. In particular the countable class **REC** of the recursive sets is not effectively meager, whence the effective category concept relativized to **REC** yields a classification schema for classes of recursive sets.

Mehlhorn has shown that every recursively presentable class of recursive sets is effectively meager but he also gave examples of effectively meager classes containing nonrecursive sets. In order to describe the strength of Mehlhorn's concept we characterize it in terms of Σ_2^0-classes.

Definition 2.2 *A class* $\mathbf{C}$ *is a* Σ_2^0-class *if there is a total recursive* 0-1-*valued functional* $\Phi: 2^\omega \times \omega \times \omega \to \{0,1\}$ *such that*

$$A \in \mathbf{C} \;\Leftrightarrow\; (\exists m)\,(\forall n)\,(\Phi^A(m,n) = 1).$$

For more details on the arithmetical hierarchy of classes we refer to Rogers [Ro67], Chapter 15.

Theorem 2.3 *For any class* $\mathbf{C}$ *the following are equivalent.*

(i) $\mathbf{C}$ *is effectively meager.*

(ii) *There is a* Σ_2^0*-class* $\mathbf{D}$ *and a recursive set* A *such that* $\mathbf{C} \subseteq \mathbf{D}$, $\mathbf{D}$ *is c.f.v., and* $A \notin \mathbf{D}$.

(iii) *There is a* Σ_2^0*-class* $\mathbf{D}$ *and a set* A *such that* $\mathbf{C} \subseteq \mathbf{D}$ *and, for any finite variant* $\hat{A}$ *of* A, $\hat{A} \notin \mathbf{D}$.

For the proof of Theorem 2.3 we need the following observations.

Proposition 2.4 *Let f be an effective system of extension functions. There is a greatest class which is meager via f, namely*

$$\mathbf{M}(f) = \{A : (\exists e)\,(\forall n)\,(f_e(A\restriction n) \not\sqsubseteq A)\}$$

Proof. Straightforward. □

Proposition 2.5 *For any effective system of extension functions f, $\mathbf{M}(f)$ is a Σ^0_2-class.*

Proof. Immediate by definition. □

Proposition 2.6 *For any effective system of extension functions f, $\mathbf{REC} \not\subseteq \mathbf{M}(f)$.*

Proof. Given f, by an effective finite extension argument inductively define a recursive set $A \notin \mathbf{M}(f)$ as follows. Enumerate $A\restriction l(s)$ in stages, where $l : \omega \to \omega$ is a strictly increasing function, and $l(s)$ and the finite extension $A\restriction l(s)$ of $A\restriction l(s-1)$ are defined at stage s: Given s and $A\restriction l(s-1)$ (where $l(-1) = 0$), let $A\restriction l(s) = f_s(A\restriction l(s-1))$. Then A is recursive and the initial segment $A\restriction l(s)$ ensures that A is not effectively nowhere dense via f_s. □

Proposition 2.7 *For any effective system of extension functions f there is an effective system of extension functions $\hat{f}$ such that $\mathbf{M}(f) \subseteq \mathbf{M}(\hat{f})$ and $\mathbf{M}(\hat{f})$ is closed under finite variants.*

Proof. Given f the function $\hat{f}$ is obtained by considering all finite variants of the extension functions of the system f in the following sense. For strings x and y such that $|x| \geq |y|$ let $var(x, y)$ be the string obtained from x by replacing it's initial segment of length $|y|$ by y. Then, for any number e and strings y and z, the extension function $\hat{f}_{\langle e,y,z\rangle}$ extends a string x with $|x| \geq |y|$ in the same way as f_e extends the string $var(x, y)$ and it maps shorter strings to an extension determined by z. To be more precise, for x with $|x| \geq |y|$, $\hat{f}_{\langle e,y,z\rangle}(x) = x\hat{\ }v$, where v is the unique string satisfying $f_e(var(x, y)) = var(x, y)\hat{\ }v$, and $\hat{f}_{\langle e,y,z\rangle}(x) = x\hat{\ }0^{|y|-|x|}\hat{\ }z$ otherwise. □

Proof of Theorem 2.3. Since the implication *(ii)* ⇒ *(iii)* is obvious, it suffices to show *(i)* ⇒ *(ii)* and *(iii)* ⇒ *(i)*.

For a proof of *(i)* ⇒ *(ii)* assume that $\mathbf{C}$ is effectively meager via f. Then, for $\hat{f}$ as in Proposition 2.7 and for $\mathbf{D} = \mathbf{M}(\hat{f})$, $\mathbf{C} \subseteq \mathbf{D}$ (by Proposition 2.4), $\mathbf{D}$ is a c.f.v. Σ^0_2- class (by Propositions 2.5 and 2.7) and there is a recursive set $A \notin \mathbf{D}$ (by Proposition 2.6).

For a proof of the implication *(iii)* ⇒ *(i)* let $\mathbf{D}$ be a Σ^0_2-class, say

$$X \in \mathbf{D} \Leftrightarrow (\exists m)\,(\forall n)\,(\Phi^X(m, n) = 1),$$

and let A be a set such that, $\hat{A} \notin \mathbf{D}$ for all finite variants of A. By Proposition 2.4, it suffices to define an effective system of extension functions f such that $\mathbf{D} \subseteq \mathbf{M}(f)$. Note that, by assumption on A, for any number m and for any string x there is a string $y \sqsupseteq x$ such that $\Phi^y(m,n) = 0$ for some number n. So, by letting $f_m(x)$ be the least such string y, f is an effective system of extension functions as required. □

The above characterization of effective meagerness easily implies the previously mentioned results of Mehlhorn on his concept.

Corollary 2.8 *(Mehlhorn [Me73])* **REC** *is not effectively meager.* □

Recall that a class $\mathbf{C}$ is *recursively presentable* (*r.p.*) if there is a recursive set $U \subseteq \omega \times \omega$ such that $\mathbf{C} = \{U_e : e \geq 0\}$. A class of recursive sets is called *bounded* if it is contained in an r.p. class.

Corollary 2.9 *(Mehlhorn [Me73]) Every bounded class of recursive sets is effectively meager.*

Proof. By Theorem 2.3 it suffices to show that every r.p. class $\mathbf{C}$ is a Σ^0_2-class. But given a recursive set U such that $\mathbf{C} = \{U_e : e \geq 0\}$,

$$X \in \mathbf{C} \Leftrightarrow (\exists e)\,(\forall n)\,(U(e,n) = X(n))$$

whence $\mathbf{C}$ is a Σ^0_2-class. □

By Corollary 2.9, the standard complexity classes are effectively meager. An interesting example of an effectively meager class containing nonrecursive sets is the following.

Corollary 2.10 *(Mehlhorn [Me73]) The class* $\mathbf{C} = \{A : \mathbf{P}^A = \mathbf{NP}^A\}$ *is effectively meager.*

Proof (Sketch). Let $\{M_e : e \geq 0\}$ and $\{N_e : e \geq 0\}$ be recursive enumerations of the deterministic respectively nondeterministic polynomial time bounded oracle Turing machines. Then, for any oracle set A, the set

$$K^A = \{\langle e, n, 2^m \rangle : N_e \text{ accepts } n \text{ in less than } m \text{ steps}\}$$

is $\mathbf{NP}^A$-complete (see Baker et al. [BGS75]). It follows that

$$\mathbf{C} = \{A : K^A \in \mathbf{P}^A\} = \{A : (\exists n)\,(\forall m)\,(K^A(m) = M_n(m))\}$$

whence $\mathbf{C}$ is a Σ^0_2-class. Moreover, for $A =^* B$, $\mathbf{P}^A = \mathbf{P}^B$ and $\mathbf{NP}^A = \mathbf{NP}^B$ whence $\mathbf{C}$ is c.f.v. Finally, Baker et al. [BGS75] have shown that there is a set $A \notin \mathbf{C}$. Hence effective meagerness of $\mathbf{C}$ follows from Theorem 2.3. □

Mayordomo [May94] pointed out some limitations of the effective category concept. For any class $\mathbf{C}$, a set A is $\mathbf{C}$-*bi-immune* if neither A nor its complement $\overline{A}$ contain any infinite set $X \in \mathbf{C}$ as a subset.

Theorem 2.11 *(Mayordomo [May94]) The class* $\overline{\mathbf{BI}(\mathbf{P})} \cap \mathbf{REC}$ *of the recursive sets which are not* **P***-bi-immune is not effectively meager though* $\overline{\mathbf{BI}(\mathbf{P})}$ *is meager in the classical sense.*

This theorem can be interpreted as follows: The effective category concept formalizes the effective finite extension method. A finite extension construction of a **P**-bi-immune set, however, cannot be effective, i.e. the finite extension method yields only nonrecursive **P**-bi-immune sets. On the other hand, however, Balcázar and Schöning [BS85] have constructed recursive **P**-bi-immune sets by a *slow diagonalization* or *wait and see argument.* The extension of Mehlhorn's category concept introduced in the next section is designed to formalize this more general diagonalization technique. For a detailed discussion of this topic we refer to Ambos-Spies [Am96].

3 The Extended Effective Category Concept

The extension of Mehlhorn's effective category concept which we will consider here is based on partial extension functions. For such a function f we say that a class **C** is nowhere dense via f if

$$(\forall A \in \mathbf{C})\,(\exists^{\infty} n)\,(f(A{\upharpoonright}n)\downarrow)$$

and

$$(\forall x)\,(f(x)\downarrow \Rightarrow \mathbf{B}_{f(x)} \cap \mathbf{C} = \emptyset).$$

Note that we can extend f to a total extension function $\hat{f}$ such that **C** is nowhere dense via $\hat{f}$ by letting $\hat{f}(x) = f(y)$ for the least $y \sqsupseteq x$ such that $f(y)\downarrow$ if such a string y exists and by letting $\hat{f}(x) = x$ otherwise. Hence, in the classical case, partial extension functions lead to the same category concept. In case of effective extensions, however, we obtain a stronger concept as we shall show below.

Definition 3.1 *(a) An* effective partial extension function f *is a partial recursive function* $f : 2^{<\omega} \to 2^{<\omega}$ *with recursive domain where* $x \sqsubseteq f(x)$ *for all strings* x *such that* $f(x)$ *is defined. An effective partial extension function* f *is* dense along *a set* A *if* $f(A{\upharpoonright}n)\downarrow$ *for infinitely many numbers* n*. A set* A meets f *if* $f(A{\upharpoonright}n) \sqsubset A$ *for some number* n *such that* $f(A{\upharpoonright}n)\downarrow$*; and* A avoids f *otherwise. A class* **C** *is* extended effectively (e.e.) nowhere dense via f *if* f *is an effective partial extension function such that* f *is dense along all sets in* **C** *and* $\mathbf{B}_{f(x)} \cap \mathbf{C} = \emptyset$ *for all strings* x *such that* $f(x)\downarrow$ *(i.e. all sets in* **C** *avoid* f*).* **C** *is* e.e. nowhere dense *if* **C** *is e.e. nowhere dense via some* f*.*

(b) *An* effective system of partial extension functions *is a partial recursive function* $f : \omega \times 2^{<\omega} \to 2^{<\omega}$ *with recursive domain such that* f_e *is an effective partial extension function for all e. A class* **C** *is* e.e. meager *via f if f is an effective system of partial extension functions and* **C** *is the union of classes* $\mathbf{C}_e$ *which are e.e. nowhere dense via* f_e*.* **C** *is* e.e. meager *if* **C** *is e.e. meager via some f, and* **C** *is* e.e. comeager *if* $\overline{\mathbf{C}}$ *is e.e. meager.*

Obviously, any class which is effectively nowhere dense via f is also e.e. nowhere dense via f, whence the above concept extends the effective category concept of Mehlhorn. As there we can show that, for every effective system f of partial extension functions, there is a largest class which is e.e. meager via f, namely

$$\mathbf{M}(f) = \{A\colon (\exists e)\,(f_e \text{ is dense along } A \,\&\, A \text{ avoids } f_e)\}.$$

Hence, to show that **REC** is not e.e. meager it suffices to show that $\mathbf{REC} \not\subseteq \mathbf{M}(f)$ for all effective systems of partial extension functions f.

Theorem 3.2 **REC** *is not e.e. meager.*

Proof. Let f be an effective system of partial extension functions. It suffices to construct a recursive set A such that $A \notin \mathbf{M}(f)$, i.e., such that A meets the requirements

$$R_e : (\exists^\infty n)\,(f_e(A{\restriction}n)\downarrow) \Rightarrow (\exists n)\,(f_e(A{\restriction}n) \sqsubseteq A)$$

for all e. The set A is constructed by a slow diagonalization where $A(n)$ is defined at stage n. At any stage of the construction action is taken for the sake of the requirement R_e with the least index e which has not yet been completely satisfied and which now can become further or completely satisfied. Formally, given $A{\restriction}n$ fix $e \le n$ and $m \le n$ minimal (in this order) such that

- there is no number $k \le n$ such that $f_e(A{\restriction}k)\downarrow$ and $f_e(A{\restriction}k) \sqsubseteq A{\restriction}n$, and
- $f_e(A{\restriction}m)\downarrow$ and $A{\restriction}n \sqsubset f_e(A{\restriction}m)$,

let $A(n) = f_e(A{\restriction}m)(n)$ and say that R_e is active at stage n (if such numbers e and m do not exist, let $A(n) = 0$). Then, by induction on e, every requirement R_e will be active at most finitely often and, if the hypothesis of R_e holds, there will be a number m such that R_e will become active at the consecutive stages $m, \ldots |f_e(A{\restriction}m)|$thereby ensuring $f_e(A{\restriction}m) \sqsubset A$. □

Remark. For the proof of Theorem 3.2 it is crucial that the domain of the system f is recursive. If we drop this requirement, we can easily define a partial recursive system f such that **REC** is meager via f. In fact, given a recursive enumeration $\{W_e : e \geq 0\}$ of the recursively enumerable sets, the class of r.e. sets will be meager via the partial recursive function f defined by $f_{2n}(x) = x\hat{\ }1^n$ for all x and $f_{2n+1}(x) = x\hat{\ }0$ if $|x| \in W_n$ and $f_{2n+1}(x)\uparrow$ otherwise. Namely, any finite set will be nowhere dense via f_{2n} for all sufficiently large n and $\{W_n\}$ will be nowhere dense via f_{2n+1} if W_n is infinite.

By Theorem 3.2 the extended effective category concept relativized to **REC** is not trivial. To show that this concept is strictly stronger than Mehlhorn's effective category, we observe that, in contrast to Theorem 2.3 the class **BI**(**C**) of the **C**-bi-immune sets is e.e. comeager for all r.p. classes **C**.

Theorem 3.3 *Let* **C** *be r.p. Then* **BI**(**C**) *is e.e. comeager.*

Proof. It suffices to define an effective system of partial extension functions f such that for every set $A \notin \mathbf{BI}(\mathbf{C})$ there is an index e such that f_e is dense along A and $f_e(A\restriction n) \not\sqsubseteq A$ whenever $f_e(A\restriction n)$ is defined. Given a recursive set U such that $\mathbf{C} = \{U_e : e \geq 0\}$ such a system f is obtained by letting $f_{2e+i}(x) = x\hat{\ }i$ if $|x| \in U_e$ and $f_{2e+i}(x)\uparrow$ otherwise ($e \geq 0,\ i \leq 1$). Namely, if $A \notin \mathbf{BI}(\mathbf{C})$ then there is an infinite set $U_e$such that $U_e \subseteq A$ or $U_e\overline{A}$ whence $\{A\}$ is e.e. nowhere dense via f_{2e} or f_{2e+1}, respectively. □

Corollary 3.4 *There is a class* **C** *of recursive sets such that* **C** *is e.e. meager but not effectively meager.*

Proof. by Theorem 2.?.By Theorems 3.3 and 2.11, $\overline{\mathbf{BI}(\mathbf{P})} \cap \mathbf{REC}$ is e.e. meager but not effectively meager. □

4 Banach-Mazur Games

Banach-Mazur games allow a game theoretic approach to the concepts of Baire category. They were originally proposed by Mazur for characterizing Baire category on the real line, but the games can be defined on any topological space [Ox57]. Here we only consider the Cantor space version.

Given a class $\mathbf{C} \subseteq 2^\omega$, the *Banach-Mazur game* $BM(\mathbf{C})$ for **C** is a two person game. In a *play* the players alternate in extending an initial segment by nonempty strings:

I	x_0		x_1		
					...
II		y_0		y_1	

The *result* or *outcome* of the play is the unique set $R \in 2^\omega$ defined by the initial segments

$$R\restriction r_n = x_0\hat{\ }y_0\hat{\ }\dots\hat{\ }x_n\hat{\ }y_n$$

constructed by the end of round n ($n \geq 0$). Player I *wins* if $R \in \mathbf{C}$. Otherwise Player II *wins*.

A *strategy* for Player I (or Player II) is some rule according to which Player I, knowing *all* his and his opponent's previous moves (classical Banach-Mazur games are games of *perfect information* [GS53]), determines his next move. A *winning strategy* is a strategy with which the player wins every play in $BM(\mathbf{C})$. If there is a winning strategy for one of the players, the game is called *determined*. Formally, strategies can be defined as *trees* [Ke95]. However, the concept of a strategy can be simplified. For the players it is sufficient to know the current initial segment $R\restriction r_n$, as defined above. The information which parts of the current initial segment were played by which player is not relevant for the strategy. So strategies for the players are just proper extension functions (i.e. extension functions f with $x \sqsubset f(x)$ for all strings x).

Definition 4.1 *Let* $\mathbf{C} \subseteq 2^\omega$. *A* strategy *for Player I or Player II in the* Banach-Mazur game $BM(\mathbf{C})$ *for* $\mathbf{C}$ *is a proper extension function. The outcome of the play of the game where the players use the strategies* g *and* h*, respectively, is the unique set* $R = R(g,h) \in 2^\omega$ *defined by the initial segments:*

$$R\restriction r_n = (h \circ g)^n(\lambda)$$

Player I wins *if* $R \in \mathbf{C}$*, otherwise Player II* wins. *The strategy* h *of Player II is a* winning strategy *if* $R(g,h) \notin \mathbf{C}$ *for all strategies* g *of Player I.*

The main theorem for Banach-Mazur games shows the connection between the determinacy of a game and Baire category.

Theorem 4.2 *(Banach and Mazur, Oxtoby) Player II has a winning strategy for* $BM(\mathbf{C})$ *if and only if* $\mathbf{C}$ *is meager.*

For a proof see e.g. [Ox57]. Mazur stated one direction of this theorem (namely, if $\mathbf{C}$ is meager, then Player II has a winning strategy) first in the *Scottish Book* (see [Mau81], Problem 43). Banach observed that the converse implication holds, too, but his proof was never published.

In order to characterize effective Baire category, the classical version of the game has to be modified. While the role of Player I is unchanged, Player II has to choose his moves effectively. In terms of strategies this can be expressed by requiring that this player uses an effective extension strategy. Lisagor [Li81] introduced this modified version of the game for the Baire space ω^ω and, subsequently, Lutz [Lu90] and Fenner [Fe91] did the same for the Cantor space.

Definition 4.3 *Given a class* $\mathbf{C} \in 2^\omega$, $BM(\mathbf{C}, \mathbf{REC})$ *denotes the* effective Banach-Mazur game *for the class* $\mathbf{C}$. *This game differs from the classical Banach-Mazur game for* $\mathbf{C}$ *in requiring Player II to use an effective extension function as a strategy.*

The main theorem on Banch-Mazur games carries over to the effective case:

Theorem 4.4 *(Lisagor [Li81], Lutz [Lu90], Fenner [Fe91]) Player II has a winning strategy for* $BM(\mathbf{C}, \mathbf{REC})$ *if and only if* $\mathbf{C}$ *is effectively meager.*

A proof can be found in [Fe91].

For the above characterization of classical and effective Baire category in terms of Banach-Mazur games it is crucial that these category concepts are based on total extension functions. Hence, for a game theoretic characterization of the extended effective category concept which is based on partial extension functions, the game has to be extended.

The basic idea for the extended Banach-Mazur game introduced below is to allow Player II to cut back the current move of Player I. Without any restrictions, however, this will give Player II complete control over the outcome of the play. So the possibility of Player II for cutting Player I's move is confined by making Player II pay for the cut: In any round where Player II does not cut he receives one token and in any round in which he chooses to cut he has to pay back one token. In particular, Player II cannot cut the first move of Player I in round 0 since his start capital is 0.

Hence a strategy for Player II does not only consist of an extension function but it also has to provide a decision procedure telling Player II whether he should cut and, if so, how much of Player I's move is to be cut. This decision should be based on the values of the possible cuts in relation to the current capital of the player. We formalize this by defining a weight function $h^w : 2^{<\omega} \to \omega$ which assigns a weight to any initial segment and, in round $n+1$, we let Player II cut the move x_{n+1} of Player I to the longest string $x \sqsubseteq x_{n+1}$ such that the weight w of $(R{\upharpoonright}r_n)\hat{}x$ is minimal provided that the current capital of Player II exeeds w. The motivation for these rules is as follows: The weight function is chosen in a way that the lower the weight of an extension the more desirable this extension is for Player II. Hence, if Player II cuts a move, he will always choose a cut with the lowest possible weight. Moreover, if in some round all possible cuts have relatively high weights, whence the advantage of a cut is relatively low, Player II will spend a token for a cut only if his capital is relatively high. This will help to avoid the situation that, in some round of the play, there is the chance for a cut of low weight, but Player II has no tokens left for performing the cut, since he spent all of them on cuts of higher weights in the preceeding rounds.

By further requiring that the strategy of Player II is effective we obtain the effective extended Banach-Mazur game which can be formally defined in terms of strategies as follows.

Definition 4.5 *For any class* $\mathbf{C}$ *the* effective extended Banach-Mazur game $EBM(\mathbf{C}, \mathbf{REC})$ *is defined as follows. A* strategy g *for Player I is a proper extension function while a* strategy h *for Player II is a pair* $h = (h^e, h^w)$

where $h^e : 2^{<\omega} \to 2^{<\omega}$ *is a proper effective extension function and* $h^w : 2^{<\omega} \to \omega$ *is a total recursive function* (weight function). *The* result *or* outcome $R = R(g,h)$ *of the play of* $EBM(\mathbf{C}, \mathbf{REC})$ *in which the players use the strategies* g *and* h*, respectively, is the unique set* R *defined by the initial segments* $R\restriction r_n$ *describing the outcomes of the play up to round* n*. Here* $R\restriction r_n$ *and the* current capital $c(n)$ *of Player II at the end of round* n *are inductively given by:*

$$R\restriction r_0 = h^e(g(\lambda)) \quad \textit{and} \quad c(0) = 1$$

and, for $n \geq 1$, $L = \{|R\restriction r_n|, \ldots, |g(R\restriction r_n)|\}$,

$$R\restriction r_{n+1} = h^e(g(R\restriction r_n)\restriction l)$$

where l *is the greatest number in* L *such that*

$$(\forall l' \in L)(h^w(g(R\restriction r_n)\restriction l) \leq h^w(g(R\restriction r_n)\restriction l')) \textit{ and}$$
$$h^w(g(R\restriction r_n)\restriction l) < c(n)$$

if such a number exists, $l = |g(R\restriction r_n)|$ *otherwise, and*

$$c(n+1) = \begin{cases} c(n) - 1 & \textit{if } l < |g(R\restriction r_n)| \\ c(n) + 1 & \textit{otherwise} \end{cases}$$

Player II wins *the play if* $R(g,h) \notin \mathbf{C}$ *and* h *is a* winning strategy *for Player II if by using strategy* h *Player II wins all plays no matter what strategy* g *Player I employs.*

Our main result shows that this extended game concept characterizes the e.e. category concept.

Theorem 4.6 *For any class* $\mathbf{C}$ *the following are equivalent.*

(i) Player II has a winning strategy for $EBM(\mathbf{C}, \mathbf{REC})$.

(ii) $\mathbf{C}$ *is extended effectively meager.*

Proof. For a proof of *(i)* $\Rightarrow$ *(ii)* let $h = (h^e, h^w)$ be a winning strategy of Player II for $EBM(\mathbf{C}, \mathbf{REC})$. It suffices to define an effective system of partial extension functions f such that $\mathbf{C} \subseteq \mathbf{M}(f)$. Let the system f consist of the following components, where $k, l \in \omega$:

$$f_l^k(x) = \begin{cases} h^e(x) & \text{if } |x| > l \,\&\, h^w(x) = k \\ \uparrow & \text{otherwise} \end{cases}$$

$$f_l^\infty(x) = \begin{cases} h^e(x) & \text{if } |x| > l \\ \uparrow & \text{otherwise} \end{cases}$$

Now, to show that $\mathbf{C} \subseteq \mathbf{M}(f)$, fix $A \notin \mathbf{M}(f)$. It suffices to define a strategy g for Player I such that $R(g,h) = A$. For the definition of g distinguish the

following two cases:
Case 1: $(\exists k)\,(\overset{\infty}{\exists}\, n)\,(h^w(A\restriction n) = k)$.
Let k_0 be the least such k and fix l_0 such that $h^w(A\restriction n) \geq k_0$ for $n \geq l_0$. Then $f_l^{k_0}$ is dense along A for all $l \geq 0$ whence, by $A \notin \mathbf{M}(f)$, A meets all these extension functions. So for $x \sqsubset A$ we may fix $n = n(x)$ minimal such that $f^{k_0}_{\max(|x|,l_0)}(A\restriction n(x)) \sqsubset A$. Note that $x \sqsubset A\restriction n(x)$, $h^w(A\restriction n(x)) = k_0$ and $h^e(A\restriction n(x)) = f^{k_0}_{\max(|x|,l_0)}(A\restriction n(x))$. Now define g as follows: For $x \sqsubset A$ let $g(x) = A\restriction n(x)$; otherwise let $g(x) = x\hat{\ }0$. To show that $R(g,h) = A$, by induction on m show that, for $R = R(g,h)$, $R\restriction r_m \sqsubset A$. Since Player II cannot cut the first move of Player I,

$$R\restriction r_0 = h^e(g(\lambda))$$

and, by definition of g,

$$h^e(g(\lambda)) = h^e(A\restriction n(\lambda)) = f_{l_0}^{k_0}(A\restriction n(\lambda)) \sqsubset A$$

and $r_0 > l_0$. For the inductive step let $x = R\restriction r_m$ and assume that $x \sqsubset A$. Then $g(R\restriction r_m) = A\restriction n(x)$, $h^w(A\restriction n(x)) = k_0$ and, by $r_m \geq r_0 > l_0$, $h^w(y) \geq k_0$ for all strings y with $R\restriction r_m \sqsubseteq y \sqsubseteq g(R\restriction r_m)$. Hence, Player II will not cut the move of Player I in round $m+1$, whence

$$R\restriction r_{m+1} = h^e(g(R\restriction r_m)) = f^{k_0}_{\max(|x|,l_0)}(A\restriction n(x)) \sqsubset A.$$

Case 2: Otherwise.
Then there is a nondecreasing function $l : \omega \to \omega$ such that, for any $k \in \omega$, $l(k) > k$ and $h^w(A\restriction n) > k$ for all $n \geq l(k)$. Since the functions f_l^∞ are dense along all sets, for $x \sqsubset A$ we can fix a number $n(x) > |x|$ such that $f^\infty_{l(|x|)}(A\restriction n(x)) \sqsubset A$. Define the extension function g by letting $g(x) = A\restriction n(x)$ if $x \sqsubset A$ and $g(x) = x\hat{\ }0$ otherwise. To show that $R(g,h) = A$, by induction on m show that, for $R = R(g,h)$, $R\restriction r_m \sqsubset A$ and $r_m > l(m)$. For $m = 0$ this follows as in Case 1. So fix m, let $x = R\restriction r_m$ and assume that $R\restriction r_m \sqsubset A$ and $r_m > l(m)$. Then $g(x) = A\restriction n(x)$ and, by $r_m > l(m)$, $h^w(y) > m$ for all y with $x \sqsubseteq y \sqsubseteq A\restriction n(x)$. So Player II cannot cut this move since $c(m) \leq m$. It follows that

$$R\restriction r_{m+1} = h^e(g(x)) = f^\infty_{l(|x|)}(A\restriction n(x)) \sqsubseteq A$$

and, since $|x| \geq r_m \geq m+1$, $r_{m+1} > n(x) > l(|x|) \geq l(m+1)$ by definition of $n(x)$.

For a proof of *(ii)* $\Rightarrow$ *(i)* assume that $\mathbf{C}$ is e.e. meager, say $\mathbf{C} \subseteq \mathbf{M}(f)$ for an effective system of partial extension functions f. It suffices to give a winning strategy $h = (h^e, h^w)$ for Player II in the game $EBM(\mathbf{M}(f), \mathbf{REC})$. Note that for $A \in \mathbf{M}(f)$ there is an index k sucht that f_k is dense along A and A avoids f_k. So, if for the outcome $R = R(g,h)$ of some play, f_k is dense along R then we have to ensure that $f_k(R\restriction n) \sqsubset R$ for some number n.

This is achieved by ensuring that if $f_k(R\upharpoonright n)$ is defined infinitely often then in some round the extension function f_k will be simulated. Here smaller indices are given higher priority i.e. will be simulated first (in case of a conflict).

Roughly speaking this is achieved by letting $h^w(x) = k$ and $h^e(x) = f_k(x)$ for the least index k such that $f_k(x)\downarrow$ and f_k is not yet met by x, i.e. $f_k(y) \not\sqsubseteq x$ for all $y \sqsubseteq x$ with $f_k(y)\downarrow$. The actual definition of h is somewhat more involved, however, since by the above procedure a strategy $f_{k'}$ with $k' < k$, which is not defined on x but on some y with $x \sqsubset y \sqsubset f_k(x)$, might be dense along R but will possibly never be met. This problem is avoided by the following inductive search for k.

Given a string x, inductively define $h_k^w(x)$ and $h_k^e(x)$ for $k = |x| + 1, |x|, \ldots, 0$ as follows and let $h^w(x) = h_0^w(x)$ and $h^e(x) = h_0^e(x)$. For $k = |x| + 1$ let $h_k^w(x) = k$ and $h_k^e(x) = x\hat{\ }0$. Given $k \leq |x|$, $h_{k+1}^w(x)$ and $h_{k+1}^e(x)$, we say that k *requires attention at* x if

$$(4.1) \qquad (\forall y \sqsubseteq x)\,(f_k(y)\downarrow \Rightarrow f_k(y) \not\sqsubseteq x)$$

$$(4.2) \qquad (\exists z)\,(x \sqsubseteq z \sqsubseteq h_{k+1}^e(x)\ \&\ f_k(z)\downarrow)$$

If k requires attention at x fix the least z as above and let $h_k^w(x) = k$ and $h_k^e(x) = f_k(z)$. Otherwise, let $h_k^w(x) = h_{k+1}^w(x)$ and $h_k^e(x) = h_{k+1}^e(x)$. This completes the definition of the strategy h.

For the remainder of the proof fix a strategy g for Player I and let $R = R(g,h)$. To show that $R \notin \mathbf{M}(f)$, we have to show that for every index k such that f_k is dense along R, R will meet f_k. This will be established by the following claims.

We say that k *requires attention in round* $m+1$ if there is a string x with $R\upharpoonright r_m \sqsubseteq x \sqsubseteq g(R\upharpoonright r_m)$ and k requires attention at x; and k *receives attention in round* $m+1$ if there is such a string x and $R\upharpoonright r_{m+1} = f_k(x) = h^e(x)$.

Claim 1. If k receives attention in round m then R meets f_k and k does not require attention after round m.

Proof. Straightforward.

Claim 2. Every k requires attention in at most finitely many rounds.

Proof. For a contradiction fix k minimal which requires attention infinitely often. Then, by Claim 1, k will never receive attention. Fix m_0 such that no $k' < k$ requires attention after round m_0 and fix $m_1 > m_0 + k$ minimal such that k requires attention in round m_1. Note that if in some round m the move of Player I is cut back to a string x with $h^w(x) = k'$ then k' receives attention in this round. So, for m with $m > m_0$, either a number $k' > k$ receives attention in round m, whence $c(m) \geq c(m-1) - 1 > k' - 1 \geq k$, or $c(m) = c(m-1) + 1$. Since $m_1 > m_0 + k$ it follows that $c(m_1 - 1) \geq k$. Hence k will receive attention in round m_1 contrary to Claim 1.

Claim 3. If f_k is dense along R then R meets f_k.

Proof. For a contradiction assume that f_k is dense along R and R does not meet f_k. By Claim 2 fix $m_0 > k$ such that no k' with $k' \leq k$ requires attention after round m_0, and, by density, fix $m > m_0$ such that $f_k(z)\downarrow$ for some string z with $R\lceil r_m \sqsubseteq z \sqsubset R\lceil r_{m+1}$. Since R avoids f_k, it follows that k requires attention at z and $h^w(z) \leq k$. On the other hand, by choice of m_0 and by definition of the strategy h, there is a number $k' > k$ and a string x with $R\lceil r_m \sqsubseteq x \sqsubseteq g(R\lceil r_m)$ such that

$$h^w(x) = h^w_{k+1}(x) = h^w_{k'}(x) = k' \text{ and } R\lceil r_{m+1} = h^e(x) = h^e_{k+1}(x) = h^e_{k'}(x).$$

Since Player II cuts back the move of a Player I to a string of lowest possible weight, it follows that

$$R\lceil r_m \sqsubseteq x \sqsubseteq g(R\lceil r_m) \sqsubset z \sqsubseteq h^e(x) = R\lceil r_{m+1}.$$

Since $h^e(x) = h^e_{k+1}(x)$, by definition of the strategy h this implies that k requires attention at x whence $h^w(x) \leq h^w_k(x) = k$ contrary to $h^w(x) = k'$. □

5 Resource-Bounded Category

Resource-bounded Baire Category concepts have been introduced and studied e.g. in [AFH88], [Lu90], [Fe91], [Fe95] and [Am96]. Most of these concepts can be defined in terms of (partial) extension functions though some of them were originally defined in terms of condition sets. In general, time and space bounds were considered and frequently the interest focused on polynomial time computable extension functions which, by $|A\lceil n| = n \approx 2^{|n|}$, correspond to diagonalizations over exponential time objects, i.e., diagonalizations over the class $\mathbf{E} = \mathbf{DTIME}(2^{lin})$. The concept of Lutz [Lu90] is the analog of Mehlhorn's concept in this setting. It is based on total extension functions comparable within the given resource bound. As Fenner [Fe91] observed this concept is quite weak. Namely, the resource bounds implicitly impose bounds on the length of the extensions. Hence, for instance polynomial-time extension functions are too short for diagonalizing over polynomial-time reductions (see [Fe91], [Am96]). Fenner [Fe91] proposed a stronger category concept overcoming these limitations in part which is based on (totally defined) extension functions which only specify a part of the extension. For obtaining a full analog of effective category in the complexity setting, however, one has to consider extension functions which are *locally* computable within the given resource bound (whence the resource bound does not bound the length of the extension). This was independently observed by Fenner [Fe95] and Ambos-Spies [Am96]. In particular, for the total locally polynomial-time computable extension functions introduced in

[Fe95], the corresponding p-category concept has a characterization by Σ_2^0-classes corresponding to Theorem 2.3: A class $\mathbf{C}$ is p-meager iff there is a Σ_2^0-class $\mathbf{D}$ and a set $A \in \mathbf{E}$ such that $\mathbf{C} \subseteq \mathbf{D}$, $\mathbf{D}$ is c.f.v. and $A \notin \mathbf{D}$ (see [Fe95] and [Am96]). In [Am96] the resource-bounded version of e.e. category based on partial extension functions is introduced. For a comparison of all these concepts and their relations to the fundamental diagonalization techniques in complexity theory we refer to [Am96].

Lutz [Lu90] and Fenner [Fe91], [Fe95] gave game theoretical characterizations of their concepts based on resource bounded variants of the Banach-Mazur games. The papers [AFH88] and [Am96], where resource-bounded category concepts based on partial extensions were introduced, focus on generic sets, i.e. sets G such that the singleton $\{G\}$ is not nowhere dense or, equivalently, not meager (in the corresponding setting).Though in [Am96] it is pointed out how meagerness can be defined in terms of genericity, meager classes were not explicitely studied. In particular no game theoretical characterization was given. Reimann [Re97] has shown, however, that the extended Banach-Mazur game introduced here can be adapted to describe the category concepts of [AFH88] and [Am96].

References

[Am96] K. Ambos-Spies, Resource-bounded genericity, in: Computability, Enumerability, Unsolvability (S.B. Cooper et al., Eds.), London Mathematical Society, LNS 224, 1996, 1-59, Cambridge University Press.

[AFH88] K. Ambos-Spies, H. Fleischhack, and H. Huwig, Diagonalizing over deterministic polynomial time, in: Proc. CSL '87, Lecture Notes Computer Science 329, 1988, 1-16, Springer Verlag.

[BGS75] T. Baker, J. Gill, and R. Solovay, Relativizations of the P=?NP question, SIAM J. Computing 5, 1975, 431-442.

[BS85] J. L. Balcázar and U. Schöning, Bi-immune sets for complexity classes, Mathematicl Systems Theory 18, 1985, 1-10.

[Fe91] S. A. Fenner, Notions of resource-bounded category and genericity, in: Proc. 6th Structure in Complexity Theory Conference, 1991, 196-212, IEEE Comput. Soc. Press.

[Fe95] S. A. Fenner, Resource-bounded Baire category: a stronger approach, in: Proc. 10th Structure in Complexity Theory Conference, 1995, 182-192, IEEE Comput. Soc. Press.

[GS53] D. Gale and F. M. Stewart, Infinite games of perfect information, Annals of Math. Studies 28,1953, 245-266.

[Ke95] A. S. Kechris, Classical Descriptive Set Theory, 1995, Springer Verlag.

[Li81] L. R. Lisagor, The Banach-Mazur Game, Mathematics of the USSR Sbornik 38, 1981, 201-216.

[Lu90] J. H. Lutz, Category and measure in complexity classes, SIAM J. Comput. 19, 1990, 1100-1131.

[Mau81] R. D. Mauldin, The Scottish Book: Mathematics from the Scottish Café, 1981, Birkhauser Verlag.

[May94] E. Mayordomo, Almost every set in exponential time is P-bi-immune, Theoretical Computer Science 136, 1994, 487-506.

[Me73] K. Mehlhorn, On the size of sets of computable functions, in: Proc. 14th IEEE Symp. on Switching and Automata Theory, 1973, 190-196.

[Od89] P. Odifreddi, Classical Recursion Theory, 1989, North-Holland.

[Ox57] J. C. Oxtoby, The Banach-Mazur game and Banach Category Theorem, in: Contributions to the theory of games, Vol. III, Annals of Math. Studies 39, 1957, 159-163.

[Ox80] J. C. Oxtoby, Measure and Category, 1980, Springer Verlag.

[Re97] J. Reimann, Topologische Spiele und resourcen-beschränkte Baire Kategorie, Diplomarbeit, Universität Heidelberg, 1997.

[Ro67] H. Rogers, Theory of Recursive Functions and Effective Computability, 1967, McGraw Hill.

Weak presentations of computable partial orderings

Marat M. Arslanov*
Kazan University, Kazan, Russia

Iskander Sh. Kalimullin†
Kazan University, Kazan, Russia

1 Introduction

A partial ordering $\mathcal{L} = \langle M, <_L \rangle$ with an underlying set M is computable (or recursive) if M is finite or $\langle M, <_L \rangle \simeq \langle \omega, <_{L'} \rangle$ with a recursive predicate L' (i. e. the following function $L'(x, y)$ is recursive: $L'(x, y) = 1$, if $x <_{L'} y$, and $L' = 0$, otherwise). Following [JSh] we define a weak presentation of a countable partial ordering $\langle M, <_L \rangle$ as a partial ordering $\langle P, <_{L'} \rangle$, $\langle M, <_L \rangle \simeq \langle P, <_{L'} \rangle$, with an underlying set $P \subset \omega$ and the *potentially recursive* predicate L' (L' is potentially recursive if it has a (total) recursive extension). In this case we also say (see [JSh]) that the set P *supports* L.

In this paper we will consider weak presentations $\langle P, <_{L'} \rangle$ of computable partial orderings $\langle M, <_L \rangle$ with r. e. underlying sets P. It is easy to see that if a partial ordering $\langle M, L \rangle$ has a weak presentation $\langle P, <_{L'} \rangle$ with an r. e. underlying set P then $\langle M, <_L \rangle$ should be a computable partial ordering. Indeed, define for all $i, j \in \omega$ $i <_{L''} j$ if and only if $a_i <_{L'} a_j$, where $a_0, a_1, \ldots$ is a recursive enumeration of P without repetition (without loss of generality we may suppose that P is infinite). Now, obviously, $\langle M, <_L \rangle \simeq \langle \omega, <_{L''} \rangle$, and L'' is a recursive predicate.

On the other hand, in Theorem 3.1 we show that there is a noncomputable partial ordering which has a weak presentation with an underlying co- r. e. set D. Therefore, in this sense the notion of weak presentation of a partial ordering generalizes the notion of computable partial ordering.

We study computable partial orderings which have or which have not weak presentations with r. e. underlying sets. It is shown that any com-

*Partially supported by Russia Foundation of Fundamental Investigations Grant 96-01-00830 and by European Network "Complexity, Logic and Recursion Theory".

†Partially supported by Russia Foundation of Fundamental Investigations Grant 96-01-00830.

putable linear ordering L is supported by any infinite r. e. set A, and there is a computable partial ordering L such that for any r. e. degree $\mathbf{a} > \mathbf{0}$ there is an r. e. set A of degree $\mathbf{a}$ which does not support L. Further, we find a sufficient condition for a computable partial ordering to have a weak presentation in each r. e. degree.

Our notation is generally standard and follows [Soare]. We sometimes identify partial orderings with their characteristic functions, thus writing for instance $L(x,y) = 1$ for $x <_L y$ and $L(x,y) = 0$ for $x \not<_L y$. Therefore, sometimes we write $\langle A, L\rangle$ instead $\langle A, <_L\rangle$. Given any set A let $A \upharpoonright x = \{y \in A : y < x\}$. For a function f and a set A $f \upharpoonright A$ means the restriction of f to the domain A.

The first author wishes to thank Saleh Aliyari (IPM, Tehran) for his contribution to an earlier version of Theorem 3.8.

2 Negative results

We begin with the following result which is similar to a Proposition from [JSh,p.200].

Proposition 2.1 *There is a recursive lattice $\mathcal{L} = \langle M, \cup, \cap\rangle$ such that no simple set supports it.*

Proof The proof uses the above mentioned Proposition from [JSh]. Let $\mathcal{L}$ be the lattice of Diagram 1 with the underlying set $M = \{a, b, c, d, e, f, x_0, \ldots\}$.

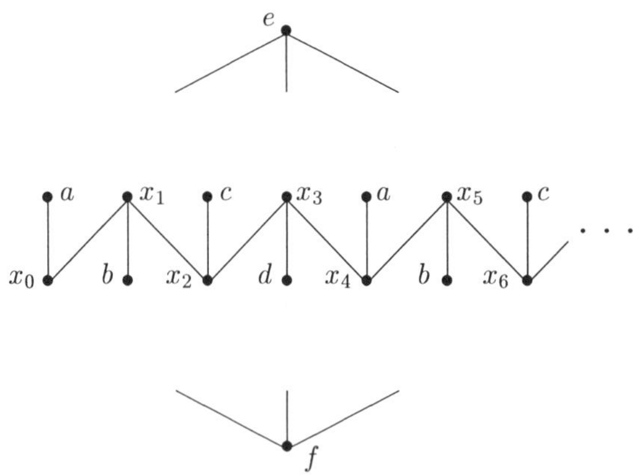

Diagram **1**

Here e is the greatest, f is the least elements of the lattice, a, b, c, d are pairwise incomparable, all other relations are shown in the Diagram.

Obviously, $\mathcal{L}$ is a recursive lattice. For any $i \geq 0$ we have

$$x_{i+1} = \begin{cases} x_i \cup b, & \text{if } a > x_i, \\ x_i \cap c, & \text{if } b < x_i, \\ x_i \cup d, & \text{if } c > x_i, \\ x_i \cap a, & \text{if } d < x_i. \end{cases}$$

Suppose some simple set S supports the lattice $\mathcal{L} = \langle M, \cup, \cap \rangle$. This means that there are recursive functions $p(x,y)$ and $q(x,y)$ such that $L \simeq \langle S, p \restriction S, q \restriction S \rangle$. Let $\tilde{a}, \tilde{b}, \tilde{c}, \tilde{d}, \tilde{e}, \tilde{f}$ members of S corresponding a, b, c, d, e, f of the lattice $\mathcal{L}$.

Let g be the following recursive function:

$$g(x) = \begin{cases} p(\tilde{b}, x), & \text{if } \tilde{a} = p(\tilde{a}, x), \tilde{b} \neq \tilde{q}(\tilde{b}, x), \tilde{c} \neq p(\tilde{c}, x), \tilde{d} \neq q(\tilde{d}, x), \\ q(\tilde{c}, x), & \text{if } \tilde{b} = q(\tilde{b}, x), \tilde{a} \neq p(\tilde{a}, x), \tilde{c} \neq p(\tilde{c}, x), \tilde{d} \neq q(\tilde{d}, x), \\ p(\tilde{d}, x), & \text{if } \tilde{c} = p(\tilde{c}, x), \tilde{a} \neq p(\tilde{a}, x), \tilde{b} \neq q(\tilde{b}, x), \tilde{d} \neq q(\tilde{d}, x), \\ q(\tilde{a}, x), & \text{if } \tilde{d} = q(\tilde{d}, x), \tilde{a} \neq p(\tilde{a}, x), \tilde{b} \neq q(\tilde{b}, x), \tilde{c} \neq p(\tilde{c}, x), \\ x, & \text{otherwise.} \end{cases}$$

It is clear that $\langle \omega, s(x) \rangle \simeq \langle \tilde{S}, g \restriction \tilde{S} \rangle$, where $\tilde{S} = S - \{\tilde{a}, \tilde{b}, \tilde{c}, \tilde{d}, \tilde{e}, \tilde{f}\}$ is a simple set. This contradicts to the Proposition from [JSh]. □

Theorem 2.2 *There is a recursive partial ordering $\mathcal{L}$ such that for any r. e. nonrecursive set A there is an r. e. set $B \equiv_{\mathrm{T}} A$ which does not support $\mathcal{L}$.*

Proof Let $\mathcal{L} = \langle M, <_L \rangle$ be the partial ordering of Diagram 2 with the underlying set $M = \{t, a_0, a_1, a_2, \ldots\}$.

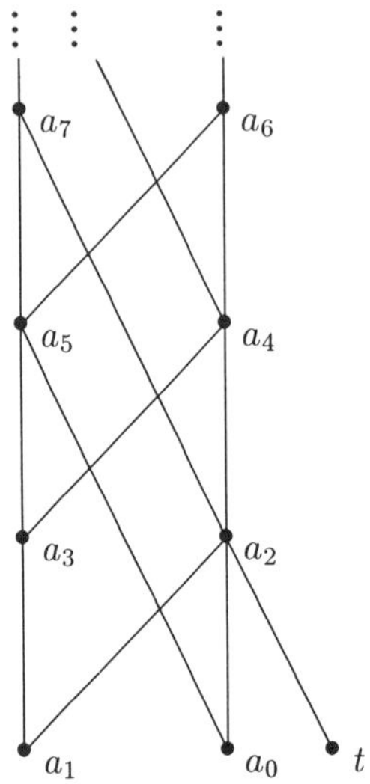

Diagram **2**

Here t, a_0, a_1 are pairwise incomparable elements of $\mathcal{L}$. All other relations are shown in the Diagram.

We need the following

Lemma 2.3 *Let $\langle S, <_P \rangle$ be a weak presentation of $\mathcal{L}$ with the underlying r. e. set S, and $\mathcal{L} \simeq \langle S, <_P \rangle$ via $1-1$-partial function $f : M \to S$. Then there is an effective enumeration $\{s_0, s_1, s_2, \ldots\}$ of S such that $s_0 = f(t)$, and $s_{i+1} = f(a_i)$ for $i > 0$. Moreover, this effective enumeration $\{s_0, s_1, s_2, \ldots\}$ of S can be found effectively by an effective enumeration of S and an index of a recursive function h, which is the extension of the partial ordering $<_P$.*

Proof Let $h : \omega^2 \to \omega$ be a (total) recursive function such that for all $x, y, (x, y) \in domP \to h(x, y) = P(x, y)$. First note that there are exactly two triples $\tau_0 = \{t, a_0, a_1\}$ and $\tau_1 = \{t, a_0, a_3\}$ of elements of $\mathcal{L}$ with pairwise incomparable elements and these two triples differ from each other by the condition $a_1 <_L a_3$. Besides, t is the only element of τ_0 which has two incomparable elements a_3 and a_5 from $\mathcal{L} - \tau_0$, and a_0 is the only element of $\tau_0 - \{t\}$ which has an incomparable element a_3 from $\mathcal{L} - \tau_0$.

Therefore, effectively enumerating elements of the set S and continously computing values of h at these enumerated elements we can determine elements $f(t), f(a_0), f(a_1)$. Further, the partial ordering $\mathcal{L}$ has following properties: for any $i \geq 0$ and any $x > 2i + 1$ $a_{2i+1} <_L a_x$, and for any a_{2i} there are only two incomparable with a_{2i} elements a_{2i+1} and a_{2i+3} from the set $\{a_x, x > 2i\}$, and also $a_{2i+1} <_L a_{2i+3}$. Further, a_{2i+2} is the unique element of the set $\{a_x, x > 2i\}$ which is incomparable with a_{2i+3}. It follows that we can determine all other elements $f(a_2), f(a_3), \ldots$ enumerating S and computing all possible values of h at these elements. □

Now we prove the Theorem 2.2. We will construct an r. e. set D so that $B = A \oplus D$ has the desired properties, namely $D \leq_{\mathrm{T}} A$, and B does not support $\mathcal{L}$.

We ensure the condition $D \leq_{\mathrm{T}} A$ by a permitting argument. To satisfy the last property we meet the following requirements for all e,

$$S_e : \Phi_e \ total \ \& \ \langle M, <_L\rangle \simeq \langle B, <_{L'}\rangle for\ some\ partial\ ordering\ <_{L'}$$

$$\rightarrow \Phi_e \ is\ not\ an\ extension\ of\ L',$$

where $\{\Phi_e\}_{e\in\omega}$ is an enumeration of all partial- recursive functions.

For each $e \in \omega$ we set aside the set of numbers $R_e = \{2\langle e, x\rangle + 1 : x \in \omega\}$ from which to draw witnesses to satisfy S_e.

The basic strategy for S_e without the requirement $B \leq_{\mathrm{T}} A$ is the following. To attack S_e we choose an unused pair of witnesses n, m from R_e and wait for a stage s such that both $\Phi_e(n,m)$ and $\Phi_e(m,n)$ are defined (if this never happens then n, m are the witnesses to the success S_e). Now there are four possible cases:

Case 1) $\Phi_e(n,m) = 1$ and $\Phi_e(m,n) = 0$ (i. e. $n <_{\Phi_e} m$).

Case 2) $\Phi_e(n,m) = 0$ and $\Phi_e(m,n) = 1$ (i. e. $m <_{\Phi_e} n$).

Case 3) $\Phi_e(n,m) = 0$ and $\Phi_e(m,n) = 0$ (i. e. $n \not<_{\Phi_e} m$ and $m \not<_{\Phi_e} n$).

Case 4) Otherwise.

In Case 4 we enumerate both numbers n and m into B and satisfy the requirement S_e since in this case the function Φ_e can not be the extension of L'.

If Case 1 holds then let $\{A_s \oplus D_s\}_{s\in\omega}$ be the enumeration of the set $A \oplus D$, chosen by Lemma 2.3 under the assumption that $\langle A \oplus D, <_{L'}\rangle$ is the weak presentation of $\langle M, <_L\rangle$, and Φ_e is an extension of L'. Then

Step 1. Enumerate m into D,

Step 2. Wait for a stage $s' > s$ such that partial ordering $\langle\{x : x \in A_{s'} \oplus D_{s'} \& x <_{\Phi_e} m\}, <_{\Phi_e}\rangle \simeq \langle\{t, a_0, a_1, \ldots a_{i_m}\}, <_L\rangle$ for some a_{i_m}

(if there is no such stage s' then the requirement S_e is obviously satisfied).

Step 3. Enumerate n into $D_{s'+1}$ and meet the requirement S_e

Both remaining cases 2 and 3 are similar.

Now we have to ensure that $D \leq_T A$ through a permitting argument. We need A to permit m to enter D at step 1 as well as to permit n to enter D at step 3.

We make infinitely many attacks on S_e by an ω-sequence of "cycles", where each cycle k proceeds as follows:

Step 1. Choose an unused pair of witnesses $m_k, n_k \in R_e$ greater than any number mentioned thus far in the construction.

Step 2. Wait for a stage s at which both $\Phi_e(n_k, m_k)$ and $\Phi_e(m_k, n_k)$ are defined.

(If this never happens then n_k, m_k are the witnesses to the success S_e).

Again there are four possible cases:

Case 1) $\Phi_e(n_k, m_k) = 1$ and $\Phi_e(m_k, n_k) = 0$ (i. e. $n_k <_{\Phi_e} m_k$).

Case 2) $\Phi_e(n_k, m_k) = 0$ and $\Phi_e(m_k, n_k) = 1$ (i. e. $m_k <_{\Phi_e} n_k$).

Case 3) $\Phi_e(n_k, m_k) = 0$ and $\Phi_e(m_k, n_k) = 0$ (i. e. $n_k \not<_{\Phi_e} m_k$ and $m_k \not<_{\Phi_e} n_k$).

Case 4) Otherwise.

Step 3. Wait for $A \restriction k$ to change (at a stage s', say) and start cycle $k+1$ to run simultaneously with cycle k.

Step 4. Stop cycles $k' > k$. If at Step 2 the Case 4 holds then enumerate m_k and n_k into A and stop.

(Now we satisfied the requirement S_e since in this case the function Φ_e can not be the extension of L'.)

In Case 1 (cases 2 and 3 are similar) let $\{A_s \oplus D_s\}_{s\in\omega}$ be the enumeration of the set $A \oplus D$, chosen by Lemma 2.3 under the assumption that $\langle A \oplus D, <_{L'}\rangle$ is the weak presentation of $\langle M, <_L\rangle$, and Φ_e is an extension of L'. Then enumerate m_k into A and

Step 5. Wait for a stage $s'' > s'$ such that the partial ordering $\langle\{x : x \in A_{s''} \oplus D_{s''} \& x <_{\Phi_e} m_k\}, <_{\Phi_e}\rangle \simeq \langle\{t, a_0, a_1, \dots a_i\}, <_L\rangle$ for some a_i.

(If there is no such stage s'' then the requirement S_e obviously satisfied).

Step 6 Wait for $A \restriction k$ to change again (at a stage $s^\star$, say) and start cycle $k+1$ to run simultaneously with cycle k.

Step 7 Enumerate n_k into $D_{s^\star+1}$ and meet the requirement S_e.

The module has the following possible outcomes:

(A) Eventually each cycle k gets stuck at steps 3 or 6 waiting for a $A \upharpoonright k$–change. In this case A is recursive, contrary to hypothesis.

(B) Some (least) cycle k_0 gets stuck at step 2 or 5. Then we satisfy S_e through cycle k.

Therefore, either we were successful in satisfying S_e through the outcome (B), or we satisfied S_e through witnesses m_k, n_k at steps 4 or 7.

The explicit construction and the remaining parts of the proof are now straightforward. So we will not give them here. □

3 Positive results

Using the method of Theorem 2.2 we can easily construct a non- recursive partial ordering which is supported by a co-r. e. set. Let $\mathcal{L} = \langle \omega, <_L \rangle$ be a partial ordering of Diagram 3 with the underlying set $M = \{r, t, a_i, b_i, c_i : i \in \omega\}$. Here $\{r, t, a_i, b_i, c_i : i \in \omega\}$ is a recursive permutation of ω.

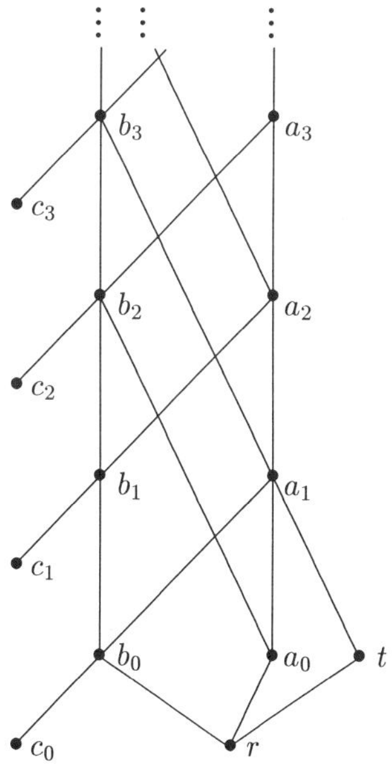

Diagram **3**

Obviously, the characteristic function L of $<_L$ is recursive.

As in Theorem 2.2 it is easy to see that if $\langle P, <_P \rangle$ is a weak presentation of $\mathcal{L}$ with the underlying r. e. set P, and $\mathcal{L} \simeq \langle P, <_P \rangle$ via $1-1$-partial function $f : M \to P$ then there is an effective enumeration $\{p_0, p_1, p_2, \ldots\}$

of P such that $p_0 = f(r), p_1 = f(t)$, and $p_{3i+2} = f(a_i), p_{3i+3} = f(b_i)$ and $p_{3i+4} = f(c_i)$ for $i \geq 0$. Again, the enumeration $\{p_0, p_1, \ldots\}$ of P can be found effectively by an effective enumeration of P and an index of a recursive function h, which is the extention of the partial ordering $<_P$.

We will construct an r. e. set A meeting for all $e \in \omega$ requirements

$$R_e : \Phi_e \ total \ \rightarrow \langle \omega, \Phi_e \rangle \not\simeq \langle \bar{A}, L \restriction \bar{A} \rangle.$$

Then, obviously, the partial ordering $\langle \bar{A}, L \restriction \bar{A} \rangle$ is not recursive but the co-r. e. set $\bar{A}$ supports it.

It will follow from the construction that we enumerate into A only elements of the set $\{c_i : i \in \omega\}$. Therefore all elements $t, r, a_i, b_i (i \in \omega)$ belong $\bar{A}$.

To meet the requirement R_e for a given e we keep c_e out of A until at a stage s we see that for some $D \subseteq \omega \restriction s$

$$\langle D, \Phi_{e,s} \restriction D \rangle \simeq \langle S_e, L \restriction S_e \rangle,$$

where $S_e = \{r, t, a_0, b_0, a_1, b_1, \ldots, a_e, b_e, c_e\}$.

If no such stage exists then R_e is automatically satisfied. If s exists then enumerate c_e into A_{s+1} satisfying R_e.

Combining this strategy by an obvious way with coding and permitting strategies we obtain the following

Theorem 3.1 *Let* $\mathbf{a} > \mathbf{0}$ *be a r. e. degree. There is an r.e. set* $A \in \mathbf{a}$ *and a noncomputable partial ordering* $\mathcal{L}$ *such that* $\omega - A$ *supports* $\mathcal{L}$.

Below in Theorem 3.8 we prove that any infinite r. e. set supports any infinite recursive linear ordering. We begin with the following

Proposition 3.2 *Let* α, β *and* γ *be recursive linear orderings,such that any infinite r. e. set supports* β. *Then any infinite r. e. set supports also* $\alpha+\beta+\gamma$.

Proof Let A be an infinite r. e. set. Without loss of generality we may suppose that A is not recursive. Let B_1, B_2 be recursive subsets of A such that $card\{B_1\} = card\{\alpha\}, card\{B_2\} = card\{\gamma\}, B_1 \cap B_2 = \emptyset$. Let $C = A - (B_1 \cup B_2)$ be an infinite r. e. set. By the condition C supports β. Therefore, there exists a recursive function g such that $\langle C, <_{g \restriction C} \rangle \simeq \beta$.

Since α, γ are recursive, there are partial recursive functions f_i, $i = 1, 2$, such that $dom f_i = B_i, i = 1, 2, \langle B_1, \leq_{f_1} \rangle \simeq \alpha, \langle B_2, \leq_{f_2} \rangle \simeq \gamma$.

We define

$$f(x,y) = \begin{cases} f_1(x,y), & \text{if } x,y \in B_1, \\ f_2(x,y), & \text{if } x,y \in B_2, \\ g(x,y), & \text{if } x,y \notin B_1 \cup B_2, \\ 1, & \text{if } x \in B_1, y \notin B_1, \\ 0, & \text{if } y \in B_1, x \notin B_1, \\ 1, & \text{if } y \in B_2, x \notin B_2, \\ 0, & \text{if } x \in B_2, y \notin B_2. \end{cases}$$

It is clear, that f is recursive and $\langle A, f \restriction A\rangle \simeq \alpha + \beta + \gamma$. □

Corollary 3.3 *Any infinite r. e. set supports $\omega + \omega^\star$.*

Proof Any infinite set supports $\langle \omega, < \rangle$. □

Corollary 3.4 *Let η be the ordering of rationals. Then any infinite r. e. set supports the ordering $\omega + (\omega^\star + \omega) \times \eta + \omega^\star$.*

Theorem 3.5 *Let $\mathcal{L} = \langle \omega, <_L \rangle$ be a recursive linear ordering without maximal (or minimal) elements. Then any infinite r. e. set A supports $\mathcal{L}$.*

Proof

Let $\mathcal{L}$ be a recursive linear ordering without maximal elements (for a recursive linear ordering without minimal elements the proof is similar).

Let A be an infinite r. e. set, $A = \cup A_s, A_s = \{a_0, a_1, \ldots, a_s\}$. We will define by stages a recursive relation $R = \bigcup_s R_s$ and a partial function $g(x)$, *dom* $g = A$, such

that $\langle \omega, <_L \rangle \simeq \langle A, R \restriction A \times A \rangle$ by the isomorphism $g : A \to \omega$. The finite part of g constructed to the end of stage s is denoted by g_s, so $g = \lim_s g_s$. We will sometimes write $x <_{R_s} y$ instead $R_s(x,y) = 1$ and $R_s(y,x) = 0$ accordingly. We say a number $x \in \omega$ is a *free number at stage s* if $x \notin$ *rang* g_s.

We define by induction numbers i_s as

$i_0 = 0$;
$i_{s+1} = \mu t\{t > i_s \ \&\ a_t > a_{i_s}\}$.

Obviously, $B = \{a_{i_s}\}_{s \in \omega}$ is an infinite recursive subset of A.

Stage $s = 0$. Define $a_0 <_{R_0} 0 <_{R_0} 1 <_{R_0} \cdots <_{R_0} a_0 - 1$ (i. e. $R_0(0,1) = 1, R_0(1,0) = 0, R_0(1,2) = 1, \ldots$ etc.).

Stage $s+1$. We consider two cases.

Case 1. $a_{s+1} \in B$. It follows $\forall t \leq s (a_t < a_{s+1})$.

1) Let x be a minimal free number at stage s. Define $g_{s+1}(a_t) = g_s(a_t)$ for all $t \leq s$, and $g_{s+1}(a_{s+1}) = x$.

2) Define $R_{s+1}(a_u, a_v) = L(g_{s+1}(a_u), g_{s+1}(a_v))$ for all $u, v \le s+1$. (Here L means the recursive characteristic function for $<_L$.)

3) For all $i, j \in (\omega - A_s) \restriction a_{s+1}$ such that $i < j$ define $i <_{R_{s+1}} j$.

4) For each $i \in (\omega - A_s) \restriction a_{s+1}$ define

   (a) $R_{s+1}(i, x) = R_{s+1}(a_t, x)$, $R_{s+1}(x, i) = R_{s+1}(x, a_t)$ for all $x \in A_{s+1} - \{a_t\}$ if there is a (least) number $a_t \notin B, t \le s$, such that $i < a_t$, and

   (b) $x <_{R_{s+1}} i$ for all $x \in A_{s+1}$ (i.e. $R_{s+1}(x, i) = 1$) if there is no such a_t.

   Note that in (a) and (b) $R_s \subseteq R_{s+1}$).

At this stage we defined $R : \omega \times \omega \to \{0, 1\}$ at $\omega \restriction a_{s+1} \times \omega \restriction a_{s+1}$.

Definitions (a) and (b) ensure transitiveness of the ordering $<_R$ on the set $\omega \restriction a_{s+1}$.

Case 2. In this case we have $a_{s+1} \notin B$. It follows $\exists t \le s(A_{s+1} \subseteq \omega \restriction a_t)$. Therefore, at stage $s+1$ the relation R is already defined at all elements of A_{s+1}.

Let R_{s+1} orders A_{s+1} by the following way:

$$(1) \qquad b_1 <_{R_{s+1}} \cdots <_{R_{s+1}} b_k <_{R_{s+1}} a_{s+1} <_{R_{s+1}} c_1 <_{R_{s+1}} \cdots <_{R_{s+1}} c_l,$$

where $b_1, \ldots, b_k, c_1, \ldots, c_l$ is a permutation of the elements of A_s.

We first choose a number x which is free at stage $s+1$ such that $g_s(c_l) <_L x$, and define

$g_{s+1}(b_i) = g_s(b_i)$ for all $i, 1 \le i \le k$, $g_{s+1}(a_{s+1}) = g_s(c_1), g_{s+1}(c_1) = c_2 \cdots g_{s+1}(c_{l-1}) = g_s(c_l), g_{s+1}(c_l) = x$.

End of the construction.

It follows from the construction that $R_s \subseteq R_{s+1}$ for all s, and $R = \bigcup_s R_s$ is defined on $\omega \times \omega$.

Let us first show, that for all $x \in A$ $\lim_s g_s(x)$ exists.
If $x = a_s$ then by the construction $\forall t \ge s\{g_t(a_s)\downarrow\}$. Let $t \ge s$ be first stage such that $A_t \restriction a_s = A \restriction a_s$. It follows from points (a) and (b) of Case 1 of the construction that $\forall u > t(a_u \notin B \to a_s <_R a_u)$. Therefore, after stage t for any $u > t$ such that $a_u \notin B$ $a_s = b_i$ for some b_i from (1), which implies $\forall u > t(g_u(a_s) = g_t(a_s))$.

It follows now that $g(x)$ is defined for any $x \in A$.

It is clear also that g is injective since so are all functions $g_s, s \in \omega$.

To prove that *rang* $g = \omega$ assume for the contradiction that for some (least) y we have $y \notin$ *rang* g.

Case 1) For all s $y \notin rang\ g_s$. Since B is infinite there is a stage s such that $a_s \in B$ and y is a least free at stage s number. Now by the construction we have $y = g_s(a_s)$. Contradiction.

Case 2) For some $s, t, t \leq s$, $y = g_s(a_t)$. Let $k > s$ be first stage such that $A \restriction a_t = A_k \restriction a_t$. It follows from the construction that there is an integer $c \in A_k$ such that $c <_R a_t$ and $y = g_k(c)$. Since $\forall u > k(a_u \notin B \to a_t <_R a_u)$, then $\forall u > k(a_u \notin B \to c <_R a_u)$. Therefore, $\forall u > k(g_u(c) = g_k(c) = y) \to g(c) = \lim_s g_s(c) = y \to y \in rang\ g$. Contradiction.

□

We now show that any recursive linear ordering has a weak presentation at any infinite r. e. set.

Definition 3.6 *An element x of the set M is an* upper (lower) condensation point *of the linear ordering $\mathcal{L} = \langle M, <_L \rangle$ if*

$$\exists y_0(y_0 <_L x) \ \&\ \forall y(y <_L x \to \exists z(y <_L z <_L x))$$

(accordingly

$$\exists y_0(x <_L y_0) \ \&\ \forall y(x <_L y \to \exists z(x <_L z <_L y)))$$

An element of M is a condensation point *if it is an upper or lower condensation point of $\mathcal{L}$.*

Corollary 3.7 *Let $\langle \omega, <_L \rangle$ be a recursive linear ordering with a condensation point. Then any infinite r. e. set supports the ordering $<_L$.*

Proof Let $x \in \omega$ be an upper condensation point (the same proof for the case when x be a lower condensation point), and let $C = \{y : y <_L x\}$.

Clearly, C is an infinite recursive set and the ordering $\langle C, <_L \rangle$ has no maximal points. It is obvious also that $\langle \bar{C}, <_L \rangle$ is a recursive ordering. It follows from Theorem 3.5 that any infinite r. e. set supports $\langle C, <_L \rangle$. Now by Proposition 3.2 any infinite r. e. set supports
$\langle \omega, <_L \rangle = \langle C, <_L \rangle + \langle \bar{C}, <_L \rangle$ □

Theorem 3.8 *Any infinite r. e. set supports any infinite linear recursive ordering α.*

Proof

By Theorem 3.5. and Corollary 3.7 it suffices to consider a linear ordering α with maximal and minimal elements and which does not contain condensation points. Therefore, for some α_1 $\alpha = \omega + \alpha_1 + \omega^\star$. If $\alpha_1 = \emptyset$, then the proof is obvious. Let x_1 is the minimal (for the natural ordering) element of α_1. Since x_1 is not a condensation point then $\alpha = \omega + \alpha_2 + \xi + \alpha_3 + \omega^\star$. Here ξ is the ordering of all integers $\omega^\star + \omega$, and $x_1 \in \xi$.

If either α_2 or α_3 is empty, then we stop the process. Otherwise, let x_2 and x_3 be minimal elements of α_2 and α_3 accordingly. (We choose x_2 and x_3 as minimal elements of α_2 and α_3 accordingly since we need all integers $x \in \omega$ to be considered.)

Again x_2 and x_3 are not condensation points. Therefore, $\alpha = \omega + \alpha_4 + \xi + \alpha_3 + \xi + \alpha_6 + \xi + \alpha_7 + \omega^\star$, and so on.

If this process is infinite, then $\alpha = \omega + \xi \times \eta + \omega^\star$ (since any linear dense ordering without endpoints is isomorphic to the ordering of rationals), and Corollary 3.4 finishes the proof.

If this process is finite, then for some i we have $\alpha_i = \emptyset$. Since $\xi = \omega^\star + \omega$ we have $\alpha = \beta + \omega + \alpha_i + \omega^\star + \gamma = \beta + \omega + \omega^\star + \gamma$ for some β and γ. Let x be a minimal (for the ordering $<_L$) element from the part ω, and y be a maximal element for the part $\omega^\star$. Then β is defined in the recursive set $\{z : z <_L x\}$ and γ is defined in $\{z : y <_L z\}$. Therefore, β and γ are recursive orderings. Now the proof of the theorem follows from Corollary 3.3 and Proposition 3.2. □

Theorem 3.9 *Let $\mathcal{L} = \langle \omega, <_L \rangle$ be a recursive partial ordering on ω such that*

$$(1) \qquad (\forall x)(\forall y)(\exists z)(x <_L z \ \& \ y <_L z).$$

Then for any r. e. degree **a** *there exists an r. e. set $A \in$* **a** *which supports $\mathcal{L}$.*

Proof

Let $\mathcal{L} = \langle \omega, <_L \rangle$ be a recursive partial ordering satisfying (1).

It follows obviously from (1) that for any finite set F there exists an integer z such that $(\forall x \in F)(x <_L z)$. Denote by L a recursive characteristic function of $<_L$, i. e. for all x, y

$$L(x,y) = \begin{cases} 1, & \text{if } x <_L y, \\ 0, & \text{otherwise.} \end{cases}$$

For any $i \in \omega$ define $[i]$ as the finite set $\{x : 2^i - 1 \leq x < 2^{i+1} - 1\}$. Then $\omega = \cup_{i \in \omega} [i]$. Let $[i] = \{a_{i,0} < a_{i,1} < \ldots < a_{i,2^i - 1}\}$.

We organize a linear ordering $<_R$ on ω using the following tree (see Diagram 4):

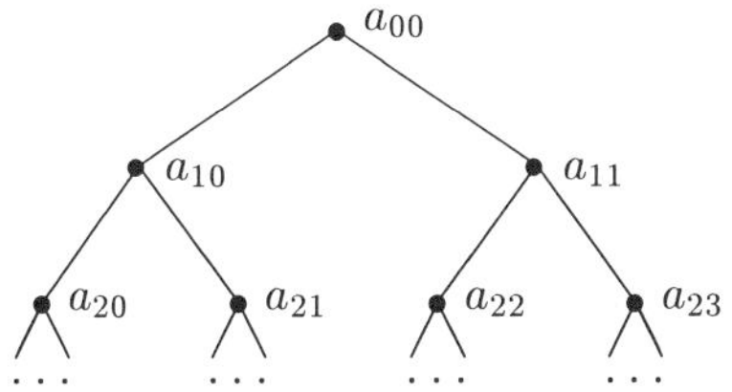

Diagram **4**

Each natural number belong to one and only one node $a_{i,j}, i \geq 0, 0 \leq j \leq 2^i - 1$, of the tree. For all $i, j \in \omega$ let $T^0_{i,j}$ and $T^1_{i,j}$ be subtrees which "grow" from the node $a_{i,j}$ of the tree (see Diagram 5).

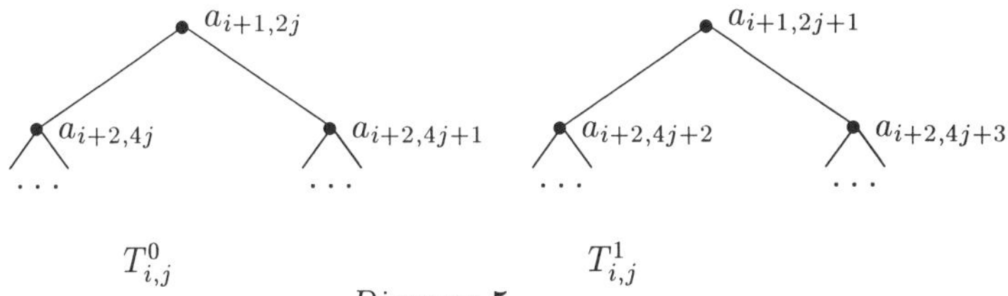

Diagram **5**

We define $x <_R a_{i,j}$ for all $x \in T^0_{i,j}$, and $a_{i,j} <_R y$ for all $y \in T^1_{i,j}$. Obviously, there is such a recursive linear ordering $<_R$ of ω. Moreover, note that for each i the set of nodes $\{a_{i,0}, a_{i,1}, \ldots, a_{i,2^i-1}\}$ are exactly the elements of $[i]$.

Let $A \in \mathbf{a}$ be an r. e. set. Fix an effective enumeration $\{A_s\}_{s\in\omega}$ of A such that for any $s \mid A_{s+1} - A_s \mid \leq 1$.

We will construct an r. e. set $B \equiv_T A$, a recursive function $P = \cup_s P_s$ and a partial-recursive function $g = \cup_s g_s$ such that g will provide an isomorphism between $\langle B \oplus \omega, P \restriction B \oplus \omega\rangle$ and $\langle \omega, <_L\rangle)$.

It will follow from the construction that

$$(2) \qquad \forall i(card\{B \cap [i]\} \leq 1)$$

For the condition $A \equiv_T B$ we will construct B meeting for all e requirements

$$S_i : i \in A \leftrightarrow [i] \cap B \neq \emptyset.$$

This immediately implies that $A \leq_T B$. To test whether $x \in B$ find i such that $x \in [i]$ and see whether $i \in A$. If $i \notin A$ then $x \notin B$ by S_i. If $i \in A$

then $[i] \cap B \neq \emptyset$. Using the enumeration of B we can obtain y such that $y \in [i] \cap B$. Now it follows from (2) that $x \in B$ iff $x = y$.

For all $x, y \in \omega$ we define $P_0(2x, 2y) = R(x, y)$. For the sake of convenience we will write $x <_P y$ instead $P(x,y) = 1$ and $P(y,x) = 0$. We say that $x \in \omega$ is *free* at stage s if $x \notin rang\ g$.

Stage $s = 0$. Define $B = \emptyset, g_0 = \emptyset$. We have $P(x,y) \downarrow \Rightarrow x, y$ even numbers.

Stage $s+1, s \geq 0$. Let x be the least free number. Let $g' = g_s \cup \{\langle 2s+1, x\rangle\}$. For all $t \in dom g_s$ define $P_{s+1}(t, 2s+1) = L(g_s(t), x), P_{s+1}(2s+1, t) = L(x, g_s(t))$. For all $2k \notin dom g_s$ define $2s+1 <_{P_{s+1}} 2k$.

If $A_{s+1} - A_s = \emptyset$ define $B_{s+1} = B_s, g_{s+1} = g'$. Otherwise, let $m \in A_{s+1} - A_s$. Let $n \in [m]$ be the $<_R$-maximal element of $[m]$ such that $\forall y < m(y \notin A_s \rightarrow \forall j \in [y](n <_R j))$.

Let $z \in \omega$ be the first integer such that $\forall k \in dom g'(g'(k) <_L z)$. Define $g_{s+1} = g' \cup \langle 2n, z\rangle, B_{s+1} = B_s \cup \{n\}$, and for all $q \in dom g', P_{s+1}(2n, q) = L(z, g'(q)), P_{s+1}(q, 2n) = L(g'(q), z)$.

In each case define $P_{s+1}(i,j) = P_s(i,j)$ for all $\langle i,j\rangle \in dom P_s$.

End of the construction.

It follows immediately from the construction that $g : B \oplus \omega \rightarrow \omega$ is a partial-recursive function (therefore, $g^{-1} : \omega \rightarrow B \oplus \omega$ is a recursive function) which provides $\langle B \oplus \omega, P \restriction B \oplus \omega\rangle \simeq \langle \omega, <_L\rangle$. Also it is obvious, that

$$\forall i(card\{B \cap [i]\} \leq 1),$$

and

$$i \in B \leftrightarrow [i] \cap B \neq \emptyset.$$

As already noticed this implies $B \equiv_T A$ and, therefore $C = B \oplus \omega \equiv_T A$. $\square$

4 Open questions

In this paper we have shown that any infinite r. e. set supports any infinite recursive linear ordering, but this is not true for recursive partial orderings: as is shown in Theorem 2.2 there exists a recursive partial ordering $<_L$ such that each r. e. degree contains an r. e. set which does not support $<_L$. On the other hand, as is shown in Theorem 3.9 if $<_L$ has the additional property $\forall x \forall y \exists z(x <_L z \& y <_L z)$ then each r. e. degree contains an r. e. set which supports $<_L$. Therefore, it is natural to ask the following

Question 4.1 *Does every infinite recursive partial ordering $<_L$ have a weak presentation in each r. e. degree (i. e., does for any infinite recursive partial ordering $<_L$ and any r. e. degree* **a** *there exist an r. e. set $A \in$* **a** *which supports $<_L$)?*

We may also ask

Question 4.2 *Do there exist a Turing degree* $\mathbf{a} > \mathbf{0}$ *such that if a partial ordering* $\mathcal{L}$ *has a weak presentation in a set* $A \in \mathbf{a}$ *then* $\mathcal{L}$ *is recursive?*

As follows from Theorem 3.1 this degree **a** (if it exists) cannot be recursively enumerable.

5 Bibliography

[Soare] Soare, R. I., *Recursively Enumerable Sets and Degrees*, Springer-Verlag, Berlin, 1987.

[JSh] Jockusch, C.G. Jr. and Shlapentokh, A., Weak presentations of computable fields, *J. Symb. Logic* **60**, 1995, 199-208.

Recursion Theory on Weak Fragments of Peano Arithmetic: A Study of Definable Cuts

C T Chong and Yue Yang

National University of Singapore

Classical recursion theory (CRT) is built on the axioms of Peano arithmetic P. Of these axioms, the induction scheme **Ind** provides both the source and the power to carry out highly sophisticated constructions (such as priority arguments), on the set of natural numbers, and to verify their correctness.

Now **Ind** may be decomposed into a hierarchy of induction schemes, in increasing order of complexity. Denote by $I\Sigma_n$ the induction scheme restricted to Σ_n formulas. Then **Ind** is equal to the union of $I\Sigma_n$ ($n = 0, 1, \ldots$). It is clear that each construction in CRT of a definable set on ω uses only $I\Sigma_n$ for some n, dictated, roughly speaking, by the number of alternating quantifiers involved therein. It follows that by assuming only fragments of **Ind**, it is possible to derive substantial number of results in CRT. From the global point of view, one is led to two general problems: (a) Classify the proof-theoretic strength of theorems and results in CRT in terms of fragments of the induction scheme; (b) Study recursion theory in non-standard models of fragments of P. It turns out that these two problems are closely related.

In this paper, by a weak fragment of Peano arithmetic, we mean an axiom system that assumes at most the $I\Sigma_2$ induction scheme. This is a weak system because it provides just the minimal environment to carry out infinite injury constructions (indeed something even less than $B\Sigma_2$ is quite sufficient if we work in a special model of $I\Sigma_1$ (See Section 2)). On the other hand it is strong enough to support a reasonable recursion theory, as we will see in the sequel.

This paper gives a survey of recursion theory on weak fragments of P. Our objectives are twofold: (a) To discuss results in CRT which are provable under a weak system, and to identify their proof-theoretic complexity wherever appropriate, and (b) To summarize the key ideas and techniques introduced into the subject over the past decade. The reader will notice that many of these had originated in ordinal recursion theory developed since the early 1960's. The fact that there is a tight link between recursion theory on ordinals and that on weak fragments of P is a fortuitous coincidence probably not anticipated by the founders of these two subjects. We believe that there is a two-way interaction between them, and expect to see more applications going in either direction.

We begin with some definitions and terminologies. Denote by P^- the Peano axioms minus the induction scheme, but adjoined with the exponential function to allow for coding of 'finite' sets. By the Σ_n bounding principle, written $B\Sigma_n$, we mean the following scheme: In a model of P^-, every Σ_n-function total on a proper initial segment (with a last point) has a bounded range. Kirby and Paris [11] showed that, for $n \geq 1$,

$$I\Sigma_{n+1} \Rightarrow B\Sigma_{n+1} \Rightarrow I\Sigma_n,$$

and that the reverse arrows fail. This shows that there is a finer hierarchy obtained by the interweaving of $B\Sigma_n$. We will see that this finer hierarchy also gives rise to 'finer' recursion theories.

The paper is organized as follows: In Section 1 we define the basic notions and notations. Section 2 concentrates on recursion theory under the hypothesis of $B\Sigma_1$, and discusses results (such as the Sacks Splitting Theorem) which are provably equivalent to $I\Sigma_1$ over this base theory. One conclusion to be drawn from results in this section is that finite injury priority arguments may be carried out essentially with only Σ_1-induction. In the absence of this, similar results (such as the Friedberg-Muchnik Theorem) may still be derived without recourse to a priority argument. In Section 3 we move up the ladder of complexity to consider recursion theory under $B\Sigma_2$. In this theory one can prove the Density Theorem. On the other hand, with $B\Sigma_2$ as the base theory, the existence of maximal sets and incomplete high r.e. degrees is equivalent to Σ_2-induction. In the final section we discuss some open problems.

1 Preliminaries

In a (non-standard) model $\mathcal{M}$ of P^-, a bounded set is *$\mathcal{M}$-finite* if it is coded (by an element). A set X in $\mathcal{M}$ is recursively enumerable (r.e.) if it is $\Sigma_1(\mathcal{M})$. X is recursive if both X and $\mathcal{M} \setminus \mathcal{X}$ are r.e. We define recursive functions (partial and total) in the obvious way. Assuming only Σ_0-induction on $\mathcal{M}$, it can be proved that there is a $\Sigma_1(\mathcal{M})$ enumeration $\{W_e | e \in \mathcal{M}\}$ of all r.e. sets. An analog of Church's thesis may be formulated in this setting as well. This will be used implicitly throughout the paper.

Given subsets A and B of $\mathcal{M}$, we say that A is *pointwise recursive in* B (written $A \leq_p B$) if there is an $e \in \mathcal{M}$ such that for all $x \in \mathcal{M}$,

$$x \in A \Leftrightarrow \exists K_1 \exists K_2 [(x, 0, K_1, K_2) \in W_e \;\&\; K_1 \subset B \;\wedge\; K_2 \cap B = \emptyset],$$

and

$$x \notin A \leftrightarrow \exists K_1 \exists K_2 [(x, 1, K_1, K_2) \in W_e \;\&\; K_1 \subset B \;\wedge\; K_2 \cap B = \emptyset],$$

where K_1 and K_2 are (codes of) $\mathcal{M}$-finite sets. We say that A is *strongly recursive in* B (written $A \leq_T B$, where 'T' stands for 'Turing') if the above holds upon replacing x by an $\mathcal{M}$-finite set. It is easy to verify that $\leq_T$ is a transitive relation, and hence gives rise to the notion of degrees of unsolvability as in CRT. However $\leq_p$ is not transitive in some model satisfying $I\Sigma_1$, as shown by Groszek and Slaman [8].

¿From here on all models $\mathcal{M}$ satisfy at least $B\Sigma_1$. A $B\Sigma_n$ *model* is a model of $B\Sigma_n$ but not $I\Sigma_n$. There is a 'topological' difference between models of $B\Sigma_n$ and models of $I\Sigma_n$, and this is exhibited through the existence of Σ_n *cuts*. A *cut* is a subset of $\mathcal{M}$ which is closed downwards and closed under the succession function (for example the set ω). For our purpose, all cuts considered are proper subsets of $\mathcal{M}$. A Σ_n-cut is a cut which is $\Sigma_n(\mathcal{M})$-definable. The following result, due to H. Friedman, characterizes $B\Sigma_n$ models.

Proposition 1.1 *Let $\mathcal{M}$ be a model of $B\Sigma_n$. Then $\mathcal{M}$ is a $B\Sigma_n$ model if and only if it has a Σ_n-cut. In such a case, there is a $\Sigma_n(\mathcal{M})$-function mapping a Σ_n-cut cofinally into $\mathcal{M}$.*

It turns out that Σ_n-cuts (which resemble limit ordinals in some sense) play a central role in our analysis of the structure theory of degrees in the current setting. The existence of such objects allows one to import techniques of ordinal recursion theory into weak fragments of P, in rather natural ways. The next result, due to Chong and Mourad [2], hints at a fine structure theory *a la* Jensen. For this, let us say that $X \subset I$, for I a cut, to be Δ_n *on* I if both X and $I \setminus X$ are $\Sigma_n(\mathcal{M})$.

Proposition 1.2 *(Coding Lemma) Let I be a Σ_n cut in a $B\Sigma_n$ model $\mathcal{M}$. If $X \subset I$ is Δ_n on I, then there is an $\mathcal{M}$-finite set K so that $K \cap I = X$.*

X is said to be *coded on* I if it satisfies the conclusion. An example of a structure with many coded sets, albeit degenerate, was given by Mytilinaios and Slaman [15], where an application of the theory of ultrapowers was made to produce a $B\Sigma_2$ model with ω as a Σ_2-cut, with the additional property that every real is coded on ω.

To strengthen the analogy with fine structure theory further, there is a result (Chong and Yang [4]) which says, roughly speaking, that $\Delta_n(\mathcal{M})$ definability is the best possible when it comes to coding sets:

Proposition 1.3 *There exist $B\Sigma_n$ models $\mathcal{M}$, such that not every $\Sigma_n(\mathcal{M})$ subset of a Σ_n cut I is coded on I.*

A refinement of this result leads to a characterization of the existence of nonlow r.e. sets in $B\Sigma_2$ models (see Section 3).

Finally, there are two important notions which make the subject different from CRT, and show its links with ordinal recursion theory:

Definition 1.1 *In a model $\mathcal{M}$ of a fragment of Peano arithmetic, a* regular set *A is one whose intersection with any $\mathcal{M}$-finite set is $\mathcal{M}$-finite.*

Cuts are examples of nonregular sets. Their existence very often results in the failure of priority arguments, and along with it theorems whose proofs use these arguments. In cases where a theorem continues to hold, different methods, of the non-priority kind, are needed.

Definition 1.2 *A set A is* hyperregular *if every function pointwise recursive in A maps a bounded set to a bounded set.*

Hyperregular sets and nonhyperregular sets are natural objects to study in connection with the jumps of r.e. degrees (Section 3). In particular, if B is a hyperregular r.e. set, then Σ_1 induction relative to B holds, and this feature is crucial in verifying that certain constructions work.

2 Σ_1 Bounding and Σ_1 Induction: The Role of a Σ_1-cut

In this section, we fix $\mathcal{M}$ to be a model of $B\Sigma_1$. When $\mathcal{M}$ is a $B\Sigma_1$ model, I is used to denote a Σ_1 cut in $\mathcal{M}$ and $f : I \to \mathcal{M}$ a Σ_1 cofinal function. We first discuss Post's Problem in a $B\Sigma_1$ model. This is contrasted with The Sacks Splitting Theorem, which is equivalent to $I\Sigma_1$ over Σ_1 bounding. An analysis of the degree of a Σ_1-cut (and its generalization) leads to the proof of the Friedberg-Muchnik Theorem in $B\Sigma_1$. To conclude this section we state some results relating to $I\Sigma_1$ but not $B\Sigma_2$ models, for example the nontransitivity of the weak reducibility relation $\leq_p$. Our objective is to highlight the crucial role played by a Σ_1 cut in recursion theory. Σ_1 cuts, whose existence in a $B\Sigma_1$ model renders many priority constructions irrelevant and ineffectual, provide a new angle for attacking the fundamental problems in a weak system. As such, constructions which are straightforward generalizations of classical ones will be largely ignored.

Let $\mathcal{M}$ be a $B\Sigma_1$ model. Then I is a nonregular r.e. set and is not recursive. Using the fact that I is downward closed, for any computation $\langle x, y, K_1, K_2\rangle$ which applies to I (meaning that $K_1 \subset I$ and $K_2 \subset \bar{I}$), we may assume that K_1 and K_2 are singleton sets. Furthermore, by choosing a different enumeration if necessary, we may ignore the positive condition K_1 in looking at computations. For convenience, we will quite often use N to denote a negative condition (which in this case is a singleton set).

For the study of r.e. sets and degrees in $\mathcal{M}$, the cut I turns out to be a natural candidate for the solution of Post's problem, as observed by Slaman and Woodin [18]:

Theorem 2.1 *In any $B\Sigma_1$ model $\mathcal{M}$, $\emptyset <_T I <_T \emptyset'$.*

Proof. It suffices to enumerate a set A such that $A \not\leq_p I$, because I is not recursive. The construction is done in I many stages. At stage i, consider all numbers m less than $f(i)$. We use $\langle m, x\rangle$ to diagonalize against Φ_m. If there is some $\langle\langle m, x\rangle, 0, N\rangle \in \Phi_{m,f(i)}$, then we choose an N_{max}, which is the maximal negative condition N among them, and for this N_{max}, choose the least x such that $\langle\langle m, x\rangle, 0, N_{max}\rangle \in \Phi_{m,f(i)}$ and x is not in $A_{m,f(i)}$. Enumerate the least x in the definition of N_{max} into A_m. This will guarantee that $A \neq \Phi_m(I)$ by the following argument. An application of the Pigeon Hole Principle produces an x_0 such that $\langle m, x_0\rangle$ is not in A. If $\Phi_m(\langle m, x_0\rangle) \neq 0$ then we are done. So let us assume that $\Phi_m(\langle m, x_0\rangle) = 0$ and is correctly computed at some stage i. At this stage, N_{max} is defined and greater than or equal to the true negative condition used in computing $\Phi_m(\langle m, x_0\rangle)$. It follows that correspondingly the least x in the definition of N_{max} is a true witness for $\Phi_m(\langle m, x\rangle) = 0 \neq A(\langle m, x\rangle)$. (Notice that unlike typical injury arguments, witnesses are not found uniformly. They are either of the form $x_0 \notin A_m$ or taken to be the least x corresponding to some N_{max}.) □

We next sketch the proof that Sacks Splitting Theorem fails in every $B\Sigma_1$ model. This was first observed by Mourad [12] using the cut I as an example for the failure of Sacks splitting.

We take a somewhat different route here, by first studying the degree (in the strong sense) of a bounded Σ_1 set X. This approach allows one to extract information on the relation between regular, nonregular r.e. sets and Σ_1 cuts in $B\Sigma_1$ models.

Let a be an upper bound of X. Let $\theta(y, u)$ be a Σ_0 formula such that $x \in X \Leftrightarrow (\exists u)\theta(x, u)$. We call the least such u the *Σ_1 witness for x.*

Theorem 2.2 *Any bounded Σ_1 set X is either $\mathcal{M}$-finite or strongly Turing equivalent to I. In particular, all Σ_1 cuts in $\mathcal{M}$ have the same Turing degree.*

Proof. If there is a uniform upper bound u_0 for all Σ_1 witnesses verifying any x to be in X, then X is a bounded Σ_0 set, hence $\mathcal{M}$-finite.

Suppose that there is no uniform upper bound for witnesses of x in X. We show that the degree of X is the same as the degree of I.

Consider the set $K \subseteq I\times \leq a$:

$$(e, y) \in K \Leftrightarrow e \in I \wedge y \leq a \wedge (\exists u < f(e))\theta(y, u).$$

In other words, $(e, y) \in K$ if and only if the Σ_1 witness of y shows up before $f(e)$.

Now K is coded on $I\times \leq a$ since it is Δ_1 on $I\times \leq a$ and so Proposition 1.2 (the Coding Lemma) applies. Let K^* be an $\mathcal{M}$-finite set such that

$K^* \cap (I\times \leq a) = K$. Using K^* and I, it is not hard to show that X is strongly Turing reducible to I. Moreover under the assumption that the Σ_1 witnesses of X are unbounded in $\mathcal{M}$, the same code K^* can be used to compute I from X as well. □

The above theorem on bounded Σ_1 sets in $B\Sigma_1$ models may be generalized to bounded Σ_n sets in $B\Sigma_n$ models for $n \geq 1$.

It is now easy to see that I is Turing reducible to any nonregular Σ_1 set, since a bounded nonregular portion is already sufficient to calculate I.

Corollary 2.1 *Let A be any nonregular Σ_1 set. Then I is strongly Turing reducible to A.*

Now the failure of Sacks Splitting Theorem follows quite easily (Mourad [12]).

Corollary 2.2 *Sacks Splitting Theorem fails in every $B\Sigma_1$ model $\mathcal{M}$.*

Proof. Let A_0 and A_1 be a splitting of the cut I. One of them, say A_0, must be nonregular. By Corollary 2.1, I is strongly recursive in A_0. □

We now give a sketch of Mytilinaios' proof [14] that Sacks Splitting Theorem is provable in $P^- + I\Sigma_1$. Together with Corollary 2.2, this will imply that Sacks Splitting Theorem is indeed equivalent to $I\Sigma_1$ over the base theory $B\Sigma_1$.

The proof follows the classical strategy of Sacks preservation. The major difficulty is that Σ_2 induction is necessary for the original construction to work. Specifically, in the absence of Σ_2 induction there may be a cut J where no uniform bound exists for the set of restraints $\{r(e)|e \in J\}$, where $r(e)$ is the restraint imposed by requirement R_e.

The solution is to use Shore's blocking method [17]. Let $\mathcal{M}$ be a model of $I\Sigma_1$ in which Σ_2 induction fails (the other case being straightforward). Divide $\mathcal{M}$ into blocks of requirements, so that requirements in the same block have the same priority. The size of each block, and the number of blocks, is not determined *a priori*, but are instead decided in a Σ_2 manner by the preservation strategy. Thus, for example, the 0th block, written $B(0)$, consists only of the requirement R_0. As the construction unfolds, the restraint $r(0,s)$ of R_0 at stage s increases. The size of the block $B(1)$ at stage s, consisting of all requirements R_e, $0 < e < r(0,s)$, will correspondingly increase, finally settling down at $r(0)$, the final position of the restraint for R_0, using Sacks preservation strategy. At each stage s, all requirements in $B(1)$ have the same priority. Collectively they impose a *block restraint* which is the supremum of the individual restraints in $B(1)$, and which we denote by $r(1,s)$. Then $r(1,s)$ in turn determines the length and size of $B(2)$ at stage s, and so on. In this way, if $B(e-1)$ has $r(e-1)$ as its final restraint, then $B(e)$ will consist (finally) of requirements with indices between $r(e-2)$ and $r(e-1)$, and under $I\Sigma_1$, the Sacks preservation strategy will ensure that $r(e)$ exists. What one obtains eventually is a Σ_2 cut J so that for each $e \in J$, $B(e)$ and $r(e)$ exist, with the additional property that each requirement belongs to one and only one block. The conclusion is that each requirement is satisfied and so Sacks splitting holds in $I\Sigma_1$.

The description above of the construction is an instance of what we call *dynamic blocking* (Chong and Yang [5]), a technique which was developed in there for infinite injury priority arguments. In the present situation we are concerned with Σ_2 dynamics since both the restraints and lengths of blocks stabilize eventually. It turns out that dynamic blocking is a natural tool for priority type constructions in models of weak fragments of Peano arithmetic, as a solution for the absence of a predetermined 'cofinality' (standard objects used in ordinal recursion theory) in the structures under study.

The degree of a Σ_1 cut, in addition to providing a solution to Post's problem in models of $P^- + B\Sigma_1$, is also the base of the cone of all nontrivial r.e. degrees. The first result that points in this direction is the theorem of Mourad [13] that I is of minimal r.e. degree.

Lemma 2.1 *If A is a regular Σ_1 set, then A is strongly recursive in I.*

Proof. Let $\exists u\theta(x, u)$ be a Σ_1 formula defining A. Define $K \subset I \times I$ by (i, j) is in K if and only if there is some element $x < f(i)$ which enters A between stages $f(j)$ and $f(j+1)$. K is coded on $I \times I$ by Proposition 1.2, as it is Δ_1 on $I \times I$. Let K^* be an $\mathcal{M}$-finite set such that $K^* \cap (I \times I) = K$. For any $\mathcal{M}$-finite set E, we find an i such that $E \subset f(i)$. Look at all pairs $(i, j) \in K^*$. By the regularity of A there is some j_0 such that (i, j_0) is in K^* and for all $j > j_0$ in I (i, j) is not in K^*. That j_0 can be found by using I. By the definition of K, $E \subset A$ if and only if $E \subset A_{f(j_0)}$ and $E \cap A = \emptyset$ if and only if $E \cap A_{f(j_0)} = \emptyset$. $\square$

Theorem 2.3 *The degree of the cut I is a minimal Σ_1 degree.*

Proof. By previous results, it suffices to show that a Σ_1 regular set is either recursive or of the same degree as I. Let A be a regular r.e. set and $\hat{h} : I \to I$ be defined by $\hat{h}(i) = k$ if $A \restriction f(i)$ settles down before stage $f(k)$. Fix an upper bound e of I. First we will find a coded approximation $h(i, j)$ of $\hat{h}(i)$. The price to pay is that $h(i, j)$ will overspill to $e \times e$. Consider the set $S \subset I \times I \times I$:

$$S = \{(i, j, k) | A_{f(k)} \restriction f(i) = A_{f(j)} \restriction f(i)\}.$$

By the Coding Lemma, S is coded on $I \times I \times I$, as it it Δ_1 on $I \times I \times I$. Define $h : e \times e \to e$ by

$$h(i, j) = \text{ the least } k \text{ such that } (i, j, k) \in S^*.$$

Then $\hat{h}(i)$ for $i \in I$ can be approximated by

$$\hat{h}(i) = \lim_{j \in I} h(i, j).$$

Observe that for each fixed i in I, the function $h(i, j)$ can be made non-decreasing for $j < e$.

Next with the help of $h(i, j)$, we can decide the degree of A based on the following two cases.

Case 1 There is a $b \in e \setminus I$ such that for all $i \in I$, $h(i, b) = \hat{h}(i)$. Then $A \upharpoonright f(i)$ settles down at stage $f(h(i, b))$, therefore A is recursive.

Case 2 For all $b \in e \setminus I$, there is an i in I such that $h(i, b) > \hat{h}(i)$. We argue that A computes I by enumerating $\overline{I}$, the complement of I, from A as follows. Consider any $b < e$. Enumerate $i \in I$, and calculate $h(i, b)$. We use A to find the value $\hat{h}(i)$. Compare $\hat{h}(i)$ and $h(i, b)$. If $\hat{h}(i)$ is less than $h(i, b)$, then b is in $\overline{I}$. □

The origin of the priority method traces back to the solution of Post's problem by Friedberg and Muchnik. In the standard construction of two r.e. sets A and B of incomparable degree (called a Friedberg-Muchnik pair), a finite injury strategy is used, such that each requirement R_e is injured at most 2^e times. Simpson had observed some years ago that using Σ_1 induction one could derive the same effective bound for each requirement, and hence obtain a Friedberg-Muchnik pair. On the other hand, in the absence of $I\Sigma_1$, an effective bound on the number of injuries for each requirement is no longer assured. With the characterization of the Sacks Splitting Theorem as being equivalent to $I\Sigma_1$, the provability of the Friedberg-Muchnik theorem in a weaker system became especially interesting. This was settled by Chong and Mourad [3]: $B\Sigma_1$ proves the Friedberg-Muchnik Theorem.

Since the cut I is actually the least nonrecursive r.e. set, any Friedberg-Muchnik pair has to consist of nonregular r.e. degrees above that of I. Nonregular r.e. degrees are typically associated with cuts. Theorem 2.2 states that all Σ_1 cuts have the same degree. Hence one has to go beyond the notion of cuts to look for a solution, and this is found in what is called the *union of cuts* . For each i in I, let L_i denote the interval $[f(i), f(i+1))$. We say A is a union of cuts if for each $i \in I$, $A \cap L_i$ is downward closed in L_i.

We sketch the injury free strategy for making $\Phi_e(A) \neq B$ in a $B\Sigma_1$ model. The Friedberg-Muchnik pair A and B will each be a union of cuts. In each interval L_i, we place two markers a_i and $\bar{a}_i$. For stages j in I, marker a_i moves up within L_i as j increases, whereas $\bar{a}_i$ moves down as j increases. At any stage, $a_i < \bar{a}_i$. The idea is that all numbers in L_i below a_i enter A, while all those above $\bar{a}_i$ in L_i are barred from A (forever). The interval L_i has to be 'large' enough (in a precise sense) to allow for the movement of these two markers throughout the construction. Correspondingly, one also has, on the B-side, markers $b_i < \bar{b_i}$ in L_i (in an actual construction, A would occupy the even intervals while B the odd intervals, in order to take care of certain technical details). The construction is arranged so that at any stage, $\bar{a_i} - a_i$ is large, and $\bar{b_i} - b_i$ is large. In addition, this property holds in the limit as well (meaning that 'largeness' is preserved after going through I-many stages). A negative condition N will now consist of an $\mathcal{M}$-finite set of points, each point belonging to one and only one interval.

We describe briefly how a typical construction works with a special case. Suppose $\Phi_{e,f(i)}(\langle x, 0, N\rangle) = 0$, with $N \in [a_2, \bar{a_2}]$ and $x \in [b_1, \bar{b_1}]$. There are two possible actions to be taken: If $N - a_2$ is large, we move $\bar{a_2}$ down to N and make this computation permanent. Find the least x with these properties (on the A-side). Denote it by x_0. If $\bar{b_1} - x_0$ is large, move marker b_1 up to capture x_0, and we achieve permanent disagreement. If $x_0 - b_1$ is large, then we move $\bar{b_1}$ downwards to x_0, and look for a new candidate for diagonalization. On the other hand, if $\bar{a_2} - N$ is large, then move a_2 up to capture N and look for another computation that allows one to perform diagonalization.

To show that the construction succeeds, suppose for the sake of contradiction that $\Phi_e(A) = B$. In L_1, choose an x not in B yet $\bar{b_1} - x$ is large at every stage of the construction (this needs to be justified). Then there is a true computation $\langle x, 0, N\rangle \in \Phi_{e,f(i)}$ at some stage i. For simplicity assume that $N \in L_2$. Being a true computation, $N - a_2$ must be large. That means that $\bar{a_2}$ is moved downwards to capture N, and either x or some number smaller is captured by b_1 as it moves up, ensuring a permanent disagreement.

Of course in general the negative condition may have components from different intervals, and so it is more complicated to handle. A careful tracking of where these components reside will be essential. A final point to note is that the requirements are arranged in such a way that for each j, the eth requirement R_e (for $e \leq f(j)$) is satisfied using less than I-many number of cuts. To do this it is essential that each interval L_i is 'large' enough to accommodate the movement of markers, in order to provide space to satisfy such requirements. One can think of the case of the limit cardinal $\aleph_\omega^L$, where for finite injury constructions of the Friedberg-Muchnik type, requirements of priority less than an $\aleph_n^L$ are satisfied by stage $\aleph_{n+1}^L$. The fact that each succeeding cardinal has size much larger than the previous cardinal allows sufficient space for markers associated with the relevant requirements to move (there is, of course, a crucial difference here: we are dealing with nonregular sets so that a straightforward cardinality argument, as is done in the case of $\aleph_\omega^L$ will not suffice).

As we have seen before, there are theorems in CRT which are equivalent to $I\Sigma_1$ over the base theory $B\Sigma_1$. In section 3, we will see theorems equivalent to $I\Sigma_2$ over $B\Sigma_2$. It is natural to ask whether a weaker base theory, say $I\Sigma_1$, is sufficient to establish the equivalence of such theorems with $B\Sigma_2$.

No example is known thus far. One of the reasons is that there is a model $\mathcal{M}$ of $I\Sigma_1$ but not $B\Sigma_2$ which behaves like the standard model ω. As such most constructions from CRT carry over with minor modification.

The particular model $\mathcal{M}$ of $P^- + I\Sigma_1 + \neg B\Sigma_2$ has a $\Delta_2(\mathcal{M})$ projection g from ω one-one onto $\mathcal{M}$. Since the projection g is Δ_2, it has a recursive approximation. This allows requirements to be indexed by ω (in a Δ_2 fashion), so that induction can be carried out over ω via g. For example, in $\mathcal{M}$ there is a maximal r.e. set (Chong [1]), and there is an incomplete high r.e. set (Yang [20]). This justifies the need for a base theory stronger than $I\Sigma_1$ to establish the equivalence of these results with Σ_2 induction.

The construction of such a model $\mathcal{M}$ was first done by Groszek and Slaman in their study of the nontransitivity of weak Turing reducibility [8]. We briefly describe the construction. Let $\mathcal{N}$ be a model of P with a nonstandard element p. The model $\mathcal{M}$ is built in ω steps. Initially let $\mathcal{M}[0] = \{p\}$. Suppose $\mathcal{M}[k] = \{m_1, ..., m_i\} \subset \mathcal{N}$. First, throw in all Σ_1 witnesses of the first k Σ_1 sentences true in $\mathcal{N}$. This will make $\mathcal{M} \prec_{\Sigma_1} \mathcal{N}$. Secondly, for each of the first k Π_1 formulas with one free variable, say $\varphi(v)$, if there is an x in $\mathcal{N}$ such that $\mathcal{N} \models \varphi(x)$ and x is less than some element of $\mathcal{M}[k]$, then we add the $\mathcal{N}$-least such x to $\mathcal{M}$. This will ensure that $\mathcal{M}$ satisfies $L\Pi_1$, which is equivalent to $I\Sigma_1$. Let $\mathcal{M} = \cup \mathcal{M}[k], k \in \omega$. Then $\mathcal{M} \models P^- + I\Sigma_1 + \neg B\Sigma_2$. And the function $g : i \mapsto$ the i-th element in the construction, is pointwise recursive in $0'$ in $\mathcal{M}$, because $0'$ is absolute in $\mathcal{N}$ and $\mathcal{M}$. Therefore, there is a recursive approximation of two variables $g(\_, s)$ of g.

3 Σ_2 Bounding and Σ_2 Induction: The role of a Σ_2-cut

We now turn our attention to infinite injury priority constructions, and more generally, constructions which seemingly require more than just Σ_1 induction. There are three theorems in CRT which interest us: the Sacks Density Theorem, the Friedberg Maximal Set Theorem, and Shoenfield's theorem on the existence of an incomplete high r.e. degree. The classical proof of each of these theorems comes with a new technique that leads to applications elsewhere. Each construction appears to require at least $I\Sigma_2$, or even $I\Sigma_3$, to succeed. Groszek and Slaman [7] have analyzed strategies in infinite priority constructions. Under their classification of complexity of outcomes of strategies, the applicability of all Π_2 constructions is equivalent (over the base theory $P^- + Exp + I\Sigma_0$) to $I\Sigma_2$. However their analysis does not provide much information on individual Π_2 constructions, for example Shoenfield's proof on the existence of high sets or Sacks' proof of the Density Theorem. As it turns out, the Density Theorem actually holds under $P^- + B\Sigma_2$, while over this as base theory, the Maximal Set Theorem and

Shoenfield's Theorem are both equivalent to $I\Sigma_2$, showing that essentially strategies of the same level of complexity are required to establish these results.

In the discussion that follows, all models satisfy $P^- + B\Sigma_2$.

We begin with the study of the Maximal Set Theorem and Shoenfield's Theorem on high r.e. sets. We will not dwell on the positive aspects of these theorems—that existence of such sets follow from Σ_2-induction. The proofs (Chong and Yang [4]) for maximal sets, and Groszek and Mytilinaios [6] for incomplete high r.e. sets) involve a careful tracking of the complexity of strategies used in the construction, and showing that $I\Sigma_2$ suffices to do the job. We concentrate instead on the other direction, that if there is either a maximal set or an incomplete high r.e. set in a model $\mathcal{M}$, then indeed Σ_2-induction holds in $\mathcal{M}$. It will be seen that the existence of a Σ_2-cut in a $B\Sigma_2$ model disengages the role of many Σ_2 or Π_2 constructions from the theorems (indeed the theorems fail completely in such models). We begin with the Maximal Set Theorem.

Definition 3.1 *A nontrivial r.e. set M is* maximal *if its complement is not $\mathcal{M}$-finite, and any r.e. set which contains M is either M or $\mathcal{M}$, save for an $\mathcal{M}$-finite difference.*

The main result of Chong and Yang [4] states:

Theorem 3.1 *Let $\mathcal{M}$ be a model of $P^- + B\Sigma_2$. Then $\mathcal{M}$ has a maximal set if and only if $\mathcal{M}$ satisfies $I\Sigma_2$. In other words, over this base theory, the existence of a maximal set is equivalent to Σ_2-induction.*

Proof. We give a sketch. Suppose that M is a maximal set. There are two cases to consider:

Case 1. There is a Σ_2-cut I and a Σ_2-function f mapping I onto the complement of M. In this case we say that M has a *small* complement. Fix an upper bound e_0 of I. It is not difficult to define a simultaneous r.e. collection of pairwise disjoint r.e. sets $\{H_d | d \leq e_0\}$ whose union is $\mathcal{M}$ with the additional property that each H_d contains at most one member not in M (the idea of generating such a collection $\{H_d\}$ originated in Lerman and Simpson [10] where he proved that there is no maximal $\aleph_\omega^L$-r.e. set). Now consider the set

$$K = \{(e,d) | f(e) \in H_d\}$$

which is a subset of $I \times I$ and Δ_2 on I. By the Coding Lemma (Proposition 1.2) K is coded on I by an $\mathcal{M}$-finite set K^*. Since K is a function on I, we may think of K^* as a function as well which we denote by f^*. Now

$$\cup\{H_{f^*(2e)} | e \leq e_0\}$$

is an unbounded r.e. set which splits the complement of M into two nontrivial pieces, contradicting the maximality of M.

Case 2. The complement of M is *large*, i.e. Case 1 does not hold. The first step is to produce a Σ_2-cut I, a simultaneous r.e. collection of pairwise disjoint r.e. sets $\{H_d|d \leq e_0\}$ (for some e_0 an upper bound of I) such that unboundedly many elements of $\mathcal{M}\backslash\mathcal{M}$ distributed nicely in $\cup\{H_d\}$ (meaning that there are cofinally many d in I which contain elements of $\bar{\mathcal{M}}$). Again let f be a Σ_2-function mapping I unboundedly into $\mathcal{M}$. Define $g(e)$ to be the first element above $f(e)$ not in M (we may arrange f nicely so that there is always a member of $\mathcal{M} \setminus \mathcal{M}$ between any two values of f). Define

$$K = \{(e,d)|g(e) \in H_d\}.$$

Again K is coded on I and so a similar argument proves that M is not maximal.

Definition 3.2 *Let* $\emptyset'' = \{(x,e)|\exists y\forall z\varphi(x,y,z,e)\}$ *where* φ *ranges over all* Σ_0 *formulas as* e *ranges over* $\mathcal{M}$. $\emptyset''$ *is called a* complete Σ_2 *set.*

It is easy to see that every Σ_2-set is recursive in $\emptyset''$.

Definition 3.3 *An incomplete r.e. set* A *is* high *if its jump* $A' = \{e|\{e\}^A(e)\downarrow\}$ *is Turing equivalent to* $\emptyset''$ *(i.e. each set is recursive in the other).*

Shoenfield's high set theorem states that there is a high r.e. set. The proof uses a Π_2-strategy which is a classic example of an infinite injury construction. Groszek and Mytilinaios [6] show that with enough care, this construction actually works in $I\Sigma_2$ and establishes Shoenfield's theorem in this system. We give here a description of how such a result must fail when only $B\Sigma_2$ is allowed. The corollary is that any construction of a high r.e. set has to have complexity at least that of the standard one. The technique used in the proof of the result also highlights the role played by a Σ_2-cut in recursion theory.

Theorem 3.2 *(Three Point Theorem) If* $\mathcal{M}$ *is a* $B\Sigma_2$ *model, then there is no high r.e. set. Indeed there exist three degrees, denoted* $\mathbf{0}'$, $\mathbf{0}^{1.5}$ *and* $\mathbf{0}''$, *which form the set of jumps of r.e. degrees. Furthermore, the jump of an incomplete r.e. degree is either* $\mathbf{0}'$ *or* $\mathbf{0}^{1.5}$.

The following is a useful observation, which is a generalization of Theorem 2.2 in Section 2.

Lemma 3.1 *In a* $B\Sigma_2$ *model, all* Σ_2*-cuts are of the same degree.*

Our analysis of the jumps of r.e. sets in $B\Sigma_2$ models relies on the notion of hyperregularity, which originated in ordinal recursion theory. The importance of hyperregularity to the subject of study revolves around the following fact.

Lemma 3.2 *In a model $\mathcal{M}$ of $P^- + B\Sigma_2$, every incomplete r.e. set A is hyperregular. Furthermore, an r.e. set A is hyperregular if and only if $\mathcal{M}$ satisfies $I\Sigma_1(A)$.*

Proof of Theorem 3.2. Let $\emptyset'$ be a complete r.e. set, and let I be a Σ_2-cut. By Lemma 3.1, the degree of $I \oplus \emptyset'$ is well-defined, which we denote by $\mathbf{0}^{1.5}$. The claim is that if A is incomplete r.e., then A' is recursive in $\emptyset' \oplus I$. By Lemma 3.2, A is hyperregular. Let f be a Σ_2-cofinal map from I into $\mathcal{M}$. Let $A'_{f(e)} = \{d | \{d\}^{A|f(e)}_{f(e)}(d) \downarrow\}$.

Then

$$K = \{(d, e) | A'_{f(e)} \upharpoonright f(d) \neq A'_{f(e-1)} \upharpoonright f(d)\}$$

is Δ_2 on I. Let K^* be an $\mathcal{M}$-finite set which codes K (Proposition 1.2). Since A is hyperregular, for each d in I there is a maximum e in I so that $(d, e) \in K^*$. This yields an algorithm to compute A' from $\emptyset' \oplus I$, using K^* as the parameter set.

The degree $\mathbf{0}^{1.5}$ turns out to be strictly between $\mathbf{0}'$ and $\mathbf{0}''$. According to Mourad [13], the degree of $\mathbf{0}^{1.5}$ is minimal over $\mathbf{0}'$ (generalizing an argument for the degree of a Σ_1 cut in a $B\Sigma_1$ model, see Section 2). The argument above shows that the jump of an r.e. degree is either $\mathbf{0}'$, $\mathbf{0}^{1.5}$, or $\mathbf{0}''$. For hyperregular r.e. sets, their jumps must be one of the first two degrees, while the jump of a nonhyperregular r.e. set automatically assumes the third degree. There exist $B\Sigma_2$ models in which only $\mathbf{0}'$ and $\mathbf{0}''$ are the only jumps of r.e. degrees (Mytilinaios and Slaman [15]), and there also exist $B\Sigma_2$ models in which all three degrees, $\mathbf{0}'$, $\mathbf{0}^{1.5}$ and $\mathbf{0}''$ are jumps of r.e. degrees (Chong and Yang [5]). Indeed a complete characterization is obtained through the notion of *tame Σ_2-coding*:

Definition 3.4 *Let $\mathcal{M}$ be a $B\Sigma_2$ model. Let I be a Σ_2 cut. A Σ_2-relation R on $I \times I$ is* tame *if for any $a \in I$, there is a $b \in I$ such that if $(x, y) \in R$ and $x < a$, then $y < b$. In particular, if R is a Σ_2 function, then R is bounded on every proper initial segment of I.*

Theorem 3.3 *([5]) Let $\mathcal{M}$ be a $B\Sigma_2$ model. Then the following are equivalent:*

(a) Every tame Σ_2-relation on a Σ_2-cut I is coded on I;
(b) $\mathbf{0}^{1.5}$ is not the jump of an r.e. degree.

Thus the degree of jumps of r.e. degrees depend on the existence/nonexistence of codes, a phenomenon which hints at a fine structure theory for models of fragments of Peano arithmetic. We believe that this is an interesting phenomenon that warrants further investigation.

The proof that (b) implies (a) in Theorem 3.3 introduces an interesting feature for infinite injury constructions on models of $B\Sigma_2$ (in the absence of Σ_2 induction) that has also surfaced (though not explicitly) in [16]. As it turns out, its finite injury counterpart for $I\Sigma_1$ was present in [14] (discussed in the previous section). This is a technique we refer to as *dynamic blocking* in [5].

We illustrate with the proof that (b) implies (a) in Theorem 3.3. Suppose for the sake of contradiction that there is a Σ_2 cut I with a tame Σ_2 subset which is not coded. We construct an r.e. set A whose jump is $\emptyset^{1.5}$. The first step to note is that there is an r.e. set B so that for each $e \in I$, $B^{[e]} = \{x|(x, e) \in B\}$ is either $\mathcal{M}$-finite or all of $\mathcal{M}$, and that $\cup_{d \leq e} B^{[d]}$ is recursive. Furthermore, there is no uniform bound for $\cup B^{[d]}$ where d ranges over those in I for which $B^{[d]}$ is $\mathcal{M}$-finite. This is an analog of piecewise recursive r.e. sets in CRT first studied by Shoenfield. The existence of B uses the noncodability of a tame Σ_2 set in I. Next one constructs a *thick subset* A of B that is not low. To do this, the technique of blocking is used. Requirements are grouped into blocks of the same priority. One argues that the limit infimum of the restraints on each block exists. Three is, however, an inherent difficulty with using I as the number of blocks. This has to do with the fact that we are dealing with a Π_2 construction (the limit infimum function on block restraint) over a Σ_2 cut, so that to argue that each block will impose a limit infimum restraint becomes impossible if we were to specify the number of blocks of requirements *a priori*.

The solution is to use the strategy of dynamic blocking. The number of blocks is not determined before hand. Rather it is decided by the construction itself. Once the size of a block (say block e) and its limit infimum restraint is known, the size of block $e + 1$ will be, roughly speaking, equal to the length of this restraint. One then argues that the limit infimum of the restraints for block $e + 1$ exists, and this determines the size of block $e + 2$, and so on. In the end, there is a subcut $J \subseteq I$ such that for each $e \in J$, the limit infimum of the restraints for block e exists, and that each requirement belongs to one of the blocks.

The r.e. set constructed will be incomplete, and its jump will be $\mathbf{0}^{1.5}$ since the noncodable tame Σ_2 set we started with is actually recursive in the jump. If A were low, that set would be $\Delta_2(\mathcal{M})$ and so coded on I, a contradiction.

The Sacks Jump Inversion Theorem states that every degree r.e. in and above $\mathbf{0}'$ is the jump of an r.e. degree.

Corollary 3.1 *The Sacks Jump Inversion Theorem is equivalent to* $I\Sigma_2$

over the base theory $B\Sigma_2$.

Proof. The construction of the Friedberg-Muchnik pair in Chong and Mourad [2] generalizes to degrees above $\mathbf{0}^{1.5}$. By Theorem 3.6, these degrees are never the jump of an r.e. degree.

Finally we sketch a proof of the Density Theorem in $B\Sigma_2$, which is done by Groszek, Mytilinaios and Slaman [16]. This is an illustration of how, in the absence of Σ_2 induction, it is still possible to carry out infinite injury constructions successfully, using all the techniques discussed thus far.

Theorem 3.4 *Suppose that* $\mathcal{M}$ *is a model of* $P^- + B\Sigma_2$. *Then the r.e. degrees are dense.*

Let A and B be r.e. sets such that $B <_T A$. Then B is incomplete, and under $B\Sigma_2$, $\mathcal{M}$ will satisfy $I\Sigma_1(B)$. We analyze once again a naive construction using the blocking technique and isolate the problems associated with satisfying various $\Sigma_1(B)$ statements. We then argue that $I\Sigma_1(B)$ is in fact sufficient for a modified construction, using dynamic blocking, to go through.

We assume that the reader is familiar with the proof of the Density Theorem in CRT cast in the language of tree constructions. We construct an r.e. set C such that $B <_T C <_T A$. We only concentrate on the parts where $B\Sigma_2$ is used. Let the notions of length of agreement, hat computation, true stage and injury set be defined as usual. There are four types of requirements. The first one—$B \leq_T C$—is done by direct coding. The second type, $C \leq_T A$, is analyzed after the construction. Let us concentrate on strategies to achieve incomparability of degrees.

$A \not\leq_T C$ is done by Sacks preservation strategy, i.e. preserving the length of agreement between $\Phi_e(C)$ and A. The worry is that the restraint reaches infinity on some cut I. The solution is to use blocking, which is similar to the proof of Sacks splitting theorem. First of all we need to argue that if we assign all the requirements in some block $\{N_e : e < b\}$ for some fixed b the same priority, then the restraint has a liminf. More precisely, suppose that the injury set is recursive in B. Then for sufficiently large B-true stages, the restraint imposed by the block will be stable. The proof is more or less a relativised version of the one used to prove the Sacks splitting theorem. $I\Sigma_1(B)$ will ensure that the restraint reaches a limit on B-true stages, as otherwise A would be recursive in B.

$C \not\leq_T B$ is obtained by using the Sacks coding strategy. Blocks of positive requirements are formed. For a block of positive requirements $\{P_e : e < b\}$, we build a Turing functional Γ such that if $\Phi_e(B)$ computes C up to x for some $e < b$, then $\Gamma(C)$ computes A up to x. The control of B on Γ is done by putting the use γ into C when we see a disagreement between $\Phi(B)$ and C or a change of use of $\Phi(B)$. We then argue that the Γ associated with any block of positive requirements cannot compute A. Fix a block $\{P_e : e < b\}$. Divide it into two groups. The first one consists of all $e < b$ such that there is some $x \in C$ and $\Phi_e(B; x) = 0$. By $I\Sigma_1(B)$, it is $\mathcal{M}$-finite, thus providing only $\mathcal{M}$-finite computations. The other group of requirements cannot produce unbounded computations at B-true stages, otherwise $\Gamma(C)$ would be total and equal to A, implying that A is recursive in B, a contradiction.

Now we perform a tree construction, similar to the usual one. Firstly each node is associated with a block of requirements. For the preservation strategy, the outcome is, roughly speaking, the collective restraint imposed by the requirements in the block. For the coding strategy, the outcome is the collective computation Γ performed by the requirements in the block. The point to note is that the size of the next block depends on the outcome of the previous node. Thus, the blocks will settle down in a Π_2 way. At non-B-true stages, it can be very large. At B-true stages, it will fall back to the true size, a phenomenon which also happens in the proof of Theorem 3.6.

We then define the true path to be the leftmost one which is accessed at cofinally many stages, as in CRT. The reason for the existence of a true path is as follows. As in the analysis of block strategies, at sufficiently large B-true stages, we will always see the true length of agreement and the true computation of Γ. Thus we will visit the same path at these stages. On the other hand, fix any node α on the true path, the nodes to the left of α are abandoned because either we see more B-true length of agreement in the preservation strategy or we see more computations Γ correctly calculate A. For a fixed α, the information required from A will be $\mathcal{M}$-finite. We can use this information as a parameter. Thus the nodes to the left of α can be enumerated from B. $I\Sigma_1(B)$ ensures that there are $\mathcal{M}$-finitely many of them, and hence these nodes can only be visited $\mathcal{M}$-finitely often.

Following a similar argument as in CRT, requirements are satisfied along the true path. The new difficulty is that the true path can be a cut. So we are required to check that every requirement falls into some block associated with a node on the true path. Fix a requirement R_n. Clearly it belongs to some block at stages larger than n. By carefully analyzing the reasons causing R_n to change to a different block, all changes can be traced back to B (again we need an $\mathcal{M}$-finite amount of A as a parameter). $I\Sigma_1(B)$ will guarantee that R_n falls into a block associated with some node on the true path.

It remains to argue that $C \leq_T A$. It suffices to enumerate $\bar{C}$ from A. The candidates for entering C are the uses γ selected by some α on the tree. Now α belongs to one of the following three regions: (I) strictly to the left of the true path, (II) on the true path, and (III) to the right of the true path. For α in region (III), there is a stage when it gets initialized. For γ chosen by some node α from region (II), we have unboundedly many stages visiting α. We can decide if γ is in $\bar{C}$, when we see enough "truth", that is, when the computations of B which could affect γ have all settled down, and $\Gamma(C, x) = A(x)$ with use γ. For α from region (I), we show that such nodes can be enumerated from A. For a fixed node β, we have argued before that the nodes to the left of β can be enumerated from B, provided that we have $\mathcal{M}$-finite amount of information on A as s parameter. Since we have A as an oracle, and A computes B, all nodes from region (I) can be enumerated from A. Hence we can decide whether γ associated with α enters C, by using A as an oracle. This completes the proof of the Density Theorem.

4 Some Open Questions

We end this paper with a list of open problems. The first two concern understanding the complexity of higher order priority constructions:

(1) Investigate the complexity of $0'''$-priority arguments. In particular, are there theorems using such constructions which are equivalent to $I\Sigma_3$ over $B\Sigma_3$ or even $B\Sigma_2$? (It is noted in Section 2 that there is a model of $I\Sigma_1$ which behaves just like the standard model ω. It is likely that certain $0'''$-priority constructions can be carried out successfully in this model.)

(2) Is there a minimal degree below $\mathbf{0}'$ in every (any) $B\Sigma_2$ model? Chong and Mourad [2] produced a model in which there is a minimal degree (the degree of a Σ_2 cut). The classical Spector construction was bypassed. This does not seem possible with the problem under discussion. Another question, which appears to be quite difficult at this point, is whether there exists a minimal pair of r.e. degrees in $B\Sigma_2$ models.[1] Phrased differently, the problem is to decide if the existence of a minimal degree below $\mathbf{0}'$ or the existence of a minimal pair of r.e. degrees is equivalent to Σ_2 induction over $B\Sigma_2$. It may be worthwhile to add that the difficulties encountered in attacking these two problems are very similar to those experienced with analogs in the case of $\aleph_\omega^L$. It is tempting to conjecture that there is a universal, model-theoretic approach to solving these problems.

The next two problems deal with understanding the degree-theoretic structure of models of weak fragments of P.

(3) In a $B\Sigma_1$ model with Σ_1 cut I, is I low ? In other words, is I' recursive in $\emptyset'$? This question extends to $B\Sigma_2$ models $\mathcal{M}$: Is every incomplete r.e. degree in $\mathcal{M}$ low$_2$? As discussed in Section 3, the jump of an incomplete r.e. degree is either $\mathbf{0}'$ or $\mathbf{0}^{1.5}$. The latter is in fact the degree of $\emptyset' \oplus I$, where I is a Σ_2 cut. One is thus asking if the jump of $\emptyset' \oplus I$ is recursive in the complete Σ_2 set. [2]

[1] *Note added in proof.* Chong, Qian, Slaman and Yang have given a negative answer to this problem, i.e., there is no r.e. minimal pair in any $B\Sigma_2$ model. However the minimal degree problem is still open.

[2] *Note added in proof.* Chong, Qian and Yang have obtained some partial results. It

(4) In a $B\Sigma_1$ model, I is of minimal r.e. degree, and is not regular. All r.e. sets above I are nonregular. It is known that there is a Friedberg-Muchnik pair above I. What is the structure of degrees above I ? Does Sacks splitting or density hold in there ? Working with nonregular sets is challenging in that very often new techniques, very different from those employed in CRT, are required for the solutions.

The following problem is admittedly rather vague. However, we feel that there are traces of evidence at hand to warrant a formal statement:

(5) Develop a general theory of computation that elucidates the phenomenon of similarity and inter-applicability of techniques and ideas between recursion theory on fragments of Peano arithmetic and ordinal recursion theory.

Finally, we pose a question that is inspired by the Coding Lemma (Proposition 1.2):

(6) Is there a fine structure theory, in the spirit of Jensen [9], for models of fragments of Peano arithmetic ?

References

[1] C. T. Chong, Maximal sets and fragments of Peano arithmetic, *Nagoya J. Math.*, **115** (1989) 165-183

[2] C. T. Chong and K. J. Mourad, The degree of a Σ_n cut, *Ann. Pure and Applied Logic*, **48**, (1990) 227-235

[3] C. T. Chong and K. J. Mourad, Σ_n Definable sets without Σ_n induction, *Trans. Am. Math. Soc.* , **334**, (1992) 349-363

[4] C. T. Chong and Y. Yang, Σ_2 induction and infinite injury priority arguments, Part I: Maximal sets and the jump operator, *J. Symbolic Logic*, to appear

[5] C. T. Chong and Y. Yang, Σ_2 induction and infinite injury priority arguments, Part II: Tame Σ_2 coding and the jump operator, it Annals of Pure and Applied Logic, Special Issue on the Recursion Theory Session, ASL 1995 Summer Meeting in Haifa. to appear

[6] M. J. Groszek and M. E. Mytilinaios, Σ_2-induction and the construction of a high degree, in: *Recursion Theory Week*, Proc. Oberwolfach 1989 (K. Ambos-Spies et al., eds), Lec. Notes Math. **1432**, Springer-Verlag, Berlin, Heidelberg, New York, Tokyo, 1990, 205-223

is now known that in some $B\Sigma_1$ model, the cut I is not low, and in some $B\Sigma_2$ model I' is not recursive in $\emptyset''$. What happens in the general $B\Sigma_2$ model remains open.

[7] M. J. Groszek and T. A. Slaman, Foundations of the priority method I: Finite and infinite injury, preprint

[8] M. J. Groszek and T. A. Slaman, On Turing reducibility, preprint

[9] R. B. Jensen, The fine structure of the constructible universe, *Annals of Math. Logic*, **4** (1972), 229–308

[10] M. Lerman and S. Simpson, Maximal sets in α-recursion theory, *Israel J. Math.*, **14** (1973), 236–247

[11] J. B. Paris and L. A. S. Kirby, Σ_n-collection schemas in arithmetic, in: *Logic Colloquium* **'77**, North Holland, Amsterdam, 1978, 199-209

[12] K. J. Mourad, *Ph.D thesis*, University of Chicago, 1988

[13] K. J. Mourad, *Unpublished manuscript*

[14] M. E. Mytilinaios, Finite injury and Σ_1-induction, *J. Sym. Logic* **54** (1989), 38-49

[15] M. E. Mytilinaios and T. A. Slaman, Σ_2-collection and the infinite injury priority method, *J. Sym. Logic*, **54** (1989), 38-49

[16] M. J. Groszek, M. E. Mytilinaios and T. A. Slaman, The Sacks density theorem and Σ_2-bounding, to appear

[17] R. A. Shore, Splitting an α-recursively enumerable set, *Trans. Amer. Math. Soc.*, **204** 65-78

[18] T. A. Slaman and W. H. Woodin, Σ_1-collection and the finite injury priority method, in: *Math. Logic. and Its Applications*, Lec. Notes Math. **1388**, Springer-Verlag, Berlin, Heidelberg, New York, Tokyo, 1989

[19] R. I. Soare, *Recursively Enumerable Sets and Degrees*, Perspectives in Mathematical Logic, Omega Series, Springer-Verlag, Berlin, Heidelberg, New York, Tokyo, 1987

[20] Y. Yang, The thickness lemma from $P^- + I\Sigma_1 + \neg B\Sigma_2$, *J. Sym. Logic*, **60** (1995)

Lattice embedding into d-r.e. degrees preserving 0 and 1 *

DING Decheng QIAN Lei
Mathematics Department, Nanjing University
Mathematics department, National University of Singapore

1 Introduction

We are used to say that a set A is r.e. if A is a domain or range of some partial recursive function. But we can also define r.e. sets by recursive approximations to their characteristic functions. A set A is r.e. iff there is a recursive function f such that for all x:

$$A(x) = \lim_s f(x, s)$$

$$f(x, 0) = 0$$

$$| \{s \mid f(x, s) \neq f(x, s+1)\} | \leq 1$$

This is to say, we begin by guessing x is not in A and we may change our mind at most once to put x into A. The natural generalization of this property is to allow the approximation to change more often. In this way we can get the notion of n-r.e. sets. A set A is n-r.e. if there is a recursive function $f(x, s)$ such that for all x:

$$f(x, 0) = 0$$

$$| \{s \mid f(x.s) \neq f(x.s+1)\} | \leq n$$

So, in particular, the 1-r.e. sets are precisely the r.e. sets. The 2-r.e. sets are known as the d-r.e. sets. d-r.e. means "difference recursively enumerable" since a set A is d-r.e. iff there are two r.e. sets B and C such

*The work is supported by Chinese National Nature Science Foundation and the Chinese 863 Project

that $A = B - C$. A degree is n-r.e. if it contains an n-r.e. set and if it is (n+1)-r.e. but not n-r.e., we say it is a properly (n+1)-r.e. degree. The existence of properly d-r.e. degrees shows that the class of d-r.e. degrees is a proper extension of r.e. degrees. The class of d-r.e. degrees is playing a more and more important role in linking the studying of r.e. degrees and Turing degrees. Interest in the d-r.e. degrees stems from the affinity of it with r.e. degrees. We have found that there were some differences and similarity between d-r.e. and r.e. degrees. These differences and similarity let us get well understanding to the structure of degrees.

One way to get an understanding to the various degree-theoretic structures is to see what lattices can be embedded into them. Lattice Embedding have been to show that such structures have undecidable theory and to show that the theory of such structures is decidable up to certain quantifier level. Studying lattice embedding has a long history. Now it is well known that all countable distributive lattices can be embedded into r.e. degrees and the situation on non-distributive lattices are more complicated: some can be and some not. Based on this results we got well understanding to the decidability of r.e. structures. Studying lattice embedding into d-r.e. degrees and comparing it with the embedding into r.e. degrees could let us to get a deep understanding to the structure of d-r.e. degrees and the structure of degrees. But, comparing with the studying of r.e. degrees embedding into r.e. degrees, the studying of d-r.e. degrees has a short history. Downey [4] proved that the diamond lattice can be embedded into the structure of d-r.e. r.e. degrees preserving both 0 and 1, which contrasts to the Lachlan's Non-Diamond Theorem of r.e. degrees. In this paper we will continue the study of embedding to the d-r.e. degrees. We improve the Downey's result to show that There are r.e. degree $\mathbf{a}$ and d.r.e. degree $\mathbf{b}$ such that $\mathbf{0} < \mathbf{a} < \mathbf{0}'$, $\mathbf{0} < \mathbf{b} < \mathbf{0}'$, $\mathbf{a} \cup \mathbf{b} = \mathbf{0}'$ and $\mathbf{a} \cap \mathbf{b} = \mathbf{0}$ and prove that M_5 and N_5 can be embedded into d-r.e. degrees preserving 0 and 1.

Our notations and terminologies are standard, which can be found in [1] and [6].

2 Diamond Lattice

Theorem 2.1 *There are r.e. degree* $\mathbf{a}$ *and d.r.e. degree* $\mathbf{b}$ *such that* $\mathbf{0} < \mathbf{a} < \mathbf{0}'$, $\mathbf{0} < \mathbf{b} < \mathbf{0}'$, $\mathbf{a} \cup \mathbf{b} = \mathbf{0}'$ *and* $\mathbf{a} \cap \mathbf{b} = \mathbf{0}$.

2.1 Strategy

To prove the above theorem, we shall construct an r.e. set A and a d.r.e. set B which satisfy the following requirements:

$$S: \qquad \Gamma(A \oplus B) = K \tag{1}$$

$$P_{2e}: \qquad A \neq \psi_e \tag{2}$$
$$P_{2e+1}: \qquad B \neq \psi_e \tag{3}$$
$$N_e: \quad \Phi_e^A = \Phi_e^B = f \text{ total} \implies f \text{ recursive} \tag{4}$$

Where $K \in \mathbf{0}'$, $\{\Phi_e\}_{e\in\omega}$ is an effective list of recursive functionals. If we can construct the sets A, B which satisfy the above requirements, then $\mathbf{a} = deg_T(A)$ and $\mathbf{b} = deg_T(B)$ should satisfy the theorem.

We use the tree method to construct sets. The tree we use is $\Lambda^{<\omega}$, where $\Lambda = \{0, 1\}$. The requirement S has the highest priority. Every node σ on the tree is used for both the positive requirement $P_{|\sigma|}$ and the negative requirement $N_{|\sigma|}$. If $\alpha <_L \beta$ or $\alpha \subset \beta$, then we say that P_α and N_α have higher priority than P_β and N_β and P_α has higher priority than N_α.

At any stage s, we guess the true path in the tree. We use the σ_s to denote our guess at stage s. If $\sigma_s \supseteq \delta$, we say s is a δ stage and 0 is a δ stage for any δ. For any δ and $s > 0$, $h(\delta, s)$ is defined to be the last δ stage before s if N_δ has not been canceled or initialized after the last δ stage; else we define it to be 0.

The strategy to satisfy a requirement P is the Friedberg-Muchnik method. The method to meet the requirement S is as follows: we define a functional Γ such that $\Gamma(A \oplus B) = K$. At first we define $\Gamma(A \oplus B; x) = 0$ with use $\gamma(x) = x$. Whenever some k enters K at some stage s_0, we put $\gamma(k)$, the use function of Γ, into B and define a new use $\gamma(x)$. At some stage $s_1 > s_0$, we may remove $\gamma(k)$ out of B, but in this case, some element $\leq \gamma(k)$ should have entered A or B between stage s_0 and s_1. We manage that $\gamma(x)$, for any x, can be changed at most finitely often. So A and B can recognize whether $x \in K$ for any x.

Now we consider the requirement N_σ. We define the length of the agreement between Φ_σ^A and Φ_σ^B to be:

$$l(\sigma, s) = \max\{x : (\forall y < x)[\Phi_{\sigma,s}^{A_{\sigma,s}}(x) = \Phi_{\sigma,s}^{B_{\sigma,s}}(x)]\}.$$

If $l(\sigma, s) > \max\{l(\sigma, t) : t < s\}$, we say s is a σ-expansion stage. In the presence of requirement S, we cannot use the minimal pair method without any modification. Our method to meet the requirement N_σ is as follows:

1. Assign a number b_σ big enough. Afterwards, if there is a number $< b_\sigma$ enters K, we cancel this requirement by canceling all parameters of this requirement **except** b_σ and restart at 2.

2. Wait for a σ-expansion stage s_1 which satisfies the following condition: there is an (least) x such that $x < l(\sigma, s_0) < l(\sigma, s_1)$, $\Phi_{s_0}^A(x) \neq \Phi_{s_1}^A(x)$ and requirement N_σ was not canceled between stage s_0 and s_1, where s_0 is the last σ-expansion stage before s_1. In this case, we say N_σ has an error at x. At stage s_1, we put $\gamma(b_\sigma)$ into A and lift $\gamma(y)$ for all $y \geq b_\sigma$ to be greater than s_0 (so the requirement S will not injure N_σ

if no element $< b_\sigma$ enters K afterwards). Now we say we are ready to get a global win on N_σ. Set a restraint for B to be s_1 and go to the next step.

3. Wait for the next σ-expansion stage s_2. Now we have that

$$B_{s_2}\lceil\phi(B_{s_1}, x, s_1) = B_{s_1}\lceil\phi(B_{s_1}, x, s_1),$$

so $\Phi^A_{s_2}(x) = \Phi^B_{s_2}(x) = \Phi^B_{s_1}(x) \neq \Phi^B_{s_0}(x)$. At stage s_2, we remove all elements which enter B between stage s_0 and stage s_2 from B. Initialize all requirements with lower priority (so $B\lceil s_0$ is preserved). Now we have $\Phi^A(x) = \Phi^A_{s_2}(x) \neq \Phi^B(x) = \Phi^B_{s_2+1}(x) = \Phi^B_{s_0}(x)$ and global win on N_σ.

Now we consider the interference between the requirements N_σ's. The main difficulty is as follows: a strategy of N_σ removes some elements out of B. This action could injure the computation of the B-side of some requirement N_α with higher priority. For instance, at a stage s_0 N_α has an error and we are ready to get a global win on it. But just before N_α takes action to get a global win, N_σ removed some numbers from B. It could injure the computations of N_α so that N_α could neither get a global win nor correct its error which appeared at the stage s_0.

However, the tree method can solve the above problem automatically. For any requirement N_α, if some requirement N_β may hurt it, then $\beta \supset \alpha$. This is because if N_β has the higher priority than N_α, then N_α is initialized and the previous action will be abandoned. If $\beta >_L \alpha$, then at any α stage, N_β is initialized. So if N_β needs to remove some elements from B, these elements must enters B between two consecutive α stages. Hence this will not interfere the action of N_α.

In the remainder what we should consider is how to make $A \oplus B \geq_T K$. In our construction, we use a mark system $\gamma(x)$. We guarantee that for any x, $\gamma(x)$ can be redefined at most finitly often and $\gamma(x)$ can increase at stage s only if some $\gamma(y)$ for $y \leq x$ enters A. If x enters K, we put $\gamma(x)$ into B. If it is removed at some later stage, then at that stage, we shall put some $\gamma(y)$ for $y < x$ into B to replace $\gamma(x)$.

2.2 Construction

A follower x of P_σ is said to be realized at stage s if $\psi_{e,s}(x) = 0$.

A requirement P_σ requires attention at stage s if it has not been satisfied and one of the following conditions is satisfied:

1. P_σ has a realized follower x.

2. P_σ has no follower.

A requirement N_σ requires attention at stage s if stage s is a σ-expansion stage and one of the following conditions is satisfied:

1. b_σ is undefined.
2. We have not found an error of N_σ since the last time when N_σ was canceled. There is an x such that $l(\sigma,s) > l(\sigma,t) > x$(where t is the last σ stage before s) and the requirement has not been canceled after t. $\Phi^A_{\sigma,t}(x) \neq \Phi^A_{\sigma,s}(x)$. In this case, we say **we have found an error at x and are ready to get a global win on N_σ.**
3. We are ready to get a global win on N_σ at stage $h(\sigma,s)$ and N_σ has not been canceled since then.

"A requirement is initialized" means we cancel all parameters of it and let it be unsatisfied. "A requirement N_σ is canceled" means we cancel all parameters of it except b_σ.

At any stage s, we define σ^s as follows:

$$\sigma^s_0 = \lambda \quad (5)$$

$$\sigma^s_{t+1} = \begin{cases} \sigma^s_t\hat{\ }0 & \text{if } s \text{ is a } \sigma^s_t \text{ expansion stage.} \\ \sigma^s_t\hat{\ }1 & \text{otherwise} \end{cases} \quad (6)$$

$$\sigma^s = \sigma^s_s \quad (7)$$

The function $h(\sigma,s)$ is defined to be the last σ-stage t, if there is such a stage and N_σ has not been canceled or initialized since the stage t; otherwise, it is defined to be 0.

Construction:

Stage 0: Set $A_0 = B_0 = \emptyset$. For all x, define $\Gamma(x) = x$, with use $\gamma(x) = x$.

Stage $s > 0$:

Stage s consists of two steps.

Step 1:

Find the requirement with the highest priority which requires attention at stage s. If no such requirement, for every $\alpha >_L \sigma_s$, initialize requirements P_α and N_α, go directly to step 2. Otherwise, adopt the appropriate case below:

(1) if this requirement is a positive requirement P_σ, do as follows:

a) if P_σ has a realized follower x, then put it into A or B according to $|\sigma|$ is even or odd. Initialize all requirements with priorities lower than P_σ.

b) if P_σ has no follower, assign a fresh number to be its follower. Initialize all requirements with priorities lower than P_σ.

(2) If this requirement is a negative requirement N_σ then do as follows:

a) if b_σ is undefined, then asign b_σ as a fresh number. Initialize all requirements with priorities lower than N_σ.

b) We are in case 2. That means we have found an error since $h(\sigma, s)$. Assume x is the least such error. We put $\gamma(b_\sigma)$ into A and, for all $x \geq b_\sigma$, redefine $\gamma(x)$ to be fresh numbers such that $\gamma(x)$ is monotone in x. Initialize all requirements with priorities lower than N_σ.

c) We are in case 3. That means we are ready to get a global win on N_σ. Remove all elements which enter B between $h_\sigma(h_\sigma(s))$ and s. let $z = \min\{y : \gamma(y) \text{ is removed at the stage } s\}$, we put $\gamma(z)$ into B. Initialize all requirements with priorities lower than N_σ.

Step 2:

Put $\gamma(k_s)$ into B. For any N_α with $b_\alpha \geq k_s$, Cancel N_α.

End of construction.

2.3 Verification

Lemma 2.2 *There exists an f such that $f = \liminf_s \sigma_s$, i.e. the true path exists.*

Proof: : Obviously.

For any n, we denote f_n to be $f\rceil n$.

Lemma 2.3 *For any n, the following statements hold:*

(i) There exists a stage s such that after s, there is no requirement R_α (R_α is P_α or N_α) with $\alpha \subseteq f_n$ acts after stage s.

(ii) Requirement P_{f_n} is met.

(iii) Requirement N_{f_n} is met.

Proof: : We prove this lemma by induction on n.

To show (i), by inductive hypothesis and the definition of f, we can find a stage s_0 such that after stage s_0, there is no requirement P_α or N_α with $\alpha \subset f_n$ or $\alpha <_L f_n$ acts after stage s_0. After stage s_0, P_{f_n} is never canceled and if P_{f_n} receives attention at some stage $s_1 > s_0$, it will not require attention again, so it can receive attention at most once after stage s_0. Let s_1 be a stage after s_0 such that after stage s_1, P_{f_n} will not require attention. At or after stage s_1, N_{f_n} must be assigned a b_{f_n} and this b_{f_n} will never be canceled (this is because b_{f_n} can be canceled only when N_{f_n} is initialized). Let $s_2 \geq s_1$ and $K\lceil b_{f_n} = K_{s_2}\lceil b_{f_n}$, then N_{f_n} will never be canceled after stage s_2 and can receive attentions at most twice after stage s_2.

To show (ii), assume s_0 as above. By our assumption of f, there must be infinitely many f_n stages. After stage s_0, if P_{f_n} is assigned a follower x, it will never be canceled by our assumption of s_0. If x is never realized, then we have $\Psi_e(x) \neq 0 = A(x)$ (for $n = 2e$) or $\Psi_e(x) \neq 0 = B(x)$ (for $n = 2e+1$). If it is realized at stage $s_3 > s_2$, then $\Psi_{e,s_3}(x) = 0$ and x enters A or B at the next f_n stage, say s_4. If $n = 2e$, then the theorem holds obviously. If $n = 2e+1$ then at stage s_4, every negative requirement with lower priority than P_{f_n} is initialized, so they can not remove x from B. Besides, by assumption, no requirement with higher priority than P_{f_n} could receive attentions after stage s_4, so they can not remove x from B too. So (ii) holds.

To show (iii), assume $s_5 > s_1$ such that after stage s_5, P_{f_n} does not receive attention and b_{f_n} is defined at stage s_5. Let $s_6 \geq s_5$ such that $K\lceil b_{f_n} = K_{s_6}\lceil b_{f_n}$.

Assume $\Phi_n^A = \Phi_n^B = g$ is total, then we must have infinitely many f_n expansion stages and $f_{n+1} = f_n\hat{}0$. In this case, we can recursively compute g as following: for any x, find an f_n expansion stage s' after s_6 such that $l(f_n, s') > x$, then we have $g(x) = \Phi_{n,s'}^A(x)$. If the above formula does not hold, then there must be two f_n expansion stages s_7, $s_8 > s_6$ such that $s_7 = h(f_n, s_8)$ and there is the least y such that $\Phi_{n,s_7}^A(y) \neq \Phi_{n,s_8}^A(y)$. Then f_n should act at stage s_8 and $\gamma(b_{f_n})$ is put into A to lift all other $\gamma(m)$ for $m \geq b_{f_n}$. At the next f_n expansion stage, say s_9, we remove all elements enter B between s_7 and s_9, put $\gamma_{s_9}(z)$, where $z = \min\{y : y \text{ is removed at stage } s_9\}$ into B and initialize all requirements with priorities lower than f_n. But at stage s_8, $\gamma(b_{f_n})$ is lift to be greater than s_8. We have $B_{s_7}\lceil s_8 = B_{s_9}\lceil s_8$. By our assumption, there is no $\gamma(m)$ such that $m \leq b_{f_n}$ enters B later, so $\Phi_{n,s_7}^B(y) = \Phi_{n,s_9+1}^B(y) = \Phi_n^B(y)$. But $\Phi_n^A(y) = \Phi_{n,s_9}^A(y) = \Phi_{n,s_9}^B(y) = \Phi_{n,s_8}^B(y) \neq \Phi_{n,s_7}^A(y)$. So $\Phi_n^A(y) \neq \Phi_n^B(y)$. This contradicts to our assumption. Thus N_{f_n} is met.

Lemma 2.4 $K <_T A \oplus B$

Proof: To calculate K via A and B, by our construction, $\gamma_{s+1}(x)$ is redefined to be different to $\gamma_s(x)$ only if $\gamma_s(x)$ or some $\gamma_s(y)$ for $y < x$ enters A. But by our construction, $\gamma_s(y)$ is put into A only if y is assigned to be b_σ for some σ and if $\gamma_s(y)$ is put into A, then no other $\gamma_t(y)$ can enter A. So for any x, $\gamma(x) = \lim_s \gamma_s(x)$ exists and the function $\lambda_x\gamma(x)$ is recursive in A. Next, if x enters K at s, we put $\gamma_s(x)$ into B, if $\gamma_s(x)$ is removed at a later stage t, there must be a $y < x$ such that $\gamma_t(y)$ enters B. So for any x, find an s such that $B\lceil\gamma(x) = B_s\lceil\gamma(x)$, then $x \in K \leftrightarrow x \in K_s$. But we have shown that $\lambda_x\gamma(x)$ is recursive in A, so $K \leq_T A \oplus B$.

3 N_5 lattice

In this section, we prove the following theorem:

Theorem 3.1 *There are r.e. degrees* $\mathbf{a}, \mathbf{b} > \mathbf{0}$ *and d.r.e. degree* $\mathbf{c} > \mathbf{0}$ *such that* $\mathbf{a} < \mathbf{b}$, $\mathbf{a} \cup \mathbf{c} = \mathbf{0}'$ *and* $\mathbf{b} \cap \mathbf{c} = \mathbf{0}$.

To prove the theorem, we shall construct d.r.e. set C and r.e. sets A, B which satisfy the following requirements:

$$
\begin{array}{rcr}
P_{2e}: & A \neq \psi_e & (8) \\
P_{2e+1}: & C \neq \psi_e & (9) \\
R_e: & B \neq \Phi_e(A) & (10) \\
N_e: & \Phi_e(B) = \Phi_e(C) = f \text{ total} \Longrightarrow f \text{ recursive} & (11) \\
S: & \exists\Gamma(\Gamma(A \oplus C) = K) & (12) \\
Q: & A \leq_T B & (13)
\end{array}
$$

Where $K \in \mathbf{0}'$ is an r.e. set and assume $\{k_s : s \in \omega\}$ is a recursive enumeration of it, $\{\Phi_e : e \in \omega\}$ is an effective enumeration of all recursive functionals and $\{\psi_e : e \in \omega\}$ is an effective enumeration of all recursive functions.

To satisfy the requirement Q, we ensure that all numbers of A are even and $B \cap \{2x : x \in \omega\} = A$.

The strategies to satisfy the requirements S, P and N are similar to the strategies used in the previous section. The only different is that we should consider the requirement R. The method to satisfy the requirement R is the Friedberg-Muchnic method.

Now we describe the construction.

We also use the tree method to construct the sets. The tree we use is same as the tree used in the above section. Each node $\sigma \in T$ is used both for requirement N, P and R. If $\alpha <_L \beta$ or $\alpha \subset \beta$ then the requirement relate to α has a higher priority than the requirement relate to β. Besides, we assume the requirement P_α has a higher priority than R_α. Requirement R_α has a higher priority than N_α.

We define the length agreement function l as follows:

$$l(\sigma, s) = \max\{x : \forall y < x(\Phi_{|\sigma|,s}(B_s, y) = \Phi_{|\sigma|,s}(C_s, y))\}$$

We define $ml(\sigma, s) = \max\{l(\sigma, t) : t < s\}$. At any σ stage s, if $l(\sigma, s) > ml(\sigma, s)$, then we say s is a σ-expansion stage. At stage s, we define σ_s recursively as follows:

$$\sigma_s^0 = \lambda \tag{14}$$

$$\sigma_s^{t+1} = \begin{cases} \sigma_s^{t\frown}0 & \text{if } s \text{ is a } \sigma_s^t \text{ expansion stage.} \\ \sigma_s^{t\frown}1 & \text{otherwise} \end{cases} \tag{15}$$

$$\sigma_s = \sigma_s^s \tag{16}$$

And the function $h(\sigma, s)$ is same as in the previous section.

A follower x of requirement P_α is said to be realized at stage s if $\psi_{e,s}(x) = 0$, where $|\alpha| = 2e + i$ for $i < 2$. A follower x of requirement R_α is said to be realized at stage s if $\Phi^A_{|\alpha|,s}(x) = 0$.

At stage s, we say requirement P_α or R_α requires attention if stage s is an α stage and one of the following conditions hold:

1) P_α has a realized follower.

2) P_α has no follower.

At stage s, we say requirement N_α requires attention if stage s is an α-expansion stage and one of the following conditions hold:

1) b_α is undefined at stage s.

2) Let t be the previous α-expansion stage. N_α has not been initialized or cancelled between stage t and stage s. There is an x such that $l(\alpha, s) > l(\alpha, t) > x$ and $\Phi_{\alpha,s}(x) \neq \Phi_{\alpha,t}(x)$. That means we have found an error at x and we say that we are ready to get a global win on N_α.

3) We are ready to get a global win on N_α at stage $h(\alpha, s)$ and N_α has not been cancel or initialized since then.

The definitions of initializing a requirement and canceling a requirement are same as that of the above section.

Construction:

Stage 0: Set $A_0 = B_0 = C_0 = \emptyset$. For all x, define $\Gamma(x) = 0$ with use $\gamma(x) = 2x$.

Stage $s > 0$:

Stage s consists of two steps.

Step 1:

Find the requirement with the highest priority which requires attention at stage s. If no such requirement, for every $\alpha >_L \sigma_s$, initialize all requirements P_α, R_α, and N_α. Directly go to step 2. Otherwise, we adopt the appropriate case below:

(1) If the requirement is P_α, then

a) If P_α has a realized follower x and $|\alpha|$ is odd, then we put x into C, else, we put x into A and B. Initialize all requirements with priorities lower than P_α

b) If P_α has no follower, then we assign an even fresh number to be a follower of P_α. Initialize all requirements with priorities lower than P_α

(2) If the requirement which requires attention with the highest priority is R_α, then do as follows:

c) If R_α has a realized follower x then we put x into B. Initialize all requirements with priorities lower than R_α.

d) If R_α has no follower, then we assign an odd fresh number to be a follower of R_α. Initialize all requirements with priorities lower than R_α

(3) If the requirement which requires attention with the highest priority is N_α, then do as follows:

e) If b_α is undefined, then we assign a fresh number to b_α. Initialize all requirements with priority lower than N_α.

f) We are in case (2). That means we have found an error at the stage $h(\alpha, s)$. Assume x is the least such error. We put $\gamma(b_\alpha)$ into A and, for all $x \geq b_\alpha$, redefine $\gamma(x)$ to be an even fresh number such that $\gamma(x)$ is increasing in x. Initialize all requirements with priorities lower than N_α. We say that we are ready to get a global win on N_α.

g) We are in case (3). That means we are ready to get a global win on N_α. In this case, remove all elements which enter C between $h_\alpha(h_\alpha(s))$. Let $x = \min\{y : y$ is removed at the stage $s\}$, put x into C and initialize all requirements with priorities lower than N_α. We say that we have got a global win on N_α.

Step 2:

Put $\gamma(k_s)$ into C. For any N_β with $b_\beta \geq k_s$, cancel N_β.

3.1 Verification

The construction of N_5 lattice is similar to that of diamond lattice. So we only discuss the difference between them.

In this construction, the requirements P and S are met as the previous section. This is because this construction and the previous construction are same for these requirements.

Let f be the true path of the construction and f_n stands for $f\lceil n$.

For a requirement R_{f_n}, if it is not injured after some stage s_0, then it acts at most twice. So its influence to other requirements is finite. Besides, after stage s_0, it must have a follower x and this follower will never be canceled. If it never receives attention after it gets a follower, then at any f_n stage t (note that there are infinitely many f_n stages), we have $\Phi_{n,t}(A, x) \neq 0 = B(x)$. So the requirement R_n is satisfied. If the requirement R_{f_n} receives attention at stage t after stage s_0 and puts its follower x into B, then we have $\Phi_{n,t}(x) = 0 \neq B_{t+1}(x) = B(x)$. But at stage t, all requirements with priorities lower than R_{f_n} are initialized. So the computation of $\Phi_{n,t}(A, x)$ could not be injured afterwards and then $\Phi_n(A, x) = \Phi_{n,t}(A, x) = 0 \neq B(x)$. Thus the requirement R_n is satisfied via R_{f_n}.

For a requirement N_{f_n}, the only difference between this construction and that of diamond lattice is that we should put some elements into B via requirements R_n. But their influence to requirements N_n is similar to that of requirements P_n. Every requirement R_{f_n} requires attention at most finitely often. Besides, if we generate a disagreement between $\Phi_{f_n}(B)$ and $\Phi_{f_n}(C)$, we initialize all requirements with lower priorities, include those R_α requirements. So they will not destroy this disagreement. This completes the proof of theorem 3.1.

4 M_5 lattice

In this section, we prove the following theorem:

Theorem 4.1 *There are r.e. degree* $\mathbf{a} > \mathbf{0}$ *and d.r.e. degrees* $\mathbf{b}, \mathbf{c} > \mathbf{0}$ *such that* $\mathbf{a} \cup \mathbf{b} = \mathbf{b} \cup \mathbf{c} = \mathbf{c} \cup \mathbf{a} = \mathbf{0}'$ *and* $\mathbf{a} \cap \mathbf{b} = \mathbf{b} \cap \mathbf{c} = \mathbf{c} \cap \mathbf{a} = \mathbf{0}$

Note that we can not force $\mathbf{b}$ or $\mathbf{c}$ to be r.e., because it contradicts to Lachlan's non-diamond theorem.

To prove this theorem we shall construct r.e. set A and d.r.e. sets B, C which satisfy the following requirements:

$$P_{3e}: \quad A \neq \psi_e \tag{17}$$
$$P_{3e+1}: \quad B \neq \psi_e \tag{18}$$
$$P_{3e+2}: \quad C \neq \psi_e \tag{19}$$
$$N_{3e}: \quad \Phi_e(B) = \Phi_e(C) = f \text{ total } \Longrightarrow f \text{ recursive} \tag{20}$$
$$N_{3e+1}: \quad \Phi_e(C) = \Phi_e(A) = f \text{ total } \Longrightarrow f \text{ recursive} \tag{21}$$
$$N_{3e+2}: \quad \Phi_e(A) = \Phi_e(B) = f \text{ total } \Longrightarrow f \text{ recursive} \tag{22}$$
$$S_0: \quad \exists \Gamma_0(\Gamma_0(B \oplus C) = K) \tag{23}$$
$$S_1: \quad \exists \Gamma_1(\Gamma_1(C \oplus A) = K) \tag{24}$$
$$S_2: \quad \exists \Gamma_2(\Gamma_2(A \oplus B) = K) \tag{25}$$

Where $K \in \mathbf{0}'$ is an r.e. set and assume $\{k_s : s \in \omega\}$ is a recursive enumeration of it. $\{\Phi_e : e \in \omega\}$ is an effective enumeration of all recursive functionals and $\{\psi_e : e \in \omega\}$ is an effective enumeration of all recursive functions.

For requirements S, at every stage s, we put $\gamma_2(k_s)$ into B and put $\gamma_0(k_s)$ and $\gamma_1(k_s)$ into C.

We also define the length agreement function l as follows($i < 3$):

$$l(3e+i, s) = \begin{cases} \max\{x : \forall y < x(\Phi_{e,s}(B, y) = \Phi_{e,s}(C, y))\} & i = 0 \\ \max\{x : \forall y < x(\Phi_{e,s}(A, y) = \Phi_{e,s}(C, y))\} & i = 1 \\ \max\{x : \forall y < x(\Phi_{e,s}(A, y) = \Phi_{e,s}(B, y))\} & i = 2 \end{cases} \tag{26}$$

the definitions of function *ml* and *expansion stage* are similar to that in the previous sections.

For requirement N_{3e+2}, we do as follows:

1.1) Assign a number b big enough. If there is a number $< b$ enters K later, we cancel this requirement by canceling all parameters of this requirement except b.

1.2) Wait for a $(3e+2)$-expansion stage s_1 such that $l(3e+2, s_1) > l(3e+2, s_0) > x$, but $\Phi_{e,s_1}(A, x) \neq \Phi_{e,s_0}(A, x)$, where s_0 is the previous

$(3e+2)$-expansion stage and the requirement was not initialized or canceled between stage s_0 and s_1. In this case, we say that we have found an error at x. At stage s_1, we put $\gamma_2(b)$ into A and lift the $\gamma_2(b)$ and all other $\gamma_2(y)$ for $y > b$ to be very big values. We say that we are ready to get a global win on N_{3e+2}.

1.3) Wait for the next $(3e+2)$-expansion stage s_2. At stage s_2 we remove all elements enter B between stage s_0 and s_2 and, let x be the least number which is removed at the stage s, put x into B. Now, if the requirement is not injured later, the requirement is satisfied permanently.

For a requirement N_{3e+1}, we do as follows:

2.1) Assign a number b big enough. If there is a number $< b$ enters K later, we cancel this requirement by cancel all parameters of this requirement except b.

2.2) Wait for a $(3e+1)$-expansion stage s_1 such that $l(3e+1, s_1) > l(3e+1, s_0) > x$, but $\Phi_{e,s_1}(A, x) \neq \Phi_{e,s_0}(A, x)$, where s_0 is the previous $(3e+1)$-expansion stage and the requirement was not initialized or canceled between stage s_0 and s_1. In this case, we say that we have found an error at x. At stage s_1, we put $\gamma_1(b)$ into A, $\gamma_0(b)$ into C and lift the $\gamma_1(y)$ and $\gamma_0(y)$ for all $y \geq b$ to be fresh values. We say that we are ready to get a global win on N_{3e+1}.

2.3) Wait for the next $(3e+1)$-expansion stage s_2. At stage s_2 we remove all elements enter C between stage s_0 and s_2 and put $\gamma_1(b)$ and $\gamma_0(b)$ into B. Now, if the requirement is not injured later, the requirement is satisfied permanently.

The strategy to meet a requirement N_{3e} is different to those of the other cases. It has two kinds of possible actions. We call them B-action and C-action. We do as follows:

3.1) Assign a number b big enough. If there is a number $< b$ enters K later, we cancel this requirement by cancel all parameters of this requirement except b.

3.2) Wait for a $3e$-expansion stage s_1 such that $l(3e, s_1) > l(3e, s_0) > x$, but $\Phi_{e,s_1}(B, x) \neq \Phi_{e,s_0}(B, x)$, where s_0 is the previous $3e$-expansion stage and the requirement has not been initialized or canceled between stage s_0 and s_1. In this case, we say we have found an error at x. At stage s_1, if $C_{s_1} \supseteq C_{s_0}$, then we do C-action, else do B-action. (We will show later that if we do B-action, then we must have $B_{s_1} \supseteq B_{s_0}$). If we do C-action, then go to 3.2a. If we do B-action, then go to 3.2b.

3.2a) Put $\gamma_1(b)$ into A, $\gamma_0(b)$ into B and lift the $\gamma_1(y)$ and $\gamma_0(y)$ for all $y \geq b$ to be fresh values. (So C is preserved) Go to 3.3a.

3.3a) Wait for the next $3e$-expansion stage s_2. At s_2, we put $\gamma_2(b)$ into A and lift the $\gamma_2(y)$ for all $y \geq b$ to be fresh values. (So the computation $\Phi_{e,s_2}(B, x) = \Phi_{e,s_1}(B, x)$ is preserved). Go to 3.4a.

3.4a) Wait for the next $3e$-expansion stage s_3. At stage s_3, we remove all elements enter C between stage s_0 and s_3 from C and put $\gamma_0(b)$ and

$\gamma_1(b)$ into C. Now if this requirement will not be initialized or canceled, then this requirement is satisfied permanently.

3.2b) For B-action, we put $\gamma_2(b)$ into A at stage s_1 and lift $\gamma_2(y)$ for all $y \geq b$ to be fresh values. (So B is preserved) We also put $\gamma_0(b)$ into C and lift $\gamma_0(y)$ for all $y \geq b$ to be fresh values. (This is because we shall put $\gamma_0(b)$ into B at next $3e$-expansion stage and this action can not destroy the current calculation.) Go to 3.3b.

3.3b) Wait for the next $3e$-expansion stage s_2. At stage s_2, we put $\gamma_0(b)$ into B and $\gamma_1(b)$ into A. We lift $\gamma_0(y)$ and $\gamma_1(y)$ for all $y \geq b$ to be fresh values. (Now, the computation of $\Phi_{e,s_2}(B,x)$ is preserved.) Go to 3.4b.

3.4b) Wait for the next $3e$-expansion stage s_3. At stage s_3, we remove all elements enters B between s_0 and s_3 from B. Now if this requirement is never initialized or canceled, then our requirement is satisfied permanently.

Now we consider interactions between requirements. At first, we have the following danger:

Let N_α have a higher priority than N_β and $|\alpha| = 3e_1 + 2$, $|\beta| = 3e_2 + 2$. Assume s_0 and s_1 are two consecutive α expansion stages and we find an error at stage s_1. If at some stage t_0 such that $s_0 < t_0 < s_1$, N_β put some $\gamma_{0,t_0}(b')$ into B to lift other $\gamma_0(y)$ for $y \geq b'$. At a stage t_1 such that $t_0 < t_1 < s_1$, y enters K, so we need put $\gamma_{0,t_1}(y)$ into C. But assume at some stage $s_2 > s_1$, $\gamma_{0,t_0}(b')$ is removed from B, so it is possible that $\gamma_{0,s_2+1}(y) = \gamma_{0,t_0}(y) < \gamma_{0,t_1}(y)$. Now, we should put $\gamma_{0,s_2+1}(y) = \gamma_{0,t_0}(y)$ into B or C again to meet the requirement S_0. However, this action will cause B or C is not d.r.e. set, so the construction should be modified. The requirement N_{3e+2} should act as following:

1.1) Assign a number b big enough. If there is a number $< b$ enters K later, we cancel this requirement by canceling all parameters of this requirement except b.

1.2) Wait for a $(3e+2)$-expansion stage s_1 such that $l(3e+2, s_1) > l(3e+2, s_0) > x$, but $\Phi_{e,s_1}(A,x) \neq \Phi_{e,s_0}(A,x)$, where s_0 is the previous $(3e+2)$-expansion stage and the requirement has not been initialized or canceled between stage s_0 and s_1. In this case, we say we have found an error at x. At stage s_1, we put $\gamma_{2,s_1}(b)$ into A and lift all $\gamma_2(y)$ for $y \geq b$ to be fresh values. *We also put* $\gamma_{0,s_0}(b)$ *into* C. (Note that $\gamma_{0,s_1}(b) = \gamma_{0,s_0}(b)$. This is because if they do not equal, then the requirement N_{3e+2} must be initialized or canceled between stage s_0 and s_1.) Lift $\gamma_{0,s+1}(b)$ to be a fresh value.

1.3) Wait for the next $(3e+2)$-expansion stage s_2. At stage s_2 we remove all elements enters B between stage s_0 and s_2 and put $\gamma_2(b)$ into B. *We also put* $\gamma_0(b)$ *into* B. Now, if the requirement is not injured later, the requirement is satisfied permanently.

Note that if there is an element $< b$ enters K between stage s_0 and s_2, then the requirement is canceled, so we can not get 1.3. If $y \geq b$ enters K at stage t between stage s_0 and s_1, then we have $\gamma_{0,s_1}(b) = \gamma_{0,s_0}(b) \leq \gamma_{0,s_0}(y)$.

But even the value of $\gamma_0(y)$ decreases, it will recover to the value at stage s_0. Actually we have put $\gamma_{0,s_1}(b)$ into C instead of $\gamma_0(y)$. If the element enters K between stage s_1 and s_2, the situation is similar. We put $\gamma_{0,s_2}(b)$ into C instead of $\gamma_0(y)$. (Because we put $\gamma_0(b)$ into C at stage s_1, the $\gamma_0(y)$ can only decrease to $\gamma_{0,s_1+1}(y) > \gamma_{0,s_2}(b)$).

For a requirement N_{3e+1}, because we have put $\gamma_{0,s_1}(b)$ into C at stage s_1 and $\gamma_{0,s_2}(b)$ into B at stage s_2, the construction needs not change. For requirement N_{3e}, either a B-action or a C-action need not change the construction due to the similar reason.

Now the last obstacle is that when we remove something from B or C, we may destroy some computations relative to B and C. But if we generate an error later, we can not make a difference by removing some elements from B or C. This obstacle can be overcome by a tree construction. For a requirement N_α, because at any $\alpha\hat{}0$-stage we initialize all N_β for $\beta >_L \alpha$, the requirement N_β for $\beta >_L \alpha\hat{}0$ can not hurt the requirement N_α. If a requirement N_β can hurt N_α, we must have $\beta \supset \alpha$. Actually, in our construction, when requirement N_β remove something from B or C(say at stage t), we never put anything into A at the same stage t. At the stage t, we initialize all requirements with lower priorities than N_β. So the A-side computation of requirement N_α can not be injured at stage t. Now at the next N_α-expansion stage, the A-side computation can not generate an error, so we will not find an error at the next N_α-expansion stage.

Next, in our construction, we never remove something from B and at same time also remove something from C. If we remove something from B at stage t, then the B-side computation of requirement N_α may be injured and we can not recover this computation. But at this case, we can not remove any element from C at this stage and all requirements with lower priorities are initialized. So for requirement N_α, we can make a disagreement by remove something from C. (This is why we use B-action and C-action for requirement N_{3e}).

Now the last danger is that $\alpha\hat{}0 \subset \beta_1$ and $\alpha \subset \beta_2$. Assume N_{β_1} remove something from B at first and N_{β_2} remove something from C later, then they may destroy the both sides' computations and both can not be recovered. If N_{β_1} has higher priority than N_{β_2}, then when N_{β_1} acts, it initializes N_{β_2}. So when N_{β_2} acts next time, we must have pass an $N_\alpha\hat{}0$ stage and the computation of B has been recovered. If N_{β_1} has a lower priority than N_{β_2}, then at the stage when N_{β_1} act, N_{β_2} can not act (or N_{β_1} can not act.). At the next time when N_{β_2} can act, this stage must be an $\alpha\hat{}0$ stage, so the B side computation also has been recovered.

Now we have finished the description of the construction. The verification is similar to the previous sections.

References

[1] S. B. Cooper & X. Yi, *Isolated d.r.e. degrees,* Preprint

[2] S. B. Cooper, H. Harrington, A. H. Lachlan, S. Lempp & R. I. Soare, *The d.r.e. degrees are not dense,* Annals of Pure & Applied Logic vol. 55 (2), 1991 pp. 125-152

[3] D. Ding & L. Qian, *Isolated d.r.e. degrees are dense in r.e. degree structure* , Arch. Math. Logic vol 36, 1996 pp. 1-10

[4] R. G. Downey, *D-recursively enumerable degrees and the nondiamond theorem,* Bull. London Math. Soc., vol 211989, pp. 43-50

[5] G. E. Sacks, *The recursively enumerable degrees are dense,* Ann. of Math. vol. 80 (2), 1964, pp.300-312

[6] R.I.Soare, *Recursively Enumerable Sets and Degrees,* Springer-Verlag, Berlin, Heidelberg, New York, London, Paris, Tokyo, 1987

On Stationary Reflection Principles

Qi Feng

§0 Introduction

It follows from the works of Gödel and Cohen that the Continuum Hypothesis can neither be proved nor disproved. In the past thirty years or so, many statements in mathematics have been proven to be independent from the axiom of set theory, ZFC. In order to settle these problems one way or the other, people have formulated many new axioms or principles of infinity. Notably, Martin's Axiom, the Diamond Principle, the Box Principles, the Proper Forcing Axioms, the Semiproper Forcing Axioms, Martin's Maximum, and the Principles of Stationary Reflection, as well as many large cardinal axioms. These powerful principles do settle many open problems and provide us with more understanding of the infinity.

In this talk, I will give a survey on the Principles of Stationary Reflection. We start with a review of stationary sets. Then in section 2, we present some interesting forms of the principles of stationary reflection. In the following sections, we will present some applications of the principles.

§1 Stationary, Projective Stationary Sets of Countable Models

Primarily, stationary sets are subsets of ordinals. Here we will be concerned a generalization of the notion of stationary sets to spaces of countable models, due to Jech [8].

Recall that if α is an ordinal and $C \subseteq \alpha$, then C is *unbounded in* α if $\alpha = \cup C$, and C is *closed in* α if $(\cup(C \cap \beta)) \in C$ for all $\beta < \alpha$. A set $S \subseteq \alpha$ is *stationary in* α if $S \cap C$ is not empty for all sets C closed and unbounded in α.

Theorem 1.1 (Fodor's Lemma) Suppose that κ is a regular uncountable cardinal.

(a) If $\langle C_\alpha \mid \alpha < \kappa \rangle$ is a sequence of sets closed and unbounded in κ, then

$$C = \{\alpha < \kappa \mid \forall\, \beta < \alpha\ \alpha \in C_\beta\,\}$$

is also closed and unbounded in κ. C is called the diagonal intersection of the sequence.

(b) Suppose that $S \subseteq \kappa$ is stationary in κ and $f : S \to \kappa$ is such that $f(\alpha) < \alpha$ for all $\alpha \in S - \{0\}$ (i.e., f is regressive on S). Then there exists a stationary $T \subseteq S$ such that f is constant on T.

Assume that κ is a regular uncountable cardinal and $\lambda \geq \kappa$ is a cardinal. Let A be a set of cardinality λ. We use $P_\kappa(A)$ to denote the set of all subsets of A of size $< \kappa$. A set $C \subseteq P_\kappa(A)$ is **closed** if for every $\subseteq$–increasing countable sequence $\langle x_\alpha \mid \alpha < \theta \rangle$ from C of length $\theta < \kappa$, ($x_\alpha \subseteq x_\beta$ for all $\alpha < \beta < \theta$), the union of the sequence $\cup\{x_\alpha \mid \alpha < \theta\}$ is also in C. C is **unbounded** if for all $x \in P_\kappa(A)$ there exists some $y \in C$ such that $x \subseteq y$. C is a **club** if C is both closed and unbounded. A set $S \subseteq P_\kappa(A)$ is **stationary** if for every club C, the intersection $S \cap C$ is not empty.

THEOREM 1.2 (JECH[8])

(1) All the clubs on $P_\kappa(A)$ generate a κ–complete normal filter. Namely, if $\langle C_a \mid a \in A \rangle$ is a sequence of sets closed and unbounded in $P_\kappa(A)$, then

$$C = \{\, x \in P_\kappa(A) \mid \forall\, a \in x\; x \in C_a \,\}$$

is also closed and unbounded in $P_\kappa(A)$. C is called the diagonal intersection of the sequence.

(2) If S is a stationary set in $P_\kappa(A)$, $f : S \to A$ satisfying that $f(x) \in x$ for all $x \in S$, then there exists a stationary $T \subseteq S$ such that f is constant on T.

Another basic fact is that given two sets A and B of the same uncountable cardinality , then there is a natural correspondence between the closed and unbounded sets and stationary sets. Namely if $f : A \to B$ is a bijection, $C \subseteq P_\kappa(A)$ is a set closed and unbounded in $P_\kappa(A)$, letting $C^* = \{f''X \mid X \in C\}$, then C^* is a set closed and unbounded in $P_\kappa(B)$ (where $f''X$ denotes the set $\{f(a) \mid a \in X\}$). If S is a set stationary in $P_\kappa(A)$, letting $S^* = \{f''X \mid X \in S\}$, then S^* is a set stationary in $P_\kappa(B)$.

Assume that $B \subseteq A$. Let C be a set closed and unbounded in $P_\kappa(A)$. Then $\{X \cap B \mid X \in C\}$ is a set closed and unbounded in $P_\kappa(B)$. Conversely, if C is a set closed and unbounded in $P_\kappa(B)$, then $\{X \in P_\kappa(A) \mid X \cap B \in C\}$ is a set closed and unbounded in $P_\kappa(A)$. Hence, if S is a set stationary in $P_\kappa(A)$, then $\{X \cap B \mid X \in S\}$ is stationary in $P_\kappa(B)$. If S is stationary in $P_\kappa(B)$, then $\{X \in P_\kappa(A) \mid X \cap B \in S\}$ is stationary in $P_\kappa(A)$.

We will be interested mainly the case that $\kappa = \omega_1$. For an uncountable set A, we use $[A]^\omega$ to denote the set of all infinite countable subsets of A. This is a set closed and unbounded in $P_{\omega_1}(A)$.

If $f : [A]^{<\omega} \to A$ and $X \subseteq A$, then we say that X is closed under f if X is nonempty and for all $e \in [X]^{<\omega}$ $f(e) \in X$. If $X \subseteq A$, the closure of X under f is the smallest $Y \supseteq X$ which is closed under f. Usually we denote the closure of X under f by $cl_f(X)$. Notice that given $f : [A]^{<\omega} \to A$, the family of all countable subsets of A which are closed under f is closed and unbounded in $[A]^\omega$. In particular, given an uncountable structure of countable language endowed with a well ordering, the set of all countable elementary submodels is closed and unbounded.

THEOREM 1.3 (KUCKER[12]) Assume that C is closed and unbounded in $[A]^\omega$. Then there exists an $f : [A]^{<\omega} \to A$ such that every countable subset X of A which is closed under f is in C. Hence $S \subseteq [A]^\omega$ is stationary if and only if for every $f : [A]^{<\omega} \to A$, there exists an $x \in S$ such that x is closed under f, i.e., $f(e) \in x$ for all $e \in [x]^{<\omega}$.

For a regular uncountable κ, $\lambda \geq \kappa$, $f : [\lambda]^{<\omega_1} \to \lambda$, the set

$$\{X \in P_\kappa(\lambda) \mid X \text{ is closed under } f\}$$

is closed and unbounded in $P_\kappa(\lambda)$. Also the set

$$\{X \in P_\kappa(\lambda) \mid X \cap \kappa \in \kappa\}$$

is closed and unbounded in $P_\kappa(\lambda)$. The following theorem is a folklore.

THEOREM 1.4 Assume that C is a set closed and unbounded in $P_\kappa(\lambda)$ (κ is regular uncountable and $\lambda \geq \kappa$ is a cardinal). Then there exists an $f : [\lambda]^{<\omega} \to \lambda$ such that if $X \in P_\kappa(\lambda)$ is closed under f and $X \cap \kappa \in \kappa$ then $X \in C$. Hence $S \subseteq P_\kappa(\lambda)$ is stationary if and only if for every $f : [\lambda]^{<\omega} \to \lambda$ there exists some $X \in S$ such that X is closed under f and $X \cap \kappa \in \kappa$.

Let κ be a regular cardinal $\geq \omega_2$. Let H_κ be the set of all sets hereditarily of size less than κ. We assume that H_κ is endowed with a well ordering. Notice that $[H_\kappa]^\omega \subseteq H_\kappa$. We are interested in such structures because $(H_\kappa, \in)$ is a model of ZFC minus the power set axiom (when κ is large, sufficient initial segment of the power set axiom will hold). Essentially, when we choose large enough κ, $(H_\kappa, \in)$ will be a good model of set theory satisfying our needs. We are interested in the stationary sets on the spaces of countable sets, primarily because there is a closed and unbounded set of countable elementary submodels of $(H_\kappa, \in)$ and theses countable models are rich enough to have many desired properties. For example, if N is a countable elementary submodel of H_κ, then $N \cap \omega_1$ is a countable ordinal. This follows from the fact that if $x \in N$ is countable then $x \subseteq N$.

We use $N \prec M$ to denote that N is an elementary submodel of M as usual.

A sequence $\langle N_\alpha \mid \alpha < \theta\rangle$ of countable submodels of H_κ is an **increasing continuous** $\in$**–chain** of length θ if for all $\alpha < \theta$, $(N_\alpha \prec H_\kappa)$ and for all $\alpha < \beta < \theta$, $(N_\alpha \in N_\beta)$, and if $\alpha < \theta$ is a limit ordinal then $N_\alpha = \bigcup_{\beta<\alpha} N_\beta$.

A sequence $\langle N_\alpha \mid \alpha < \theta\rangle$ of countable submodels of H_κ is an **increasing continuous $\subseteq$–chain** of length θ if for all $\alpha < \theta$, $(\alpha \subseteq N_\alpha \prec H_\kappa)$ and for all $\alpha < \beta < \theta$, $(N_\alpha \subseteq N_\beta)$, and if $\alpha < \theta$ is a limit ordinal then $N_\alpha = \bigcup_{\beta<\alpha} N_\beta$.

It follows that every increasing continuous $\in$–chain is an increasing continuous $\subseteq$–chain.

An interesting fact about the stationary sets of countable models is that if $S \subseteq [H_\kappa]^\omega$ is stationary then S contains an increasing continuous $\in$–chain of length α for every countable ordinal α.

This is a generalization of the following lemma (see [9], Exercise 7.12, Page 60).

LEMMA 1.1 If $S \subseteq \omega_1$ is stationary, then for any countable ordinal α, for any countable ordinal γ, there exists an $x \subseteq S - \gamma$ such that x is closed and has order type $\alpha + 1$.

The proof of this lemma is by induction on α. We will use this lemma to prove the following:

THEOREM 1.5 If $S \subseteq [H_\kappa]^\omega$ is stationary, then for every countable ordinal α, there exists an increasing continuous $\in$–chain $\langle N_\gamma \mid \gamma \leq \alpha\rangle$ of length $\alpha + 1$ such that $N_\gamma \in S$ for all $\gamma \leq \alpha$.

Proof Let $S \subseteq [H_\kappa]^\omega$ be stationary. Let $\alpha < \omega_1$ be fixed. We will show that in some generic extension by a σ–closed forcing the conclusion of the lemma holds for this S and α. That will be enough.

Consider the following forcing P.

A condition p is an increasing continuous $\in$–chain of length $\theta + 1$ for some $\theta < \omega_1$ such that $p(\beta) \prec H_\kappa$ is countable for all $\beta < \theta + 1$.

The ordering is by extension.

Claim P is σ–closed.

To see this, let $\langle p_n \mid n < \omega\rangle$ be a decreasing sequence of conditions. Let θ_n be the largest countable ordinal in the domain of p_n. Let $\theta = \bigcup_{n<\omega} \theta_n$ and let $N = \bigcup_{n<\omega} p_n(\theta_n)$. Then $N \prec H_\kappa$ is a countable elementary submodel. Define $q(\theta) = N$ and $q(\beta) = p_n(\beta)$ for $\beta < \theta$ with n the least such that $\beta \leq \theta_n$. q is then a condition extending all the p_n.

Hence P is a σ–closed forcing.

Let $G \subseteq P$ be a generic filter over V. A general fact says that σ–closed forcing preserves stationary subsets of $[A]^{\aleph_0}$. It follows that S is stationary in $V[G]$. In $V[G]$, let $f = \bigcup G$. Then f is an increasing continuous $\in$–chain of length ω_1 which enumerates a closed unbounded subset of $([H_\kappa]^\omega)^V$. Since S remains stationary in the generic extension, the following set is a stationary subset of ω_1:

$$T = \{\beta < \omega_1 \mid f(\beta) \in S\}.$$

By the quoted lemma above, let $x \subseteq T$ be a closed set of order type $\alpha + 1$. Let $x = \{\gamma_\beta \mid \beta \leq \alpha\}$ be the canonical enumeration. Let $N_\beta = f(\gamma_\beta)$ for $\beta \leq \alpha$. Then $\langle N_\beta \mid \beta \leq \alpha \rangle \in V$ is the desired increasing continuous $\in$–chain of length $\alpha + 1$.

This proves the theorem. $\square$

At this point, one would like to ask the following question: Which stationary sets of $[H_\kappa]^\omega$ contain an increasing continuous $\in$–chain of length ω_1? This turns out to be a very interesting and nontrivial question.

Suppose $S \subseteq [H_\kappa]^\omega$ is stationary and S contains an increasing continuous $\in$–chain of length ω_1. Then necessarily the set $\{N \cap \omega_1 \mid N \in S\}$ contains a closed and unbounded set in ω_1.

This motives the following concept of projective stationary set, introduced in [4].

DEFINITION 1.1 The **projection** of a set $S \subseteq [H_\kappa]^\omega$ is the set $\text{Proj}(S) = \{X \cap \omega_1 : X \in S\}$. We say that $S \subseteq [H_\kappa]^\omega$ is **projective stationary** if for every club $C \subseteq [H_\kappa]^\omega$, $\text{Proj}(S \cap C)$ contains a club in ω_1. Equivalently, S is projective stationary if and only if the set $S_T = \{X \in S \mid X \cap \omega_1 \in T\}$ is stationary for every stationary subset T of ω_1.

It follows from the definition that every projective stationary set is stationary. But the converse is not true. For example, let S, T be two disjoint stationary subsets of ω_1. Then both

$$\{N \in [H_\kappa]^\omega \mid N \cap \omega_1 \in S\}$$

and

$$\{N \in [H_\kappa]^\omega \mid N \cap \omega_1 \in T\}$$

are stationary sets but none of it is projective stationary.

THEOREM 1.6 Let $A \subseteq \kappa$ be stationary such that all $\alpha \in A$ are of cofinality ω. Then $\{N \in [H_\kappa]^\omega \mid sup(N \cap \kappa) \in A\}$ is projective stationary.

Proof Let $T \subseteq \omega_1$ be stationary. Let $f : [H_\kappa]^{<\omega} \to H_\kappa$. Let $\omega_1 \subseteq X \subseteq H_\kappa$ be such that $X \cap \kappa \in A$ and X is closed under f. Let $\langle \alpha_n \mid n < \omega \rangle$ be a countable sequence from X which is cofinal in $sup(X \cap \kappa)$. Let Y be the closure of $\omega_1 \cup \{\alpha_n \mid n < \omega\}$. We may assume that $Y \prec X \prec H_\kappa$. Now choose $N \prec Y$ so that N is countable , $N \cap \omega_1 \in T$ and $\alpha_n \in N$ for all $n < \omega$. This finishes the proof. $\square$

Hence, there are disjoint projective stationary sets. For example, let A, B be disjoint stationary subsets of κ such that every $\alpha \in A \cup B$ is of cofinality ω. Then

$$\{N \in [H_\kappa]^\omega \mid sup(N \cap \kappa) \in A\}$$

and

$$\{N \in [H_\kappa]^\omega \mid sup(N \cap \kappa) \in B\}$$

are disjoint projective stationary sets.

§2 PRINCIPLES OF STATIONARY REFLECTION

We can now ask if every projective stationary subset of $[H_\kappa]^\omega$ can contain an increasing continuous $\in$–chain of length ω_1 for every regular $\kappa \geq \omega_2$. The answer is independent of the axiom of set theory. And the positive answer has many interesting consequences.

We now present the positive answer as the following Projective Stationary Reflection principle (**PSRP**) [4]:

Projective Stationary Reflection: *For every regular cardinal $\kappa \geq \omega_2$, if $S \subseteq [H_\kappa]^\omega$ is a projective stationary set, then there exists an increasing continuous $\in$–chain $\langle N_\alpha \mid \alpha < \omega_1\rangle$ of countable elementary submodels of H_κ of length ω_1 such that $N_\alpha \in S$ for all $\alpha < \omega_1$.*

We now show that the Projective Stationary Reflection Principle follows from Martin's Maximum.

Let us recall Martin's Maximum:

A forcing P is ω_1–stationary preserving if every stationary subset of ω_1 in the ground model remains stationary in the generic extension. Martin's Maximum is the following statement:

If P is a ω_1–stationary preserving partially ordered set, if $\{D_\alpha \mid \alpha < \omega_1\}$ is a sequence of dense subsets of P of size $\aleph_1$, then there exists a filter $G \subseteq P$ which meets every D_α.

It is proved in [6] that if the existence of a supercompact cardinal is consistent with the axioms of set theory, then Martin's Maximum is also consistent with the axioms of set theory. It is shown also in [6] that Martin's Maximum has many interesting consequences.

THEOREM 2.1 (FENG–JECH [4]) Assume Martin's Maximum. Then the Projective Stationary Reflection principle holds. In fact, if $S \subseteq [H_\kappa]^\omega$ is projective stationary, then there exists a strongly increasing continuous $\in$–chain of length ω_1 through S.

Proof Let $S \subseteq [H_\kappa]^\omega$ be projective stationary.

The idea is to shoot a strongly increasing continuous $\in$–chain of length ω_1 through S. Thus, a condition $p = \langle N_\alpha \mid \alpha \leq \theta\rangle$ is a strongly increasing continuous $\in$–chain of length $\theta + 1$ for some $\theta < \omega_1$ such that for all $\alpha \leq \theta$ $(N_\alpha \in S)$. The ordering is by extension.

Let P be the set of all conditions. We are going to show that forcing with P preserves stationary subsets of ω_1.

For $\alpha < \omega_1$, let $D_\alpha = \{p \in P \mid \alpha \in dom(p)\}$. For $x \in H_\kappa$, let $D_x = \{p \in P \mid \exists \alpha \in dom(p)\ (x \in p(\alpha))\}$.

We claim that each D_α and each D_x is dense. To see this, let $\alpha < \omega_1$ and $x \in H_\kappa$. Let $p = \langle N_\beta \mid \beta \leq \theta \rangle$ be a condition.

Then $S_1 = \{N \in S \mid \{p, x\} \subseteq N\}$ is stationary. Applying the lemma above, let $\langle M_\gamma \mid \gamma \leq \alpha \rangle$ be a strongly increasing continuous $\in$–chain from S_1. We then define $q(\beta) = M_\beta$ for all $\beta \leq \alpha$ such that $\beta > \theta$ and define $q(\beta) = p(\beta)$ for all $\beta \leq \theta$. It follows that $q \leq p$ and $q \in D_\alpha \cap D_x$.

We can now show that the forcing preserves stationary subsets of ω_1.

Let $T \subseteq \omega_1$ be stationary. Let $\dot{C}$ be a name for a club subset of ω_1. We would like to show that $\Vdash \check{T} \cap \dot{C} \neq \emptyset$.

Let $p \in P$ be a condition.

Let $S_T = \{N \in S \mid N \cap \omega_1 \in T\}$. Then S_T is stationary.

Let $\lambda \geq (2^{2^{|P|}})^+$ be a regular cardinal. Consider the structure

$$\mathcal{H} = \langle H_\lambda, \in, \triangle, P, \dot{C}, S_T, \cdots \rangle,$$

where $\triangle$ is a well ordering of H_λ.

Let $N \prec \mathcal{H}$ be countable such that $N \cap H_\kappa \in S_T$, $N \cap H_\kappa \prec H_\kappa$, and $\{\dot{C}, S, T, S_T, p\} \subseteq N$. Let $\delta = N \cap \omega_1$. Let $\langle D_n \mid n < \omega \rangle$ be a list of all dense subsets of P which are in N. Let $p_0 = p$. Inductively, pick $p_{n+1} \in D_n \cap N$ so that $p_{n+1} \leq p_n$. Let θ_n be the largest countable ordinal in $dom(p_n)$. By elementarity and a density argument, we have $\delta = \bigcup_{n<\omega} \theta_n$ and $N \cap H_\kappa = \bigcup_{n<\omega} p_n(\theta_n)$. Therefore,

$$q = \bigcup_{n<\omega} p_n \cup \{(\delta, N \cap H_\kappa)\}$$

is a condition stronger than all p_n. Since $\dot{C} \in N$, for each $\beta < \delta$, there is a name $\dot{\gamma} \in N$ such that

$$\Vdash \dot{\gamma} \in \dot{C} \ \&\ \check{\beta} < \dot{\gamma}.$$

Each such name corresponds to a dense subset in N. Hence, $q \Vdash \dot{\gamma} \in \check{\delta}$. It follows that $q \Vdash$ "$\dot{C} \cap \check{\delta}$ is unbounded in $\check{\delta}$.". Thus, $q \Vdash \check{\delta} \in \dot{C}$. Therefore, $q \Vdash \dot{C} \cap \check{T} \neq \emptyset$.

It follows then that forcing with P preserves stationary subsets of ω_1.

Now let $G \subseteq P$ be a filter meeting all the dense subsets D_α for $\alpha < \omega_1$ defined above. Let

$$\langle N_\alpha \mid \alpha < \omega_1 \rangle = \bigcup G.$$

Then this is a strongly increasing continuous $\in$–chain of length ω_1 with each $N_\alpha \in S$.

This finishes the proof. □

From the proof above, we have seen that for a given stationary $S \subseteq [H_\kappa]^\omega$, there is a natural forcing notion P_S associated with S to shoot an increasing continuous $\in$–chain of length ω_1 through S.

COROLLARY 2.1 $S \subseteq [H_\kappa]^\omega$ is projective stationary if and only if the forcing notion P_S preserves stationary sets of ω_1.

Proof The harder direction has been proved in the above. To see the other direction, assume that S is not projective stationary. Let $T \subseteq \omega_1$ be stationary such that $S_T = \{N \in S \mid N \cap \omega_1 \in T\}$ is not stationary. Then T is not stationary in the extension by P_S. □

After the work of Foreman, Magidor and Shelah [6], Todorcevic in a circulated hand written note [16] in September 1987 formulated the following Strong Reflection Principle (SRP) (See also [1], page 57–60):

Strong Reflection Principle: *For every κ, every $S \subseteq [\kappa]^\omega$ and for every regular $\theta > \kappa$ there is an increasing continuous $\in$–chain $\{N_\alpha \mid \alpha < \omega_1\}$ of countable elementary models of H_θ (with N_0 containing a prescribed element of H_θ) such that for all $\alpha < \omega_1$, $N_\alpha \cap \kappa \in S$ if and only if there exists a countable elementary submodel M of H_θ such that $N_\alpha \subseteq M$, $M \cap \omega_1 = N_\alpha \cap \omega_1$, and $M \cap \kappa \in S$.*

In the following, we show that the Strong Reflection Principle and the Projective Stationary Reflection principle are equivalent.

THEOREM 2.2 (FENG–JECH [4]) The Strong Reflection Principle holds if and only if the Projective Stationary Reflection principle holds.

Proof First we show that the Strong Reflection Principle implies the Projective Reflection principle.

Let $S \subseteq [H_\kappa]^\omega$ be a projective stationary set. Let $\{N_\alpha \mid \alpha < \omega_1\}$ be a continuous $\in$–chain of countable elementary submodels of H_θ with $\theta > \kappa$, given by SRP for S.

We would like to show that $\{\alpha < \omega \mid N_\alpha \cap H_\kappa \in S\}$ contains a club in ω_1.

Assume not. Let $T = \{\alpha < \omega_1 \mid N_\alpha \cap H_\kappa \notin S \text{ and } N_\alpha \cap \omega_1 = \alpha\}$. Then T is stationary.

Define D to be the following set

$$D = \{N \in [H_\theta]^\omega \mid H_\kappa \in N \;\&\; \forall \beta \in N \cap \omega_1 \; N_\beta \in N\}.$$

By normality, D contains a club on $[H_\theta]^\omega$.

Since S is projective stationary, there exists an $N \in D$ such that $N \cap \omega_1 \in T$, $N \cap H_\kappa \in S$ and N is an elementary submodel of H_θ. Let $\alpha = N \cap \omega_1$.

Since if $\beta < \alpha$ then $N_\beta \subseteq N$, $N_\alpha \cap \omega_1 = \alpha = N \cap \omega_1$ and $N_\alpha \subseteq N$. Hence $N_\alpha \cap H_\kappa \in S$ by the property of N_α. This is a contradiction.

This proves that $\{\alpha < \omega_1 \mid N_\alpha \cap H_\kappa \in S\}$ contains a club in ω_1.

We prove now that the Projective Stationary Reflection principle implies the Strong Reflection Principle.

Let $\kappa \geq \omega_1$ and $\theta > \kappa$. Assume that $S \subseteq [\kappa]^\omega$ and θ is regular. Define S^* to be the following set: for $N \in [H_\theta]^\omega$, let $N \in S^*$ if and only if $N \prec (H_\theta, \in, \triangle)$ and that there exists a countable $M \prec (H_\theta, \in, \triangle)$ such that $N \subseteq M$, $N \cap \omega_1 = M \cap \omega_1$ and $M \cap \kappa \in S$ implies that $N \cap \kappa \in S$.

Claim S^* is projective stationary in $[H_\theta]^\omega$.

Let $g : [H_\theta]^{<\omega} \to H_\theta$ and $T \subseteq \omega_1$ be stationary in ω_1.

Let λ be a regular cardinal larger than the cardinality of H_θ.

Let $N' \prec (H_\lambda, \in, \triangle)$ be countable such that $N' \cap \omega_1 \in T$ and

$$\{\kappa, \theta, S, g\} \subseteq N'.$$

Assume that there exists a countable $M \prec (H_\theta, \in, \triangle)$ such that $M \cap \omega_1 = N' \cap \omega_1$, $N' \cap H_\theta \subseteq M$ and $M \cap \kappa \in S$. (If there are no such M, then $N' \cap H_\theta \in S^*$. We have what we want.) Let N be the skolem hull of $N' \cup (M \cap \kappa)$ in the structure $(H_\lambda, \in, \triangle)$.

We claim that $N \cap \kappa = M \cap \kappa$. Hence $N \cap H_\theta \in S^*$ and we finish the proof.

Let $\alpha \in N \cap \kappa$. Let τ be a skolem term. Let $a \in N'$ and $\alpha_1, \cdots, \alpha_m \in M \cap \kappa$ be such that $\alpha = \tau(a, \alpha_1, \cdots, \alpha_m)$.

Define $h : [\kappa]^m \to \kappa$ by

$$h(x_1, \cdots, x_m) = \begin{cases} \tau(a, x_1, \cdots, x_m), & \text{if } \tau(a, x_1, \cdots, x_m) < \kappa, \\ 0, & \text{otherwise.} \end{cases}$$

Then $h \in N'$. Hence $h \in N' \cap H_\theta \subseteq M$. Therefore, $\alpha = h(\alpha_1, \cdots, \alpha_m) \in M \cap \kappa$.

This finishes the proof that S^* is projective stationary in $[H_\theta]^\omega$.

Applying the Projective Stationary Reflection principle, let $\langle N_\alpha \mid \alpha < \omega_1 \rangle$ be an increasing continuous $\in$–chain of length ω_1 such that $N_\alpha \in S^*$ for all $\alpha < \omega_1$. Certainly, this is what the Strong Reflection Principle needs for the given S. □

There are some weaker forms of stationary reflection principles known before Todorcevic's Strong Reflection principle was presented. In the following, we give some examples.

The Weak Stationary Reflection principle

For all A of cardinality at least $\aleph_2$, for all stationary $S \subseteq [A]^\omega$, there exists a subset $X \subseteq A$ of cardinality $\aleph_1$ such that $S \cap [X]^\omega$ is stationary in $[X]^\omega$.

THEOREM 2.3 (FENG–JECH [3]) Let A be a set of cardinality at least $\aleph_2$. Then the following are equivalent:

(1) For all stationary $S \subseteq [A]^\omega$, there exists an $X \in [A]^{\aleph_1}$ such that $S \cap [X]^\omega$ is stationary in $[X]^\omega$.

(2) For all stationary $S \subseteq [A]^\omega$, for all $f : [A]^{<\omega} \to A$, there exists some $X \in [A]^{\omega_1}$ such that X is closed under f and $S \cap [X]^\omega$ is stationary in $[X]^\omega$.

The Stationary Reflection principle

For all A of cardinality at least $\aleph_2$, for all stationary $S \subseteq [A]^\omega$, for all $Y \in [A]^{\omega_1}$, there exists a subset $X \subseteq A$ of cardinality $\aleph_1$ such that $Y \subseteq X$ and $S \cap [X]^\omega$ is stationary in $[X]^\omega$.

It is still a question whether the weak Stationary reflection principle is really weaker than the stationary reflection principle above. In [3], it is shown that on ω_2, these two principles are the same.

THEOREM 2.4 (FENG–JECH [3]) Let A be a set of cardinality at least $\aleph_2$. Then the following are equivalent:

(1) For all stationary $S \subseteq [A]^\omega$, for all $Y \in [A]^{\omega_1}$, there exists an $X \in [A]^{\aleph_1}$ such that $Y \subseteq X$ and $S \cap [X]^\omega$ is stationary in $[X]^\omega$.

(2) For all stationary $S \subseteq [A]^\omega$, for all closed and unbounded $C \subseteq [A]^{\omega_1}$, there exists some $X \in C$ such that $S \cap [X]^\omega$ is stationary in $[X]^\omega$.

(3) There exists a $Y \in [A]^{\omega_1}$ such that for all stationary $S \subseteq [A]^\omega$, there exists some $X \in [A]^{\omega_1}$ such that $Y \subseteq X$ and $S \cap [X]^\omega$ is stationary in $[X]^\omega$.

A subset $F \subseteq [H_\kappa]^{\aleph_1}$ is ω_1–closed if for every $\subseteq$–increasing sequence $\{X_\alpha \mid \alpha < \omega\}$ of length ω_1 from F, the union $\cup_\alpha X_\alpha$ is in F. F is unbounded if for every $X \in [H_\kappa]^{\aleph_1}$ there is some $Y \in F$ such that $X \subseteq Y$. Fleissner [5] introduced the following stationary reflection principle, called **Axiom R.**

If $S \subseteq [H_\kappa]^\omega$ is stationary, $T \subseteq [H_\kappa]^{\aleph_1}$ is ω_1–closed and unbounded, then there exists some $X \in T$ such that $S \cap [X]^\omega$ is stationary in $[X]^\omega$.

THEOREM 2.5 (FENG–JECH [4]) Assume the Projective Stationary Reflection Principle. If $S \subseteq [H_\kappa]^\omega$ is stationary and $E \subseteq [H_\kappa]^{\aleph_1}$is ω_1–closed and unbounded in $[H_\kappa]^{\aleph_1}$, then there exists an increasing continuous $\in$–chain $\langle N_\alpha \mid \alpha < \omega_1\rangle$ such that $\{\alpha < \omega_1 \mid N_\alpha \in S\}$ is stationary in ω_1, and $\cup_\alpha N_\alpha$ is in E. Hence the Stationary Reflection Principle holds.

Notice that the Stationary Reflection Principle and the Axiom R are strictly weaker than the Strong Reflection Principle, because they do not imply that the nonstationary ideal on ω_1 is saturated.

Proof Add E to our language as a unary predicate. Let $S \subseteq [H_\kappa]^\omega$. There is another projective stationary set $p(S) \subseteq [H_\kappa]^\omega$ naturally associated with S.

Let $T \subseteq \omega_1$. Define $S_T = \{N \in S \mid N \cap \omega_1 \in T\}$.

Let $F \subseteq \{T \subseteq \omega_1 \mid T$ is stationary and S_T is not stationary $\}$ be a maximal antichain of the smallest cardinality.

Define $p(S)$ to be the following set:

$$\{N \prec H_\kappa \mid N \in [H_\kappa]^\omega \text{ and } (\exists\, A \in N \cap F\, (N \cap \omega_1 \in A) \iff N \notin S)\}.$$

Then $p(S)$ is projective stationary in $[H_\kappa]^\omega$.

If we assume that $\langle N_\alpha \mid \alpha < \omega_1 \rangle$ is an increasing continuous $\in$–chain from $p(S)$, then S is stationary if and only if $\{\alpha < \omega_1 \mid N_\alpha \in S\}$ is stationary.

Let us assume that S is stationary. Toward a contradiction, let us assume that $\{\alpha < \omega_1 \mid N_\alpha \in S\}$ is not stationary. Then the complement of this set contains a club. It follows that for every $A \in F$ there is some $\alpha < \omega_1$ such that $A \in N_\alpha$. Let $f : \omega_1 \to F$ be a surjective mapping. For each $\alpha < \omega_1$ let $C_\alpha \subseteq [H_\kappa]^\omega$ be a club to witness that $f(\alpha) \in F$. Let $C = \Delta_{\alpha<\omega_1} C_\alpha$. Then $C \cap S$ is stationary and the following set T is stationary in ω_1:

$$T = \{N \cap \omega_1 \mid N \in C \cap S\}.$$

Let $\alpha \in T$ be such that for all $A \in N_\alpha \cap F$ there is a $\beta < \alpha$ with $A = f(\beta)$, for all $\beta < \alpha$, $f(\beta) \in N_\alpha$, and $\alpha = N_\alpha \cap \omega_1$. Then $N_\alpha \in S$. This is a contradiction. Hence $\{\alpha < \omega_1 \mid N_\alpha \in S\}$ is stationary in ω_1.

To see the other direction, assume that S is not stationary. By minimality, the cardinality of F must be one and the member of F must contain a club. By elementarity, there must be such a club in N_0. Therefore, for all $\alpha < \omega_1$, $N_\alpha \cap \omega_1$ must be in this club. Hence no N_α will be in S.

We now check that the union of the N_α's is in E.

Let $\alpha < \omega_1$. Since $N_\alpha \in N_{\alpha+1}$, let $\{X_n \mid n < \omega\} \in N_{\alpha+1}$ be an enumeration of $N_\alpha \cap E$. Let $X = \cup_n X_n$. Then $X \in N_{\alpha+1}$. Since it is true that "there is some $Y \in E$ such that $X \subseteq Y$", $N_{\alpha+1}$ must satisfy this sentence. Let $X_\alpha \in N_{\alpha+1}$ be a witness. Then $X \subseteq X_\alpha$ by elementarity. It follows that the sequence $\{X_\alpha \mid \alpha < \omega_1\}$ is a $\subseteq$–increasing sequence from E. Hence $X^* = \cup_\alpha X_\alpha$ is in E. If $a \in N_\alpha$, then it is true that "there is some $Y \in E$ such that $a \in Y$", N_α satisfies this sentence. Hence $N_\alpha \subseteq X^*$ for all $\alpha < \omega_1$. Since each $X_\alpha \in N_{\alpha+1}$, $X_\alpha \in Y^* = \cup_\alpha N_\alpha$. It is true that $|X_\alpha| = \aleph_1$. Therefore, Y^* satisfies this. Let $f \in Y^*$ be a bijection. Since $\omega_1 \subseteq Y^*$, by elementarity, the range of f must be a subset of Y^*. Hence $X_\alpha \subseteq Y^*$. It follows that $X^* = Y^*$.

This shows that the Projective Stationary Reflection principle implies Fleissner's Axiom R as well as Velickovic's form of stationary reflection principle [18]. $\square$

Let us consider another type of stationary reflection.

Let E_κ be the following statement: *κ is a regular cardinal $\kappa > \omega_1$, there exists a stationary subset $E \subseteq \kappa$ such that every member of E has cofinality ω and for every $\delta < \kappa$, $E \cap \delta$ is nonstationary in δ.*

In [11] Jensen showed that for every $\kappa > \omega$, E_{κ^+} holds in L.

We will see that by the works of Foreman, Magidor and Shelah, and Todorcevic, when Martin's Maximum, or the Projective Stationary Reflection principle holds, the negation of E_κ hold in a very strong sense. This is also related to a question of Friedman.

In [7], Friedman asked if every stationary $E \subseteq \kappa$ satisfying that if $\alpha \in E$ then $cf(\alpha) = \omega$ can contain a closed copy of ω_1, for regular $\kappa \geq \omega_2$. A positive answer follows from the Martin's Axiom, as proved by Foreman, Magidor and Shelah in [6]. Todorcevic showed in [16] that the Strong Reflection principle suffices.

THEOREM 2.6 (TODORCEVIC [16])

Assume the Projective Stationary Reflection. Let $\kappa \geq \omega_2$. Let E be a stationary subset of κ such that every $\alpha \in E$ has cofinality ω. Then E contains a closed copy of ω_1.

Proof Fix a regular cardinal $\kappa \geq \omega_2$.

Let E be a stationary subset of κ such that every $\alpha \in E$ has cofinality ω.

Consider the following set:

$$S = \{N \in [H_\kappa]^\omega \mid N \prec H_\kappa, \text{and } \sup(N \cap \kappa) \in E\}.$$

We have seen previously that S is projective stationary.

Applying the Strong Reflection Principle, let $\{N_\alpha \mid \alpha < \omega_1\}$ be an increasing continuous $\in$–chain such that $N_\alpha \in S$ for $\alpha \in \omega_1$. Then for each $\alpha < \omega_1$, $\sup(N_\alpha \cap \kappa) \in E$. We are done. □

COROLLARY 2.2 Assume the Projective Stationary Reflection principle. Then the principle E_κ fails.

§3 SATURATION OF NS_{ω_1}

In this section, let us consider the saturation of the nonstationary ideal on ω_1.

Let F be a set of stationary sets on ω_1. F is an antichain if $A \cap B$ is nonstationary for $A \neq B$ in F. F is a maximal antichain if F is an antichain and for every stationary subset $T \subseteq \omega_1$ there exists some $A \in F$ such that $T \cap A$ is stationary. The nonstationary ideal on ω_1 is saturated if every maximal antichain has size at most $\aleph_1$.

In [14], Steel and Van Wesep, later improved by Woodin in [20], showed that the nonstationary ideal may be saturated using determinancy. In [6], Foreman, Magidor and Shelah showed that Martin's Maximum implies that the nonstationary ideal on ω_1 is saturated. In [16], Todorcevic showed that the Strong Reflection Principle is sufficient.

THEOREM 3.1 (TODORCEVIC [16]) Assume the Projective Stationary Reflection principle. Then the nonstationary ideal on ω_1 is saturated.

Proof Let F be a maximal antichain of stationary subsets of ω_1. Consider the following set:

$$S = \{N \in [H_{\omega_2}]^\omega \mid N \prec H_{\omega_2}, F \in N, \text{and } \exists\, A \in F \cap N\ (N \cap \omega_1 \in A)\}.$$

Claim S is projective stationary.

Let $T \subseteq \omega_1$ be stationary. Let $A \in F$ be such that $A \cap T$ is stationary. Let C be a club on $[H_{\omega_2}]^\omega$. Then we can find an elementary submodel N of H_{ω_2} with the property that $N \cap \omega_1 \in A \cap T$ and $N \in C$. Hence S is projective stationary.

By the Projective Stationary Reflection Principle, let $\{N_\alpha \mid \alpha < \omega_1\}$ be an increasing continuous $\subseteq$-chain of elementary submodels of H_{ω_2} such that $\alpha \subseteq N_\alpha$ for every α and there exists a club $C \subseteq \omega_1$ such that if $\alpha \in C$ then $\alpha = N_\alpha \cap \omega_1$ and $N_\alpha \in S$.

We proceed to check that $F \subseteq X = \bigcup\{N_\alpha \mid \alpha < \omega_1\}$.

Let $A \in F$. Assume that this $A \notin X$. Let Y be the skolem hull of $X \cup \{A\}$. Let M_α be the skolem hull of $N_\alpha \cup \{A\}$. Let $D \subseteq C$ be a club such that for every $\alpha \in D$ we have that $M_\alpha \cap \omega_1 = N_\alpha \cap \omega_1 = \alpha$.

Let $\alpha \in D \cap A$. Then $\alpha = N_\alpha \cap \omega_1 = M_\alpha \cap \omega_1$. There must be some $B \in F \cap N_\alpha$ such that $\alpha \in B$. This B must be different from A. Hence $A \cap B$ must be nonstationary. But by elementarity, as both A and B are in M_α, there must be some closed and unbounded subset $E \in M_\alpha$ of ω_1 such that $E \cap A \cap B$ is empty. We then have a contradiction because $E \in M_\alpha$ implies that $M_\alpha \cap \omega_1 \in E$ and $M_\alpha \cap \omega_1 \in A \cap B$.

Therefore $F \subseteq X$. Hence the size of F is at most $\aleph_1$. $\square$

We now take a closer look at the saturation of the nonstationary ideal on ω_1. It turns out that it itself is a kind of reflection property.

To start with, let us define first certain filters. For a regular cardinal $\kappa \geq \omega_2$, for $X \subseteq [H_\kappa]^\omega$, let X be in $\mathcal{F}_\kappa$ if and only if for every stationary subset $A \subseteq \omega_1$ there exist a stationary subset $B \subseteq A$ and a closed and unbounded subset $C \subseteq [H_\kappa]^\omega$ such that $\{N \in C \mid N \cap \omega_1 \in B\} \subseteq X$.

THEOREM 3.2 (FENG–JECH [4]) The following are equivalent:

(1) The nonstationary ideal NS_{ω_1} on ω_1 is saturated.

(2) For every regular cardinal $\kappa \geq \omega_2$, for every stationary set $S \subseteq [H_\kappa]^\omega$, there exists a stationary set $A \subseteq \omega_1$ such that S is A–projective

stationary (i.e., for every stationary $B \subseteq A$, the set $\{N \in S \mid N \cap \omega_1 \in B\}$ is stationary).

(3) For every regular cardinal $\kappa \geq \omega_2$, for every $X \subseteq [H_\kappa]^\omega$, $X \in \mathcal{F}_\kappa$ if and only if X contains a closed and unbounded subset. Namely, the filter $\mathcal{F}_\kappa$ is just the club filter on $[H_\kappa]^\omega$.

(4) For every regular cardinal $\kappa \geq \omega_2$, for every $X \in \mathcal{F}_\kappa$, there exists an increasing continuous $\in$–chain $\langle N_\alpha \mid \alpha < \omega_1\rangle$ of countable elementary submodels of H_κ of length ω_1 such that $N_\alpha \in X$ for all $\alpha < \omega_1$.

Proof (1) $\Rightarrow$ (2)

Let $S \subseteq [H_\kappa]^\omega$ be a stationary set. If S is projective stationary, then there is nothing needed to be proved. So we assume that S is not projective stationary.

For a stationary $T \subseteq \omega_1$, we let $T \in F$ if and only if there exists a club $C \subseteq [H_\kappa]^\omega$ such that

$$T \cap \{ N \cap \omega_1 \mid N \in C \cap S\} = \emptyset.$$

Since S is not projective stationary, F is not empty.

Because the nonstationary ideal on ω_1 is saturated, we can take a sequence $\{A_\alpha \mid \alpha < \omega_1\}$ from F so that for every $A \in F$, $A - \nabla_{\alpha<\omega_1} A_\alpha$ is nonstationary, where $\nabla_{\alpha<\omega_1} A_\alpha$ is the diagonal union of the sequence $\{A_\alpha \mid \alpha < \omega_1\}$ defined by

$$\nabla_{\alpha<\omega_1} A_\alpha = \{\beta < \omega \mid \exists\, \alpha < \beta\, (\beta \in A_\alpha)\}.$$

For each $\alpha < \omega$, we let $C_\alpha \subseteq [H_\kappa]^\omega$ be a witness for A_α to be in F.

Let $\Delta_{\alpha<\omega_1} C_\alpha$ be the diagonal intersection of the sequence $\{C_\alpha \mid \alpha < \omega_1\}$ defined by

$$\Delta_{\alpha<\omega_1} C_\alpha = \{N \in [H_\kappa]^\omega \mid \forall \alpha \in N \cap \omega_1\; N \in C_\alpha\}.$$

Let $A = \omega - \nabla_{\alpha<\omega_1} A_\alpha$. We claim that S is A–projective stationary.

First notice that A is stationary. This is because $\Delta_{\alpha<\omega_1} C_\alpha \cap S$ is stationary. Now let $T \subseteq A$ be stationary and let $C \subseteq [H_\kappa]^\omega$ be a club.

Since $T \cap \nabla_{\alpha<\omega_1} A_\alpha$ is not stationary, $T \notin F$. Therefore the following intersection is not empty:

$$T \cap \{N \cap \omega_1 \mid N \in C \cap \Delta_{\alpha<\omega_1} C_\alpha \cap S\}.$$

Hence $\{N \in S \mid N \cap \omega_1 \in T\}$ is stationary.

(2) $\Rightarrow$ (3)

Since every closed and unbounded subset of $[H_\kappa]^\omega$ is in the filter $\mathcal{F}_\kappa$, and (2) implies that every stationary subset of $[H_\kappa]^\omega$ intersects every member of the filter $\mathcal{F}_\kappa$, the filter $\mathcal{F}_\kappa$ and the club filter on $[H_\kappa]^\omega$ are the same filter.

(3) implies (4) is trivial.

(4) $\Rightarrow$ (1)

Let F be a maximal antichain of stationary subsets of ω_1. Let $S_F = \{N \in [H_\kappa]^\omega \mid \exists\, A \in F \cap N\ (N \cap \omega_1 \in A)\}$. Then $S_F \in \mathcal{F}_\kappa$. Namely, for a given stationary subset $A \subseteq \omega_1$, let $T \in F$ be such that $B = A \cap T$ is stationary. Let C be the club of all countable elementary submodels of H_κ which contains T. Then if $N \in C$ and $N \cap \omega_1 \in B$ then $N \in S_F$. Now by the argument of Example 2.1, assuming (4), F has cardinality at most $\aleph_1$.

This completes the proof. □

The next thing we want to show is that the presaturation of the nonstationary ideal on ω_1 is also a kind of reflection property, which in turn is equivalent to the filter $\mathcal{F}_\kappa$ being countably closed.

First let us recall that the nonstationary ideal on ω_1 is **presaturated** if for every countable sequence $\{F_n \mid n < \omega\}$ of maximal antichains of the nonstationary ideal on ω_1, for every stationary subset T, there exists a stationary subset $B \subseteq T$ such that for each n the set $\{A \in F_n \mid A \cap B$ is stationary $\}$ has cardinality at most $\aleph_1$ (see [6]).

THEOREM 3.3 (FENG–JECH [4]) The following are equivalent:

(1) The nonstationary ideal NS_{ω_1} on ω_1 is presaturated.

(2) For every regular $\kappa \geq \omega_2$, the filter $\mathcal{F}_\kappa$ is σ–closed.

(3) For every regular cardinal $\kappa \geq \omega_2$, for every countable sequence $\langle X_n \mid n < \omega\rangle$ from $\mathcal{F}_\kappa$, for every stationary subset $T \subseteq \omega_1$, there exists an $M \prec H_\kappa$ such that $\omega_1 \subseteq M$, M has cardinality $\aleph_1$ and $\{N \in [M]^\omega \cap \bigcap_{n<\omega} X_n \mid N \cap \omega_1 \in T\}$ is stationary in $[M]^\omega$.

Proof (1) $\Rightarrow$ (2)

Let $X_n \in \mathcal{F}_\kappa$ for $n < \omega$. Let $X = \bigcap_{n<\omega} X_n$. We need to show that $X \in \mathcal{F}_\kappa$.

For each $n < \omega$, let $\langle B^n_\alpha, C^n_\alpha \mid \alpha < \theta_n\rangle$ be such that all the B^n_α's form a maximal antichain of stationary sets and for each $\alpha < \theta_n$, $N \in C^n_\alpha$ and $N \cap \omega_1 \in B^n_\alpha$ imply that $N \in X_n$.

Let $A \subseteq \omega_1$ be stationary. Applying the presaturation property of the nonstationary ideal, let $B \subseteq A$ be stationary such that for each $n < \omega$, the set $I_n = \{\alpha < \theta_n \mid B \cap B^n_\alpha$ is stationary $\}$ has cardinality at most $\aleph_1$. To simplify the notation, we assume that $I_n = \omega_1$. Let

$$T = B \cap \bigcap_{n<\omega} \nabla_{\alpha<\omega_1} B^n_\alpha$$

and let

$$C = \bigcap_{n<\omega} \Delta_{\alpha<\omega_1} C^n_\alpha.$$

It follows that if $N \in C$ and $N \cap \omega_1 \in T$ then $N \in X$.

Hence the filter $\mathcal{F}_\kappa$ is σ–closed.

(2) $\Rightarrow$ (3)

It will be sufficient to show that every $X \in \mathcal{F}_\kappa$ reflects.

Let $X \in \mathcal{F}_\kappa$. Let T be a stationary subset of ω_1. Let $B \subseteq T$ be stationary and let $C \subseteq [H_\kappa]^\omega$ be a closed and unbounded subset such that for all $N \in C$, $N \cap \omega_1 \in B$ implies that $N \in X$.

Let $\langle N_\alpha \mid \alpha < \omega_1 \rangle$ be an increasing continuous $\in$–chain from C. Let $M = \bigcup_{\alpha<\omega_1} N_\alpha$. Then $\{N_\alpha \mid \alpha = N_\alpha \cap \omega_1 \in B\}$ is stationary in $[M]^\omega$.

(3) $\Rightarrow$ (1)

Fix a sequence $\{F_n \mid n < \omega\}$ of maximal antichains of the nonstationary ideal on ω_1. For each $n < \omega$, let

$$S_n = \{N \prec H_\kappa \mid N \text{ is countable and } \exists\, A \in F_n \cap N\ (N \cap \omega_1 \in A)\}.$$

Then every $S_n \in \mathcal{F}_\kappa$.

Fix a stationary subset T of ω_1. Applying (3), let $X \prec H_{\omega_2}$ be such that $\omega_1 \subseteq X$, X has cardinality $\aleph_1$ and $\{N \in [X]^\omega \cap \bigcap_{n<\omega} S_n \mid N \cap \omega_1 \in T\}$ is stationary in $[X]^\omega$. Write $X = \bigcup_{\alpha<\omega_1} N_\alpha$, the union of an increasing continuous $\subseteq$–chain of countable elementary submodels with $\alpha \subseteq N_\alpha$. Then

$$B = \{\alpha \in T \mid \alpha = N_\alpha \cap \omega_1\ \&\ N_\alpha \in \bigcap_{n<\omega} S_n\}$$

is a stationary subset of T.

We claim that for every n the set $\{\ A \in F_n \mid A \cap B \text{ is stationary }\ \}$ has cardinality at most $\aleph_1$.

Assume not. Let n be the least counterexample. Let $A \in F_n - X$ be such that $A \cap B$ is stationary. Let M_α be the skolem hull of $N_\alpha \cup \{A\}$ for each $\alpha < \omega_1$. Then

$$C = \{\alpha < \omega_1 \mid \alpha = N_\alpha \cap \omega_1 = M_\alpha \cap \omega_1\}$$

is a club. Let $\alpha \in C \cap A \cap B$. Let $Z \in N_\alpha \cap F_n$ be such that $\alpha \in Z$. Since both A and Z are in M_α which is an elementary submodel of H_κ, and $M_\alpha \cap \omega_1 = \alpha \in A \cap Z$, we conclude that $A \cap Z$ is stationary. This is a contradiction.

Hence for every $n < \omega$, for every $A \in F_n$, if $A \cap B$ is stationary, then $A \in X$. We are done since X has cardinality $\aleph_1$. $\square$

To close this section, let us present the proof of Foreman, Magidor and Shelah of the fact that Stationary Reflection principle implies the presaturation of the nonstationary ideal on ω_1.

THEOREM 3.4 (FOREMAN, MAGIDOR AND SHELAH [6])

Assume the Stationary Reflection Principle. Then the nonstationary ideal NS_{ω_1} is presaturated.

Proof We prove the theorem by proving two lemmas. Let κ be a sufficiently large regular cardinal.

LEMMA 3.1 Let $T \subseteq \omega_1$ be stationary. Let $\{X_\beta \mid \beta < \theta\}$ be a maximal antichain of stationary sets below T. Then there is a club $C \subseteq [H_\kappa]^\omega$ such that each $N \in C$ is an elementary submodel and if $N \in C$, $N \cap \omega_1 \in T$, then there is some β such that $N \cap \omega_1 \in X_\beta$ and the skolem hull of $N \cup \{\beta\}$ contains no new countable ordinals.

Proof of the lemma. Assume otherwise. Let S denote the set of countable elementary submodels N such that $N \cap \omega_1 \in T$ and for every β, either $N \cap \omega_1 \notin X_\beta$ or $N \cap \omega_1$ is in the skolem hull of $N \cup \{\beta\}$. Then S is stationary in $[H_\kappa]^\omega$.

By Reflection, let X be an elementary submodel of cardinality $\aleph_1$ such that $\omega_1 \subseteq X$ and $S \cap [X]^\omega$ is stationary in $[X]^\omega$. Let X be the union of a $\subseteq$-increasing sequence $\{N_\alpha \mid \alpha < \omega_1\}$ of countable models of length ω_1. Let $S^* = \{N_\alpha \cap \omega_1 \mid N_\alpha \in S\}$. Then $S^* \subseteq T$ is stationary. Now if $\beta \in X$, then $S^* \cap X_\beta$ is bounded. Hence there is some $\beta \notin X$ such that $S^* \cap X_\beta$ is stationary. Let β be the least such. Let Y be the skolem hull of $X \cup \{\beta\}$ and N'_α be the skolem hull of $N_\alpha \cup \{\beta\}$. Let $\alpha < \omega_1$ be such that $N'_\alpha \cap \omega_1 = N_\alpha \cap \omega_1 \in S^* \cap X_\beta$. Then we get a contradiction.

This finishes the proof of the lemma.

Let T be stationary in ω_1. Let $\{X_n \mid n < \omega\}$ be a sequences of maximal antichains of stationary sets below T. Each $X_n = \{X^n_\beta \mid \beta < \theta_n\}$. Let

$$S = \{N \in [H_\kappa]^\omega \mid \forall\, n < \omega\ \exists\, \beta \in N\ (N \cap \omega_1 \in X^n_\beta)\}.$$

LEMMA 3.2 S is stationary in $[H_\kappa]^\omega$.

Proof of the lemma. Let D be a club in $[H_\kappa]^\omega$. Applying the previous lemma to each $n < \omega$, we get a club C_n. Let C be the intersection of all the C_n's. Then C is a club. Using the club $C \cap D$, by induction, we get an $N \in D \cap S$.

To finish the proof of the theorem, we apply the Stationary Reflection principle to the stationary set S. Let X be of cardinality $\aleph_1$ such that X is an elementary submodel, $\omega_1 \subseteq X$, and $S \cap [X]^\omega$ is stationary in $[X]^\omega$. Let X be the union of a $\subseteq$-increasing sequence $\{N_\alpha \mid \alpha < \omega_1\}$ of countable elementary submodels. Let $T^* = \{N_\alpha \cap \omega_1 \mid N_\alpha \in S\}$. Then $T^* \subseteq T$ is stationary. This is a desired stationary set. □

§4 ON CARDINAL ARITHMETIC

The first consequence on cardinal arithmetic of the stationary reflection principles was proved by Todorcevic in [15]. Then Velickovic in [18] proved SCH from a slightly stronger version of the Stationary Reflection principle.

THEOREM 4.1 (TODORCEVIC [15])
Assume that every stationary set $S \subseteq [\omega_2]^\omega$ reflects. Then $2^\omega \leq \omega_2$.

Proof For $\alpha < \omega_2$, let $e_\alpha : \alpha \to |\alpha|$ be a bijection. Let C be the set of all countable $A \subseteq \omega_2$ such that for all $\alpha \in A$, A is closed under e_α and e_α^{-1}, and $\omega \subseteq A$. Then C is a club in $[\omega_2]^\omega$.

Let us consider the following set S:

$$S = \{A \in C \mid \forall\, \alpha < \omega_1\ \forall \beta < \omega_2\ A \neq e''_\beta \alpha\ \}.$$

Claim Let $X \in [\omega_2]^{\aleph_1}$. If $\omega \subseteq X$ and for all $\alpha \in X$, X is closed under e_α and e_α^{-1}, then $[X]^\omega \cap S$ is nonstationary in $[X]^\omega$.

First notice that if $\alpha \in X \cap \omega_1$ then $\alpha \subseteq X$. If $\omega_1 \subseteq X$, then $X \in \omega_2$. In this case, the set

$$\{e''_X \alpha \mid \alpha < \omega_1\}$$

is a club in $[X]^\omega$, disjoint from S. If $X \cap \omega_1 \neq \omega_1$, then $X \cap \omega_1 \in \omega_1$ and the order type of X is ω_1. In this case, let $\delta = X \cap \omega_1$. Then the set

$$\{e''_\alpha \delta \mid \alpha \in X - \omega_1\}$$

is a club in $[X]^\omega$, disjoint from S.

By the assumption, S is not stationary in $[\omega_2]^\omega$. Hence $C - S$ contains a club in $[\omega_2]^\omega$. By a theorem of Baumgartner and Taylor [0], $|C - S| = 2^{\aleph_0}$. Hence $2^{\aleph_0} \leq \omega_2$. □

THEOREM 4.2 (VELICKOVIC [18]) Assume that for every stationary $S \subseteq [H_\kappa]^\omega$, there is a continuous $\in$–increasing chain of countable models $\{N_\alpha \mid \alpha < \omega_1\}$ such that $\{\alpha < \omega_1 \mid N_\alpha \in S\}$ is stationary in ω_1. Then for all regular $\kappa \geq \omega_2$, $\kappa^{\aleph_0} = \kappa$.

Although the Stationary Reflection principle implies SCH, it does not solve the Continuum problem. It does not decide which of the two values $2^{\aleph_0}$ must take. Namely, it is consistent that the Stationary Reflection principles plus CH; it is also consistent the Stationary Reflection principles plus $2^{\aleph_0} = \aleph_2$, assuming large cardinals are consistent. However, the Strong Reflection Principle does solve the Continuum problem. Recently, Woodin in [19] has shown that the Strong Reflection principle implies that $2^{\aleph_0} = \aleph_2$. In fact, Woodin has proved that it implies the second uniform indicernible is ω_2.

THEOREM 4.3 (WOODIN [19]) Assume the Projective Stationary Reflection principle. Then $2^{\aleph_0} = \aleph_2$.

THEOREM 4.4 (TODORCEVIC [16])
Assume the Projective Stationary Reflection principle. Then for all regular $\kappa \geq \omega_2$, $\kappa^{\aleph_1} = \kappa$. Hence the Singular Cardinal Hypothesis holds.

Proof
Let $E = \{\alpha < \kappa \mid cf(\alpha) = \omega\}$. Partition E into κ many pairwise disjoint stationary subsets, say $E = \cup\{E_\alpha \mid \alpha < \kappa\}$.

Let $\{T_\alpha \mid \alpha < \omega_1\}$ be a partition of ω_1 so that each T_α is stationary for $\alpha < \omega_1$ and for every stationary subset $X \subseteq \omega_1$, there is some $\alpha < \omega_1$ such that $X \cap T_\alpha$ is stationary.

Let $f : \omega_1 \to \kappa$ be a strictly increasing function. Then the following set S_f is projective stationary:

$$S = \{N \in [H_\kappa]^\omega \mid \forall \alpha < N \cap \omega_1 \; N \cap \omega_1 \in T_\alpha \Rightarrow \sup(N \cap \kappa) \in E_{f(\alpha)}\}.$$

To see this, let $T \subseteq \omega_1$ be stationary. Let α be the least countable ordinal such that $T \cap T_\alpha$ is stationary. Then the following set is stationary:

$$\{N \in [H_\kappa]^\omega \mid \alpha < N \cap \omega_1 \in T \cap T_\alpha \,\&\, \sup(N \cap \kappa) \in E_{f(\alpha)}\}.$$

Let $\{N_\alpha \mid \alpha < \omega_1\}$ be an increasing continuous $\in$–chain from S_f. Let

$$\theta_f = sup\{sup(N_\alpha \cap \kappa) \mid \alpha < \omega_1\}.$$

It follows that $E_\alpha \cap \theta_f$ is stationary in θ_f if and only if there is some $\beta < \omega_1$ such that $\alpha = f(\beta)$.

Now the mapping from f to θ_f is a one–to–one mapping. Hence $\kappa^{\omega_1} = \kappa$. By Silver's theorem [13], this in turn implies the Singular Cardinal Hypothesis holds:

For every singular cardinal λ, if $2^{cf(\lambda)} < \lambda$, then $\lambda^{cf(\lambda)} = \lambda^+$. □

§5 ON CHANG'S CONJECTURE AND THE BOX PRINCIPLE

The first indication that stationary reflection principles are very strong properties is probably that Chang's conjecture follows from the weak Stationary Reflection principle, as proved by Magidor in an unpublished work.

THEOREM 5.1 (MAGIDOR)

Assume the weak Stationary Reflection Principle. Then for λ sufficiently large, the following form of strong Chang's conjecture holds.

(*) For any expansion of $H_\lambda, \mathcal{A} = \langle H_\lambda, \in, f_i \rangle_{i<\omega}$, of countable language, there is a club $C \subseteq [H_\lambda]^\omega$ such that

(1) for each $N \in C$ $N \prec \mathcal{A}$,

(2) for each $N \in C \, \exists \, \alpha \in (\omega_2 - N)$ such that $Sk^{\mathcal{A}}(N \cup \{\alpha\}) \cap \omega_1 = N \cap \omega_1$.

Proof Assume otherwise. Define S so that for $N \in [H_\lambda]^\omega$

$$N \in S \iff N \prec \mathcal{A} \,\&\, \forall \alpha \in \omega_2 - N \; (Sk^{\mathcal{A}}(N \cup \{\alpha\}) \cap \omega_1 \neq N \cap \omega_1).$$

Then S is stationary.

Let $X \prec H_\lambda$ be such that $|X| = |X \cap \omega_2| = \aleph_1$ and $X \prec \mathcal{A}, S \cap [X]^\omega$ is stationary in $[X]^\omega$.

Case 1. $X \cap \omega_1 \neq \omega_1$. We may assume that $X \cap \omega_1 \in \omega_1$. Now it is easy to build a normal chain of countable elementary submodels so that $X = \bigcup_{\alpha<\omega_1} N_\alpha$ and $N_\alpha \cap \omega_1 = X \cap \omega_1$ for each $\alpha < \omega_1$. Let $N_\alpha \in S$.

Since $|X \cap \omega_2| = \aleph_1$, can find a $\beta \in X \cap \omega_2 - N_\alpha$. But then

$$Sk^{\mathcal{A}}(N_\alpha \cup \{\beta\}) \cap \omega_1 = N_\alpha \cap \omega_1.$$

This is a contradiction.

Case 2. $\omega_1 \subseteq X$.

Build up an elementary chain $X = \bigcup_{\alpha<\omega_1} N_\alpha$.

Then $C_1 = \{N_\alpha \cap \omega_1 : \alpha < \omega_1\}$ is a club in ω_1. Let $\beta \in \omega_2 - X$. Let $Y = SK^{\mathcal{A}}(X \cup \{\beta\})$. Then $N'_\alpha = Sk^{\mathcal{A}}(N_\alpha \cup \{\beta\})$ $(\alpha < \omega_1)$ form an elementary chain whose union is Y. Let $C_2 = \{N'_\alpha \cap \omega_1 : \alpha < \omega_1\}$. Then C_2 is a club. Let $N_\alpha \in S$ be such that

$$N_\alpha \cap \omega_1 = N'_\alpha \cap \omega_1 \in C_1 \cap C_2.$$

Then $N'_\alpha = Sk^{\mathcal{A}}(N_\alpha \cup \{\beta\})$.

This is a contradiction. □

Todorcevic in [17] proved that this type of Chang's conjecture implies certain stationary reflection principle.

THEOREM 5.2 (TODORCEVIC [17])

Let λ be some large regular cardinal such that $(*)$ of the Magidor's theorem above holds. Then every stationary subset of $[\omega_2]^\omega$ reflects.

Proof Assume that $S \subseteq [\omega_2]^\omega$ is stationary but $S \cap [\alpha]^\omega$ is not stationary in $[\alpha]^\omega$ for all uncountable $\alpha < \omega_2$.

For each uncountable $\alpha < \omega_2$, let $e_\alpha : \omega_1 \to \alpha$ be a bijection and $C_\alpha \subseteq \omega_1$ be a club in ω_1 such that for all $\beta \in C$, $e''_\alpha\beta \notin S$. Since S is stationary in $[\omega_2]^\omega$, and by $(*)$, let M and M^* be two countable elementary submodels of $(H_\lambda, \in)$ such that S and $\{e_\alpha, C_\alpha \mid \omega_1 \leq \alpha < \omega_2\}$ are in M and $M \cap \omega_2 \in S$ and such that $M \cap \omega_1 = M^* \cap \omega_1$, $M \subseteq M^*$, and $M \cap \omega_2 \neq M^* \cap \omega_2$. Let α be the least ordinal in $M^* - M$. Notice that $M^* \cap \alpha = M \cap \omega_2$. Then e_α, C_α are in M^*. Hence $M \cap \omega_1 \in C_\alpha$ and $M \cap \omega_2 = e''_\alpha(M \cap \omega_1)$. This is a contradiction. □

We have seen previously that the □-principle fails. Here we want to prove something stronger.

Let C be a sequence defined on all the limit ordinals $< \kappa$ such that for all limit ordinals $\alpha < \kappa$, $C_\alpha \subseteq \alpha$ is a club in α and for all limit ordinals $\beta < \alpha$, if $\beta \in C_\alpha$ is a limit point of C_α, then $C_\beta = \beta \cap C_\alpha$. C is a $\square^*_\kappa$ sequence if there is no club $D \subseteq \kappa$ such that for every limit point α of D, $C_\alpha = \alpha \cap D$.

Let S_C be the set of all countable elementary submodels N of H_κ such that, letting $\alpha = \sup(N \cap \kappa)$, $C_\alpha \cap N$ is bounded in α.

The following theorem is essentially due to Velickovic [18].

THEOREM 5.3 The following are equivalent:

(1) S_C is projective stationary.

(2) S_C is stationary.

(3) There is no club $D \subseteq \kappa$ such that for all limit points α of D, $C_\alpha = \alpha \cap D$. That is, C is a $\square^*_\kappa$ sequence.

Proof

(2) $\Rightarrow$ (3). Assume that (3) is false. Let D be a witness. Let $\theta > \kappa$ be a sufficiently large regular cardinal. Let $N \prec H_\theta$ be a countable elementary submodel containing all the relevant objects. Then $N \cap H_\kappa \notin S_C$.

(3) $\Rightarrow$ (1). This is essentially due to Velickovic [18].

For an $f : [H_\kappa]^{<\omega} \to H_\kappa$, and a countable ordinal α, let us consider the following game $G(\alpha, f)$, due to Velickovic [10]. Two players, Player One and Player Two, take in turn playing in ω many steps. Player One plays an interval $I_n \subseteq \kappa$ of size less than κ and an ordinal $\alpha_n \in I_n$ at his nth move. Player Two plays an ordinal β_n in his responding move. At the end of the play, Player One wins the play if and only if for every $n < \omega$, $\beta_n < \inf(I_{n+1})$ and the elementary submodel N of H_κ, which is generated by $\alpha \cup \{\alpha_n \mid n < \omega\}$ and which is also closed under f, has the property that $N \cap \omega_1 = \alpha$ and $N \cap \kappa \subseteq \bigcup_{n<\omega} I_n$.

Notice that if Player One loses a play, Player One loses at some stage even before the game is over. This game is closed (for Player One) and hence determined.

LEMMA 5.1 (Velickovic) There exists a club $E_f \subseteq \omega_1$ such that for every $\alpha \in E_f$, Player One has a winning strategy σ_f^α for the game $G(\alpha, f)$.

Let $f : [H_\kappa]^{<\omega} \to H_\kappa$. Let $T \subseteq \omega_1$ be stationary.

Assume (3), we need to find a countable elementary submodel $N \prec H_\kappa$ with the property that N is closed under f, $N \cap \omega_1 \in T$ and $N \in S_C$.

Let $\alpha \in T \cap E_f$. Let σ be a winning strategy for Player One in the game $G(\alpha, f)$.

Let $\theta = (2^\kappa)^+$. Let $\langle M_\alpha \mid \alpha < \kappa \rangle$ be an increasing continuous $\in$–chain of elementary submodels of H_θ of size $< \kappa$ such that all the relevant objects are in M_0 and $M_\alpha \cap \kappa \in \kappa$ for all $\alpha < \kappa$. Let $D = \{M_\alpha \cap \kappa \mid \alpha < \kappa\}$. Then D is a club in κ.

Let $M \prec H_\theta$ be an elementary submodel of size less than κ, containing all the relevant objects, such that $M \cap \kappa$ is an ordinal of cofinality ω_1. Let $\delta = M \cap \kappa$.

Assume that there is a $X \in M$ such that $\delta \cap X \subseteq C_\delta$ and $M \models$ "X is a club in κ ". Let Y be the set of all the limit points of X. Then $Y \in M$. Let $\gamma < \beta$ be in M such that both are in Y. Then γ and β are limit points of C_δ. Hence $C_\gamma = \gamma \cap C_\beta$. Therefore, for every pair $\gamma < \beta$ from Y, we have $C_\gamma = \gamma \cap C_\beta$. This contradicts to our assumption (3).

Thus, for every club $X \subseteq \kappa$, if $X \in M$ then there is some ordinal $\gamma \in X \cap \delta$ which is not in C_δ. It follows that for every $\gamma < \delta$, there is some $\xi < \delta$ such that $\eta_\xi = M_\xi \cap \kappa$ is not in C_δ and $\eta_\xi > \gamma$.

Inductively pick ξ_n so that $\eta_{\xi_n} \notin C_\delta$ and there is some ordinal $\beta \in C_\delta$ with $\eta_{\xi_n} < \beta < \eta_{\xi_{n+1}}$.

Let $\beta = \bigcup\{\eta_{\xi_n} \mid n < \omega\}$. Then $C_\beta = \beta \cap C_\delta$.

Consider the following play of the game $G(\alpha, f)$. Player One follows his winning strategy σ which is in M_0. Player Two plays $\mu_n = \max(\eta_{\xi_n} \cap C_\delta)$, which is in M_{ξ_n}, in his nth move. Let I_n and β_n be the nth move of Player One following his winning strategy σ. Then $\beta_n \in I_n \subseteq (\mu_n, \eta_{\xi_n})$ for all $n < \omega$. Let $N \prec M$ be generated by $\alpha \cup \{\beta_n \mid n < \omega\}$ which is closed under f. Then $N \cap \omega_1 \in T$ and $\sup N \cap \kappa = \beta$ and $N \cap C_\beta \subseteq I_0$. Hence $N \cap H_\kappa \in S_C$.

This shows that (3) implies (1). □

COROLLARY 5.1 If for every stationary $S \subseteq [H_\kappa]^\omega$ there is an increasing continuous $\in$–chain $\langle N_\alpha \mid \alpha < \omega_1 \rangle$ such that $\{\alpha < \omega_1 \mid N_\alpha \in S\}$ is stationary, then there is a club $C \subseteq \kappa$ such that for every limit point $\alpha \in C$, $C_\alpha = \alpha \cap C$. In particular, there is no $\square_\kappa^*$ sequence.

Proof By the previous theorem, we need only to prove that S_C is not stationary.

Toward a contradiction, let us assume that S_C is stationary. Let $\langle N_\alpha \mid \alpha < \omega_1 \rangle$ be an increasing continuous $\in$–chain be such that $T = \{\alpha < \omega_1 \mid N_\alpha \in S_C\}$ is stationary. Let $\gamma_\alpha = \sup(N_\alpha \cap \kappa)$ for $\alpha < \omega_1$. Let

$\delta = \sup\{\gamma_\alpha \mid \alpha < \omega_1\}$. Let $\alpha \in T$ be such that $\gamma_\alpha \in C_\delta$ and γ_α is a limit point of $C_\delta \cap \{\gamma_\beta \mid \beta < \omega_1\}$. We get a contradiction, since $\alpha \in T$ implies that $N_\alpha \in S_C$ and $C_{\gamma_\alpha} = \gamma_\alpha \cap C_\delta$ whose intersection with N_α is not bounded in γ_α. □

REFERENCES

[0] J. BAUMGARTNER AND A. TAYLOR *Saturation properties of ideals in generic extensions II* **Trans. Amer. Math. Soc.** 271 (1982) 587–609

[1] M. BEKKALI **Topics in Set Theory** Lecture Notes in Mathematics, Vol. 1476, Springer–Verlag, Berlin, New York, 1991

[2] Q. FENG AND M. MAGIDOR *On reflection of stationary sets* **Fund. Math.** 140 (1992) 175–181

[3] Q. FENG AND T. JECH *Local Clubs, Reflection, and Preserving Stationary Sets* **Proc. London Math. Soc.** (3) 58 (1989) 237–257

[4] Q. FENG AND T. JECH *Projective Stationary sets and Strong Reflection Principle* **J. London Math. Soc.** (to appear)

[5] W. FLEISSNER *Left–separated spaces with point–countable bases*

[6] M. FOREMAN, M. MAGIDOR AND S. SHELAH *Martin's Maximum, Saturated Ideals, and Nonregular Ultrafilters. Part I* **Annals of Mathematics** 127 (1988) 1–47

[7] H. FRIEDMAN *One Hundred and two problems in mathematical logic* **J. Symb. Logic** 40 (1975) 113–129

[8] T. JECH *Some combinatorial problems concerning uncountable cardinals* **Ann. Math. Logic** 5 (1973) 165–198

[9] T. JECH **Set Theory** Academic Press, New York 1978

[10] T. JECH **Multiple Forcing** Cambridge Tracts in Mathematics, Cambridge University Press, 1986

[11] R. B. JENSEN *The fine structure of the constructible hierarchy* **Ann. Math. Logic** 4 (1972) 229–308

[12] D. KUEKER *Countable Approximations and Löwenheim–Skolem Theorem* **Annals of Mathematical Logic** 11 (1977) 57–103

[13] J. SILVER *On the Singular Cardinals Problem* **Proc. Internat'l Cong. Math.** Vancouver, B. C., 1974, Vol. 1, 265–268

[14] J. STEEL AND R. VAN WESEP *Two Consequences of Determinancy Consistent with Choice* **Trans. A. M. S.** 272(1) (1982) 67–85

[15] S. TODORCEVIC *Reflecting Stationary Sets I* Hand Written Notes, 1985

[16] S. TODORCEVIC *Strong Reflections* Hand Written Notes, 1987

[17] S. TODORCEVIC *Conjectures of Rado and Chang and Cardinal Arithmetic* Preprint

[18] B. VELICKOVIC *Forcing Axioms and Stationary Sets* **Advances in Mathematics** 94 (1992) 256–284

[19] H. WOODIN *The axiom of determinacy, forcing axiom and the non-stationary ideal* preprint

[20] H. WOODIN *Some consistency results in ZFC using AD* **Cabal Seminar 79–81** Lecture Notes in Mathematics, Vol. 1019, 172–198

Department of Mathematics, National University of Singapore, Singapore 0511

Email: feng@math.nus.sg

Definable Sets of Real Numbers, Infinite Games and Core Model Theory

Kai Hauser *

Department of Mathematics, University of California, Berkeley, CA 94720, USA
Email: hauser@math.berkeley.edu

Abstract

Applying Steel's Σ^1_3 correctness theorem for the core model for one Woodin cardinal [St2], it is shown that $\mathbf{\Delta}^1_2$ determinacy follows from the assumptions that every projective set is Lebesgue measurable and has the property of Baire together with the uniformization property for $\mathbf{\Pi}^1_3$ relations.

1. Introduction

General questions about arbitrary sets of real numbers - like for instance Cantor's continuum hypothesis CH (asserting that an uncountable set of real numbers is equinumerous with the whole set of real numbers) - have a partial solution within the context of definability. For example a classical theorem of descriptive set theory due to Suslin (cf. [Lu1]) states that no *analytic* set (i.e., continuous image of a Borel set) can give a counterexample to CH. The analytic sets posess further properties indicative of well-behavedness: They are *Lebesgue measurable* [Lu1] and have the *property of Baire* [LuSi]. (This is the dual notion of measurability in terms of category: A set has the property of Baire if its symmetric difference with an open set is small in the topological sense that it can be covered by a countable union of closed sets with empty interior.)

Another regularity property of fundamental interest to descriptive set theorists is the following choice-like principle introduced in [Lu2]: Given a

*The author is a Heisenberg Fellow of the Deutsche Forschungsgemeinschaft. Travel support from the Ministry of Foreign Affairs of the Federal Republic of Germany to attend the 6th Asian Logic Conference is gratefully acknowledged.

subset A in the plane, a partial function F is called a *uniformization* if it picks for each x in the projection of A onto the real line a real number y putting (x, y) into A. In ordinary set theory the uniformization problem is solved trivially with the help of the axiom of choice, whereas it becomes highly non-trivial when definable uniformizations are sought, even if the given set is of a simple form. Also in this regard definable sets exhibit well-behavedness. Kondo [Ko] proved that the *coanalytic* sets (the complements of analytic sets) have the uniformization property: Any coanalytic subset of the real plane can be uniformized by a function with coanalytic graph.

Naturally a great deal of work was invested in descriptive set theory towards extending these regularity results to more complex sets of real numbers. A suitable framework for this investigation is provided by the *projective sets,* i.e., the smallest collection containing all Borel sets and closed under forming complements and (pointwise) continuous images. This can be stratified into a natural hierarchy: $\mathbf{\Sigma}^1_1$ is the class of all analytic sets; For $n \geq 1$, a set is $\mathbf{\Pi}^1_n$ iff its complement is $\mathbf{\Sigma}^1_n$, it is $\mathbf{\Sigma}^1_{n+1}$ iff it is the continuous image of a $\mathbf{\Pi}^1_n$ set and $\mathbf{\Delta}^1_n$ iff it is both $\mathbf{\Sigma}^1_n$ and $\mathbf{\Pi}^1_n$. Finally a set is projective iff it is $\mathbf{\Sigma}^1_n$ for some $n \geq 1$. Thus $\mathbf{\Delta}^1_1$ is precisely the Borel sets [Su] , and any projective set can be generated in finitely many steps by the above two operations. These are the sets occuring in ordinary mathematical practise, and so one would like to have a unified theory guaranteeing the well-behavedness of as many projective sets as possible.

There is, however, an insurmountable principal barrier, despite the fact that projective sets somehow seem to be simply definable: Through the work of Gödel [Go2] and Cohen [Co] (and many others after them) all the central questions are now known to be unanswerable on the basis of the standard axiomatic system for set theory ZFC (Zermelo-Fraenkel with choice). Historically, the first attempt to overcome this limitation was in the use of *large cardinal axioms* (postulates implying the existence of sets of large size).

In the mid 60s Solovay (cf. 42.11 in [Jc]) showed that the existence of a *measurable cardinal* (a prominent large cardinal axiom) implies all sets in the second level of the projective hierachy (i.e., all continuous images of coanalytic sets) are Lebesgue measurable and have the property of Baire. This important theorem led to the hope that eventually sufficiently strong large cardinal axioms may decide all open questions in this area. Quite unexpectedly that was achieved with the help of a different class of axioms: *determinacy axioms.*

Let ω denote the set of all natural numbers and ω^ω the set of all infinite sequences of natural numbers. From a topological viewpoint the space ω^ω (with its natural topology) is indistinguishable from the subspace of the Euclidian space, $\mathbb{R}$, consisting of the irrationals, and its elements are referred to as *reals* in descriptive set theory. The advantage of this identification is

ω^ω accomodates in a more natural way the notion of a *game:* With each set $A \subset \omega^\omega$ a game G_A is associated in which two players cooperatively produce a real by picking alternatively natural numbers in ω many rounds. Player I wins the game G_A if and only if this real belongs to the payoff set A. A *strategy* for one of the players is an algorithm telling him which number to pick at each of his turns. Such a strategy is *winning* for a player if it leads to a win for that player whenever he follows it regardless of the other player's moves. The game is *determined* if one of the two players has a winning strategy.

Simple games (for instance the ones with open payoff sets) are determined [GaSt]. The same is true for games with a Borel payoff set [Ma], yet additional hypotheses stronger than ZFC are needed for more complicated games. The *Axiom of Determinacy AD* [MySt] settles this issue by decree in that it asserts that for all $A \subset \omega^\omega$, G_A is determined. Its consequences are rather striking (among them Lebesgue measurability [MySw] and the property of Baire (Banach-Mazur, cf. [Ml]) for *any* set of reals, but unfortunately it contradicts the axiom of choice. The axiom of choice - as usual - fails to produce "constructible" non-determined games, and this motivates the introduction of axioms of *definable determinacy* (as first suggested in [MySt] and independently by R. Solovay and G. Takeuti) which allow mathematicians to keep the axiom of choice and with it standard tools of mathematics (like the Hahn-Banach theorem for example) while all sets arising "in practice" enjoy the usual regularity properties. A central role in this regard is played by the axiom of *projective determinacy* PD which states that every game with projective payoff set is determined. It answers all descriptive set theoretic questions about projective sets and extends the structure theory (including a strong form of uniformization) provided by ZFC for the coanalytic sets to all further levels of the projective hierarchy (cf. ch.6 of [Mo]). A natural question therefore arises:

How much of PD is "really" needed?

In 1981 Woodin [Wo1] formulated the following conjecture to the effect that *all* of PD is needed:

(ZFC) Assume

(1) Every projective set of reals is Lebesgue measurable ("projective measurability") and has the property of Baire ("projective category").

(2) Every projectiove set can be uniformized by a function with projective graph ("projective uniformization").

Then PD holds.

Subsequently this question has been included as the last Delfino problem in the expanded list of [KeMaSt]. It is to be regarded as a crucial test for the naturalness of the hypothesis PD, and rephrased in metamathematical terms, a positive answer would in effect mean that PD is the only way of extending the structure theory of the coanalytic sets provided by ZFC to the entire projective hierarchy. In terms of heuristics, Woodin's conjecture expresses the belief that PD *is* the underlying structure enabling the conflicting statements (1) (amounting to a frailure of the axiom of choice at the projective level) and (2) (which retains some form of choice in the projective hierarchy) to hold simultaneously.

Woodin [Wo1] supplies some evidence for this ambitious claim: It is shown there that (1) and (2) imply $\mathbf{\Pi}^1_1$ determinacy. The theorem proved below pushes this one level higher in the projective hierarchy (recall that on general grounds, $\mathbf{\Delta}^1_2$ determinacy is (in ZFC) equivalent to $\mathbf{\Sigma}^1_2$ determinacy by a theorem of Martin cf. [WoMaHa])

Theorem *(ZFC) Assume*

(1) Projective measure and category

(2) $\mathbf{\Pi}^1_3$ *uniformization.*

Then $\mathbf{\Delta}^1_2$ *determinacy holds.*

This may seem a tiny step toward the solution of the Delfino problem. It is, however, worth noting that $\mathbf{\Delta}^1_2$ determinacy represents a definite cut point in the large cardinal hierarchy with far reaching global consequences for the structure of the set theoretical universe. (This is in analogy with the appearance of sharps for reals.)

The above theorem falls into the area best described as *reverse descriptive set theory.* Here the aim is to show that propostions asserting regularity properties for higher order projective sets which are derivable from strong set theoretic principles (such as axioms of determinacy) conversely imply those principles. The astonishing feature of these theorems is that despite being phrased in purely descriptive set theoretic terms, their only known proofs rely on *core model theory.* (More examples of problems dealing exclusively with definable sets of reals that have been solved with core model machinery can be found in [Hj] and [St1, sec. 7].) Core models occupy a special place in the *inner model program* where the guiding idea is to exhibit for each large cardinal axiom A a canonical model M_A that is in some sense minimal and whose structure theory can be analyzed in detail. This is not a consistency proof for A as one must assume A in order to conclude that A holds in M_A. Nevertheless it is to be expected that any hidden inconsistency in A should emerge quickly in the detailed structure theory of M_A.

The prototype of an inner model is the universe of constructible sets L introduced by Gödel [Go1] to establish the consistency of the generalized continuum hypothesis and the axiom of choice with the original axioms of set theory. Gödel's analysis of L was immensely refined by R. Jensen who pioneered *fine structure* theory (a detailed study of how individual sets enter into an inner model) in the early 70s (cf. [Je]). Also discovered in Jensen's fine structure laboratory was a new important class of models, *core models.* Their name refers to the fact that the simplest one, K_{DJ}, introduced in collaboration with A.J. Dodd (cf. [Do]) in the 70s, can be characterized as the "core" of all models with measurable cardinals. Unlike L which comes out the same no matter in which world it is computed in, the shape of K_{DJ} varies with the large cardinal situation in the surrounding universe, and it mirrors that situation below measurability. The core models considered in connection with the last Delfino problem are of a much higher order: they approximate inner models with *Woodin cardinals* (this large cardinal notion of far greater strength than measurability is intimately related to projective determinacy, cf. [MaSt] and [WoMaHa]).

More specifically, the proof of the above theorem rests on a recent theorem of Steel [St2] about the correctness for $\mathbf{\Pi}^1_3$ truth as computed in the the core model for one Woodin cardinal, K. This model is the contructible closure of the ordinals under a predicate consisting of a sequence of so called *extenders.* These are directed systems of measures coding up set sized restrictions of an embedding of the universe into a transitive class. Woodinness is defined in terms of the existence of embeddings satisfying additional conditions, and the extenders appearing on the sequence of K are attempts to approximate such embeddings. The construction of K was invented by Steel [St1] and relies on previous work of Michell and Steel [MiSt] who built a fine structural inner model with a Woodin cardinal (provided *some* inner model with a Woodin cardinal exists). In order for the K- construction to produce a model with the desired properties (among which is a strong form of the generalized continuum hypothesis, certain fine structural features and the ability to reflect the large cardinal situation in the universe below one Woodin cardinal) one needs background assumptions. The existence of a measurable cardinal suffices (in fact less, cf. [St1]) if there is no inner class model containing a Woodin cardinal. In the presence of sharps for all sets this is enshured by the failure of $\mathbf{\Delta}^1_2$ determinacy via a theorem of Woodin (cf. [WoMaHa]). Thus the proof of the theorem takes the form of an indirect argument: Assuming a failure of $\mathbf{\Delta}^1_2$ determinacy one builds a model of ZFC in which K exists and is correct about $\mathbf{\Pi}^1_3$ truth and derives a contradiction. The construction of that model uses the descriptive set theoretic assumptions of the theorem and is essentially a refinement of the arguments in [Wo1].

In some sense the proof of the theorem is more important than its statement: Modulo the appropriate correctness of higher order versions of K and the closure of the universe under the sharp operation for the appropriate higher order analogues of L, the proof given in the next section generalizes to other levels of the projective hierarchy. This issue and related topics are discussed in section 3.

2 The Proof

The reader who is not familiar with the inner workings of core models can think of K (or rather its relativization K_z to some real z) as a generalization of the constructible universe L. As mentioned in the previous section the proof rests on the fact that (under the appropriate background assumptions) K correctly computes Σ^1_3 truth about its reals as proved by Steel [St2].

Theorem. *(ZFC) Suppose*

(1) Every projective set is Lebesgue measurable (PM) and has the property of Baire (PB).

(2) $\mathbf{\Pi}^1_3$ uniformization.

Then $\mathbf{\Delta}^1_2$ determinacy holds.

Proof. The argument proceeds in two steps: First a transitive model M of ZFC is constructed which is both Σ^1_4 correct and Σ^1_4 absolute in the sense of [Wo1]. Then one establishes that $\mathbf{\Delta}^1_2$ determinacy holds inside M from which the theorem follows because $\mathbf{\Delta}^1_2$ determinacy is a Π^1_4 sentence.

Construction of M. This uses a minor modification of the the construction in [Wo1]. Notice that PB of (1) and (2) clearly imply $**\mathbf{\Pi}^1_3$ uniformization (as defined in [Wo1]). This yields as in lemma 2 of [Wo1]

(3) For any real c Cohen over V, V is Σ^1_5 correct in $V[c]$.

Next, using (2), fix a Π^1_3 definition from a parameter x_0 for a function uniformizing the complete Π^1_3 set in $\mathbb{R} \times \mathbb{R}$. Let $Unif(\Pi^1_3, x_0)$ be the $\Pi^1_5(x_0)$ formalization of this situation. Thus by (3),

(4) For any real c Cohen over V, $Unif(\Pi^1_3, x_0)$ holds in $V[c]$.

Similiarly one obtains

(5) For any real c Cohen over V, $*\mathbf{\Pi}^1_2$ uniformization holds in $V[c]$

where $*\mathbf{\Pi}^1_2$ uniformization is $**\mathbf{\Pi}^1_2$ uniformization restricted to $\mathbf{\Pi}^1_2$ subsets of the plane whose projection is the whole real line. This restriction seems necessary to arrive at a Π^1_5 notion. On the other hand, the restricted version suffices to run the argument in lemma 2 of [Wo1] for Σ^1_4, and with (3) it follows:

(6) Whenever c_1, c_2 are mutually Cohen generic over V then $V[c_1]$ is Σ^1_4 correct inside $V[c_1][c_2]$.

Armed with these facts, one can now proceed exactly as in the proof of theorem 1 of [Wo1] building a Σ^1_4 correct and Σ^1_4 absolute model M of the form $L_{\omega_1}(x_0, F)$. Essentially, F is a partial function of $HC \longrightarrow HC$ assigning to a given tuple of terms $\vec{\tau}$ for reals (in the forcing language associated to a poset belonging to M) a canonical term for a witness for a Σ^1_4 formula in those terms using a $\Pi^1_3(x_0)$ uniformization as in (4), provided such a witness exists. By a theorem of Shelah [Sl], PM implies that there is no uncountable projective sequence of distinct reals. (The proof of Shelah's theorem given in [Ra] actually shows that whenever S is an uncountable sequence of distinct reals, there is a non Lebesgue measurable set projective in S.) This implies that the power set axiom holds in M, i.e., M is a model of ZFC.

$\boldsymbol{\Delta}^1_2$ determinacy in *M*. Notice that in M there still is a $\Pi^1_3(x_0)$ definable uniformization of the complete Π^1_3 set since $\Pi^1_5(x_0)$ statements relativize to M by Σ^1_4 correctness. The rest of the proof takes place inside the model M.

Assume toward a contradiction that $\boldsymbol{\Delta}^1_2$ deteminacy fails. Fix a real z such that $\Delta^1_2(z)$ determinacy fails. Without loss of generality z is Turing above x_0.

Claim 1. There is no proper class inner model with a Woodin cardinal containing z.

Proof of claim 1. Otherwise, the fact that every set has a sharp (by Σ^1_3 absoluteness) allows to find a countable ordinal δ which is Woodin in some inner model N of the form $L(V^N_\delta)$ containing z. By a theorem of Woodin [WoMaHa], $\Delta^1_2(z)$ determinacy holds in $N^{Coll(\omega,\delta)}$. Since δ is countable N generics for $Coll(\omega, \delta)$ exist. But $\Delta^1_2(z)$ determinacy is upward absolute for proper class models , a contradiction.
□ Claim 1.

Via covering over the Dodd-Jensen core model for two measurables, Σ^1_4 absoluteness implies the existence of $x^{\dagger\dagger}$ for any set x. Now go to an inner model containing all the reals and two measurable cardinals $\kappa < \Omega$ and build K_z, the core model for one Woodin cardinal (cf. [St1]) relativized to the real z up to height Ω. By a theorem of Steel [St2] K_z is Σ^1_3 correct.

Notice that $\mathbb{R} \cap K_z$ is $\Sigma^1_3(z)$ since for any real x

$$x \in K_z \text{ iff } \exists \mathcal{M}(\mathcal{M} \text{ is a countable 1-small iterable premouse built over } z \text{ and } x \in \mathcal{M}).$$

The displayed formula is indeed $\Sigma^1_3(z)$ because iterability for countable premice built over z is a Π^1_2 condition as there is no proper class inner model with a Woodin cardinal containing z. The above equivalence holds because for $x \in K_z$ one can take the first level of K_z where x appears for $\mathcal{M}$. The other direction comes from comparing such an $\mathcal{M}$ with K_z which must win the comparison by universality (cf. ch. 5 of [St1]).

It follows that with $A \subset \mathbb{R} \times \mathbb{R}$ defined as

$$(x, y) \in A \text{ iff } y \text{ is Cohen generic over } \bigcup\{\mathcal{M} : \mathcal{M} \text{ is a countable 1-small } \Pi^1_2\text{-iterable premouse built over } x\}$$

A is a Π^1_3 relation such that for all c

$$(z, c) \in A \text{ iff } c \text{ is Cohen over } K_z.$$

Claim 2. $\mathbb{R} \cap K_z$ is countable.

Proof of claim 2. Define $S \subset \mathbb{R} \times \mathbb{R}$ by

$$(x, y) \in S \text{ iff } x \text{ codes some countable ordinal } \xi_x \text{ such that } y \text{ is the } \xi_x\text{-th real in the order of construction of } K_z.$$

If the claim fails then S defines an uncountable $\Sigma^1_3(z)$ sequence of distinct reals. But this is a $\Pi^1_4(z)$ statement which would contradict Shelah's theorem [Sl] in the outside universe where M is Σ^1_4 correct. □ Claim 2.

So there is a real Cohen over K_z, i.e., $\exists y (z, y) \in A$. Now let A^* be a $\Pi^1_3(z)$ uniformization of A. Fix a real c such that $(z, c) \in A^*$. By Shoenfield absoluteness $K_z[c] \models (z, c) \in A^*$. Pick a Cohen condition p such that, over K_z, $p \Vdash (z, \dot{c}) \in A^*$ and such that $p \subset c$. Changing a finite amount of information about c, one obtains c' extending p so that $K_z[c'] \models (z, c') \in A^*$.

An inspection of Steel's proof [St2] for the Σ^1_3 correctness of the 1-Woodin core model reveals that in actual fact $K_z[c']$ must be Σ^1_3 correct. (The version of the Martin-Solovay tree for Π^1_2 shown by Steel to exist inside the core model survives "small" forcing.) It follows that $(z, c') \in A^*$ contradicting $c \neq c'$. □ theorem

3 Concluding Remarks

Once suitable strengthenings of Σ^1_3 correctness for the core models for more Woodin cardinals are at hand, certain parts of the above proof can be lifted. This generates the hope to prove the following weakened version of the Delfino Problem (which should be compared with the conjecture made in the remarks just after 4.1 of [Ha1]).

(ZFC) Assume

(1) Every projective set is Lebesgue measurable and has the property of Baire

(2) $\mathbf{\Pi}^1_{2n+1}$ *uniformization for each* $n \geq 0$.

Then PD holds.

One way of obtaining higher order versions of the one Woodin K proceeds via definable singleton operations coming from inner model theory. [These singletons appear naturally in the course of generalizing the structure theory of the first level of the projective hierarchy to its higher levels (cf. [Ke1], [Ke2]), and their connection with inner models for large cardinals related to determinacy was conjectured early on (cf. [KeMaSo]).] In this approach the least failure of an instance of PD (i.e., the least level n at which $\mathbf{\Sigma}^1_n$ determinacy fails) corresponds to a first occurence of one of these singleton operations being non-total. This can be expressed in terms of a smallness condition about the sequence of the preliminary K^c which in turn yields an iteration strategy for K^c. But now one is in the position to simply repeat the process of extracting K as a certain Skolem hull of K^c just as in the one Woodin setting of [St1]. (This method is essentially an initial segment of the far more complex procedure of obtaining a core model from a least failure of $AD^{L(\mathbb{R})}$ which was developed by Woodin in his solution of the $L(\mathbb{R})$ version of the last Delfino Problem.)

In the first step beyond one Woodin cardinal of the above construction the core model is built inside the smallest model of set theory closed under the y_0 operation and containing a rank initial segment of the universe. [For any set (of ordinals) a, $y_0(a)$ denotes the sharp of a fully iterable fine structure model with one Woodin cardinal "built above the set a". This is a Π^1_3 singleton in any real (generically) coding a.] More precisely: The smallest model of set theory closed under the y_0 operation and containing a given rank initial segment of the universe can be re-arranged as a fine structure model in the sense of [MiSt]. Σ^1_5 absoluteness implies the failure of a weak form of the covering lemma over any such model. Woodin (cf. [SchWo]) observed that this yields indiscernibles for such models, i.e., normal, countably complete ultrafilters that measure all subsets of some ordinal belonging to the model and are weakly amenable to the model. The construction of K^c in [St1] can be carried out with such an indiscernible playing the role of a measurable cardinal. Under the above mentioned smallness condition, K^c satisfies a Π^1_3 iterability condition and one can extract "true" K^2 as it will be called here just as in [St1] from K^c. Otherwise one obtains that the universe is actually closed under the next singleton operation (corresponding to the sharp of a fully iterable two-Woodin model). Then - under the smallness condition that the universe fails to be closed under the three-Woodin-sharp operation - one builds a core model inside the smallest model of set theory closed under the two-Woodin-sharp operation above some rank. This core model - called K^3 in the sequel - will be closed under the two-Woodin-sharp operation and satisfy a Π^1_4 iterability condition. [In this numerology K^1 would denote the one Woodin K and K^0 corresponds to L.]

One would now like to work with the core model K^3 imitating the proof of the previous section. However, in order to do so one must first get past the core model K^2, i.e., the closure of the universe under the two-Woodin-sharp operation has to be established. This corresponds to closing the reals under sharps (i.e., $\mathbf{\Pi}^1_1$ determinacy) in the proof of the previous section. [Results of Woodin [Wo2] when combined with the transfer theorem in [Ne] show that the closure of the reals under the two-Woodin-sharp operation is equivalent to $\mathbf{\Pi}^1_3$ determinacy.] In that proof already the Σ^1_3 absoluteness combined with the covering lemma for the core model L yielded the desired closure. Now the fact that the construction of the various core models takes place inside a universe satisfying Σ^1_6 absoluteness by itself will not suffice to derive the necessary closure as illustrated by the following example: Cut M_2 (the minimal fully iterable fine structure model with two Woodin cardinals) at its bottom Woodin cardinal. This gives a model of ZFC + global $\mathbf{\Delta}^1_2$ determinacy with unboundedly many strong cardinals (because M_2 is closed under the y_0 operation below its bottom Woodin). By a theorem of Woodin [Wo2], collapsing sufficiently many of the strong cardinals of that model to ω with finite conditions results in a model N where any prespecified amount of projective forcing absoluteness is available and global $\mathbf{\Delta}^1_2$ determinacy holds, yet Π^1_3 determinacy fails.

In contrast to this counterexample the universe in the situation where one tries to get past the core model K^2 does satisfy a fair amount of uniformization (namely, $\mathbf{\Pi}^1_5$ uniformization). This suggests the following approach to getting past K^2: First establish the correctness with respect to a lightface fragment of projective truth in V of generic collapses (of a suitable number) of strong cardinals of K^2 assuming that K^2 exists. [By the appropriate generalization of theorem 3.10 of [Ha1] strong cardinals exist in K^2 by projective absoluteness.] Then prove a negative uniformization result for generic extensions of K^2 of the above form. This would contradict that uniformization holds in V and thus K^2 did not exist in the first place, i.e., the universe is closed under the next singleton operation. Thus the stage is set for the construction of K^3. [Note that in the above mentioned counterexample, N is itself the generic extension of K^2 via the collapse of strong cardinals because we cut M_2 below its bottom Woodin cardinal (i.e., theorem 2.6 of [Ha2] applies) and by the the forcing absoluteness of K^2. Thus presumably Π^1_5 uniformization fails in N.]

Unfortunately, the approach sketched in the previous paragraph would suggest that the solution of the restricted Delfino problem (assuming level by level uniformization) is no easier than that of the full Delfino problem (cf. the remarks made in the end of section 3 of [HaHj]). This is yet another indication that getting past K^2 is a serious issue.

Provided that K^3 is Σ^1_5 correct and global $\mathbf{\Pi^1_3}$ determinacy can be derived, the proof of the previous section lifts and one obtains:

Assume

(1) Projective measure and category

(2) $\mathbf{\Pi^1_3}$ *uniformization and* $\mathbf{\Pi^1_5}$ *uniformization.*

Then $\mathbf{\Delta^1_4}$ *determinacy holds.*

As a warm-up for showing Σ^1_5 correctness of K^3 one may try and establish Σ^1_4 correctness for K^2 first. This would follow if it could be shown that cofinally in a suitable Π^1_3 norm on the complete Π^1_3 set witnesses in K^2 exist for Π^1_3 statements true in V about parameters from K^2. [The closure of the universe under the y_0 operation implies global $\mathbf{\Delta^1_2}$ determinacy. Thus the pointclass Π^1_3 is normed. Recall that the initial segments of a Π^1_3 norm are uniformly $\mathbf{\Delta^1_3}$ whence the above cofinality condition transforms Σ^1_3 correctness into Σ^1_4 correctness.]

The Σ^1_4 correctness at this level is related to earlier attempts of Woodin to derive the scale property for Π^1_3 combinatorially, i.e., from combinatorial properties of the corresponding inner model. Towards this end Woodin [Wo2] has created a scenario requiring a plausible hypothesis about the behavior of K (at the appropriate level) under coarse iterations of of its surrounding universe. This hypothesis says (under the necessary background assumptions for the construction of the core model):

> Suppose $\mathcal{T}$ is a coarse iteration tree built on a coarse premouse M with last model M_θ and associated embedding $i : M \longrightarrow M_\theta$. If K denotes the core model at the appropriate level, $K^{M_\theta} = i(K^M)$ is the final model of a fine-structural iteration tree based on K^M.

Despite all this the ultimate goal is still to settle the Delfino Problem in its original form. A positive solution will have the added advantage of showing that among all the conceivable patterns leading to projective uniformization a particular one stands out: No matter which universe of set theory we live in, as long as any projective set is Lebesgue measurable and has the property of Baire, projective uniformization can only be achieved by having the uniformization property oscillate between the odd levels on the Π side and the even levels on the Σ side of the projective hierarchy.

References

[Co] P.J. Cohen, *The Independence Of The Continuum Hypothesis, I, II,* Proc. Natl. Acad. Sci. USA **50** (1963), pp. 1143-1148, **51** (1964) pp. 105-110.

[Do] A.J. Dodd, *The Core Model*, London Math. Soc. Lec. Note Series **61** (1984), xxxviii + 229.

[GaSt] D. Gale and F.M. Stewart, *Infinite Games With Perfect Information*, Ann. Math. Studies **28** (1953), pp. 245-266.

[Go1] K. Gödel, *The Consistency Of The Axiom Of Choice And The Generalized Continuum Hypothesis*, Proc. Natl. Acad. Sci. USA **24** (1938), pp. 556-557.

[Go2] K. Gödel, *The Consistency Of The Axiom Of Choice And Of The Generalized Continuum Hypothesis With The Axioms Of Set Theory,* Annals of Math. Studies, Study 3 (1940), Princeton University Press, Princeton, New Jersey

[Ha1] K. Hauser, *The Consistency Strength Of Projective Absoluteness,* Annals of Pure and Applied Logic **74** (1995), pp. 245-295.

[Ha2] K. Hauser, *A Minimal Counterexample to Universal Baireness*, submitted for publication.

[HaHj] K. Hauser and G. Hjorth, *Strong Cardinals in the Core Model*, in Annals of Pure and Applied Logic **83** (1997), pp. 165-198.

[Hj] G. Hjorth, Two Applications of Inner Model Theory to the Study of $\mathbf{\Sigma}^1_2$ *sets*, Bulletin of Symbolic Logic **2** (1996), pp. 94 - 107.

[Jc] T.Jech *Set Theory*, Academic Press (1978), xi + 621.

[Je] R.B. Jensen *The Fine Structure Of The Constructible Hierarchy*, Annals of Math. Log. **4** (1972), pp. 229-308.

[Ke1] A.S. Kechris, *The Theory of Countable Analytical Sets,* Trans. Amer. Math. Soc. **202** (1975), 259-297.

[Ke2] A.S. Kechris, *Recent Advances in the Theory of Higher Level Projective Sets*, in *The Kleene Symposium,* J. Barwise, H.J. Keisler and K. Kunen (eds.), North Holland (1980), 149-166.

[KeMaSo] A.S. Kechris, D.A. Martin and R.M. Solovay, *Introduction to Q-Theory,* in *Cabal Seminar 79-81*, Springer Lecture Notes in Math. **1019** (1983), 199-282.

[KeMaSt] A.S. Kechris, D.A. Martin and J.R. Steel (eds.), *Appendix: Victoria Delfino Problems II,* in *Cabal Seminar 81-85*, Springer Lecture Notes in Math. **1333** (1988), pp.221-224.

[Ko] M. Kondo, *Sur L'Uniformization Des Complementaires Analytiques Et Les Ensembles Projectifs De La Second Classe*, Jap. J. Math **15** (1938), pp. 197-230.

[Lu1] N. Lusin, *Sur La Classification De M. Baire*, Comptes Rendus Acad. Sci. Paris **164** (1917), pp. 91-94.

[Lu2] N. Lusin, *Sur Le Problème De M. J. Hadamard D'Uniformisation Des Ensembles*, Comptes Rendus Acad. Sci. Paris **190** (1930), pp. 95-96.

[LuSi] N. Lusin and W. Sierpinski, *Sur Un Ensemble Non Measurable B*, Journal de Mathématiques, 9e serie, 2 (1923), pp. 53-72.

[Ma] D.A. Martin, *Borel Determinacy*, Annals of Math. **102** (1975), pp. 363-371.

[MaSt] D.A. Martin and J.. Steel, *Projective Determinacy*, Proc. Natl. Acad. Sci. USA **85** (1988), pp. 6582-6586.

[MiSt] W.J. Mitchell and J.R. Steel, *Fine Structure And Iteration Trees*, Lecture Notes in Logic **3**, Springer Verlag (1994), 130 pages.

[Ml] R.D. Mauldin (ed.), *The Scottish Book,* Birkhäuser (1981).

[Mo] Y.N. Moschovakis, *Descriptive Set Theory*, North-Holland (1980), xii + 637.

[My] J. Mycielski, *On The Axiom Of Determinateness I,II*, Fund. Math. **53** (1964), pp. 205-224, **59** (1966), pp. 203-212.

[MySt] J. Mycielski and H. Steinhaus, *A Mathematical Axiom Contradicting The Axiom Of Choice*, Bull. Pol. Acad. **10** (1962), pp. 1-3.

[MySw] J. Mycielski and S. Swierczkowski, *On The Lebesgue Measurability And The Axiom Of Determinateness*, Fund. Math. **54** (1964), pp. 67-71.

[Ne] I. Neeman, *Optimal Proofs of Determinacy*, Bull. Symb. Logic **1** (1995), 327-339.

[Ra] J. Raisonnier, *A Mathematical Proof Of S. Shelah's Theorem On The Measure Problem And Related Results,* Isr. J. Math **48 (1)** (1984), pp. 48- 56.

[SchWo] , *The Jensen Covering Property*, manuscript (1996).

[Sl] S. Shelah, *Can You Take Solovay's Inaccessible Away?,* Isr. J. Math. **48 (1)** (1984), pp. 1-47.

[St1] J.R. Steel, *The Core Model Iterability Problem*, Lecture Notes in Logic **8**, Springer Verlag (1996), 112 pages.

[St2] J.R. Steel, *The Σ^1_3 correctness for the 1-Woodin K*, manuscript (1993).

[Su] M. Suslin, *Sur une definition des ensembles mesurables B sans nombres transfinis,* Comptes Rendus Acad. Science, Paris **164** (1917), pp. 88-91.

[Wo1] W.H. Woodin, *On The Consistency Strength Of Projective Uniformization*, in *Logic Colloquium '81*, J. Stern (ed.), North-Holland (1982), pp. 365-383.

[Wo2] W.H. Woodin, personal communication.

[WoMaHa] W.H. Woodin, A.R.D. Mathias and K. Hauser *The Axiom of Determinacy*, monograph in preparation.

THE DESCRIPTIVE CLASSIFICATION OF SOME CLASSES OF C*-ALGEBRAS

Alexander S. Kechris[1)]
Department of Mathematics
California Institute of Technology
Pasadena, California 91125

We introduce here a parametrization of separable C^*-algebras by a standard Borel space and study the descriptive complexity of various canonical classes of C^*-algebras in this parametrization. This can be viewed as providing an analog of the corresponding classification of classes of von Neumann algebras (acting on a fixed separable Hilbert space) in the Effros Borel space of von Neumann algebras (see for example Nielsen [13]). However, in contrast with the von Neumann case, where most interesting classes (like: factors, type I, II, III, hyperfinite) turn out to be Borel, in the C^*-algebra case many important classes turn out to be co-analytic but *not* Borel. This makes the situation more interesting from the set-theoretic point of view, and leads to further questions, like the construction of canonical co-analytic norms, which we also address here. Finally, we relate the above to work of Sutherland [17] on parametrization of Polish groups and indicate how one might be able to show that various familiar classes of (second countable) locally compact groups are non-Borel as well.

We assume that the reader is familiar with the basics of descriptive set theory (see Moschovakis [12] and Kechris [11]) and C^*-algebras (see Dixmier [4] and Fillmore [5]).

Acknowledgement. We would like to thank E. Effros for many valuable conversations about this subject.

[1)] Research and preparation of this paper were partially supported by NSF Grants DMS–9317509 and 9619880.

§1. Parametrizing separable C^*-algebras

Let H be a fixed infinite dimensional Hilbert space, say ℓ^2, and $\mathcal{B}(H)$ the C^*-algebra of bounded operators on H. Every separable C^*-algebra is isomorphic to a C^*-subalgebra of $\mathcal{B}(H)$. We can parametrize or "encode" a separable C^*-subalgebra A of $\mathcal{B}(H)$ by giving a sequence of generators for A. So our parameter space will be

$$\Gamma := \mathcal{B}(H)^{\mathbb{N}}.$$

To each $\gamma = (\gamma_n) \in \Gamma$ we associate the C^*-subalgebra of $\mathcal{B}(H)$ given by

$$A(\gamma) := \text{ the } C^* - \text{subalgebra of } \mathcal{B}(H) \text{ generated by } \{\gamma_n : n \in \mathbb{N}\}.$$

We endow $\mathcal{B}(H)$ with the Borel structure of the weak topology (which is the same as the Borel structure of the σ-weak, strong and σ-strong topologies). This is a standard Borel space. We give Γ the product Borel structure, so this is again a standard Borel space.

DEFINITION 1.1. Let $\mathcal{A}$ be a class of separable C^*-algebras. We say that $\mathcal{A}$ is **Borel, analytic, co-analytic**, etc., if

$$\tilde{\mathcal{A}} = \{\gamma \in \Gamma : A(\gamma) \in \mathcal{A}\}$$

is Borel, analytic, co-analytic, etc., in Γ.

Denote by N the class of separable **nuclear** C^*-algebras, by GCR the class of separable **postliminal** (or **type** I) C^*-algebras, by CCR the class of separable **liminal** C^*-algebras, by GTC the class of separable C^*-algebras with **generalized continuous trace**, by TC the class of separable C^*-algebras with **continuous trace**, by FD the class of separable C^*-algebras all of whose irreducible representations are **finite-dimensional** and by BD the class of separable C^*-algebras all of whose irreducible representations have **bounded finite** dimension. Finally, let NGCR be the class of **antiliminal** separable C^*-algebras.

We then have the following inclusions

$$\mathrm{BD} \subseteq \left.\begin{matrix}\mathrm{FD}\\ \mathrm{TC}\end{matrix}\right\} \subseteq \mathrm{GTC} \subseteq \mathrm{CCR} \subseteq \mathrm{GCR} \subseteq \mathrm{N}.$$

THEOREM 1.2. *The classes* BD, TC, N *and* NGCR *are Borel. (The result about* N *is due to Effros).*

THEOREM 1.3. *The classes* FD, CCR, GCR *are co-analytic but not Borel. In fact they are complete co-analytic (i.e., every co-analytic set* *in a standard Borel space is the inverse image of each one of them by a Borel function). Moreover there is no analytic class of separable C^*- algebras $\mathcal{A}$ such that*

$$\text{FD} \subseteq \mathcal{A} \subseteq \text{GCR}.$$

(So if an analytic property holds for all separable C^*-algebras in FD, it must also hold for a non-type I separable C^*-algebra).

PROBLEMS. Let H be the class of separable C^*-algebras with **Hausdorff spectrum** and AF the class of **approximately finite** separable C^*-algebras. Then H is co-analytic and AF is analytic. Are they Borel? Finally, is the class GTC co-analytic? (It is not analytic by Theorem 1.3).

§2. Encoding pure states

Let A be a separable C^*-algebra. Denote by $B_1(A^*)$ the unit ball of the dual of A and by $B_1^+(A^*)$ the set of positive elements of $B_1(A^*)$. Let $P(A) \subseteq B_1^+(A^*)$ be the set of pure states of A. Thus $0 \notin P(A)$ and $P(A) \cup \{0\}$ is the set of extreme points of $B_1^+(A^*)$.

Enumerate in a sequence $P_1, P_2, \cdots$ all polynomials in non-commuting variables $x_1, \cdots, x_n, x_1^*, \cdots, x_n^*$ with complex rational coefficients; say $P_i = P_i(x_1, \cdots, x_{n_i}, x_1^*, \cdots, x_{n_i}^*)$. If $\gamma \in \Gamma$, let $\overline{\gamma}_i = P_i(\gamma_1, \cdots, \gamma_{n_i}, \gamma_1^*, \cdots, \gamma_{n_i}^*)$. So again $\overline{\gamma} \in \Gamma$ and $A(\gamma) = A(\overline{\gamma})$. Moreover, $\{\overline{\gamma}_n : n \in \mathbb{N}\}$ is a dense "$\mathbb{Q} + i\ \mathbb{Q}$-$*$-subalgebra" of $A(\gamma)$. If $f \in B_1(A(\gamma)^*)$, then $|f(\overline{\gamma}_n)| \leq ||\overline{\gamma}_n||$, so we can view f, which is clearly completely determined by $n \mapsto f(\overline{\gamma}_n)$, as a member of $\prod_n \Delta(||\overline{\gamma}_n||)$, where $\Delta(a) = \{x \in \mathbb{C} : |x| \leq a\})$, i.e., we have $B_1(A(\gamma)^*) \subseteq \prod_n \Delta(||\overline{\gamma}_n||)$. Moreover $B_1(A(\gamma)^*)$ consists of all the members of $\prod_n \Delta(||\overline{\gamma}_n||)$ which satisfy the linearity conditions $f(q_1\overline{\gamma}_{n_1} + q_2\overline{\gamma}_{n_2}) = q_1 f(\overline{\gamma}_{n_1}) + q_2 f(\overline{\gamma}_{n_2})$ for all $n_1, n_2 \in \mathbb{N}$, $q_1, q_2 \in \mathbb{Q}+i\ \mathbb{Q}$, so it forms a compact convex set in $\prod_n \Delta(||\overline{\gamma}_n||)$. Further we can identify $\prod_n \Delta(||\overline{\gamma}_n||)$ with $\Delta^{\mathbb{N}}$ (= the unit ball of $\ell^\infty(\mathbb{N})$ with the weak*-topology), where $\Delta = \Delta(1)$, under the obvious scaling in each coordinate. Say $h_\gamma^{-1} : \prod_n \Delta(||\overline{\gamma}_n||) \to \Delta^{\mathbb{N}}$ is the corresponding isomorphism. Put $h_\gamma^{-1}(B_1(A(\gamma)^*)) = \mathbf{B}_1(A(\gamma)^*)$, a compact convex set in $\Delta^{\mathbb{N}}$. The positive part $B_1^+(A(\gamma)^*)$ clearly forms again a compact convex set in $\prod_n \Delta(||\overline{\gamma}_n||)$ as it is defined by the positivity conditions: $f(\overline{\gamma}_n\overline{\gamma}_n^*) \geq 0$. Let $h_\gamma^{-1}(B_1^+(A(\gamma)^*)) = \mathbf{B}_1^+(A(\gamma)^*)$, again a compact convex set in $\Delta^{\mathbb{N}}$. Finally, let $h_\gamma^{-1}(P(A(\gamma))) = \mathbf{P}(A(\gamma))$. Then $\{0\} \cup \mathbf{P}(A(\gamma))$ is the set of extreme points of $\mathbf{B}_1^+(A(\gamma)^*)$. Our first task will be to choose in a Borel way a dense subset in $\mathbf{P}(A(\gamma))$ (in the weak*-topology as usual). This follows from the following lemmas:

LEMMA 2.1. *The map $\gamma \in \Gamma \mapsto \mathbf{B}_1^+(A(\gamma)^*)$ is a Borel map from Γ into $K_c(\Delta^{\mathbb{N}})$, the compact, metrizable space of compact convex non-$\emptyset$ subsets of $\Delta^{\mathbb{N}}$ (with the standard Hausdorff metric).*

LEMMA 2.2. *Let X be a separable Banach space, $B_1(X^*)$ the unit ball of its dual with the weak*-topology, $K_c(B_1(X^*))$ the space of compact non-$\emptyset$ convex subsets of $B_1(X^*)$ with the Hausdorff metric. Then there is a Borel map $F : K_c(B_1(X^*)) \to B_1(X^*)^{\mathbb{N}}$ such that for $K \in K_c(B_1(X^*))$, $F(K) = \{x^*_{n,K}\}_{n\in\mathbb{N}}$ is a dense sequence in the set of extreme points* $\mathrm{Ext}(K)$ *of* K.

The proof of 2.1 is routine so we will omit it.

PROOF OF 2.2. This is based on an argument of Choquet, given in Dixmier [4], p. 395.

First note that we can find in a Borel way for each $K \in K_c(B_1(X^*))$ a point $x_K \in \mathrm{Ext}(K)$. Indeed, fix a dense set $x_1, x_2, \cdots$ in X. Define then inductively

$$\begin{aligned} K_1 &= \{x^* \in K : \text{ Re } \langle x_1, x^*\rangle \text{ is maximum}\},\\ K_2 &= \{x^* \in K_1 : \text{ Re } \langle x_2, x^*\rangle \text{ is maximum}\},\\ &\ \ \cdots\\ K_{n+1} &= \{x^* \in K_n : \text{ Re } \langle x_{n+1}, x^*\rangle \text{ is maximum}\}\\ &\ \ \cdots \end{aligned}$$

Then $\bigcap_n K_n$ is a singleton, $\{x_K\}$, and $x_K \in \mathrm{Ext}(K)$. Easily $K \mapsto x_K$ is Borel.

Fix now an open basis $\{U_n\}$ for $B_1(X^*)$ (in the weak*-topology). It will be enough to show how to find for each $K \in K_c(B_1(X^*))$ and each U_n with $U_n \cap \mathrm{Ext}(K) \neq \emptyset$, a point $\mathbf{x}^*_{n,K} \in U_n \cap \mathrm{Ext}(K)$ such that $\mathbf{x}^*_{n,K} \in \Delta^1_1(K, z)$, where z is a real "encoding" the compact metric space $B_1(X^*)$. This will show that the set $C = \{(n, K) : U_n \cap \mathrm{Ext}(K) \neq \emptyset\}$ is Borel and will provide a Borel function $G : C \to B_1(X^*)$ such that $G(n, K) \in U_n \cap \mathrm{Ext}(K)$. From this it is easy to define F.

Fix $K \in K_c(B_1(X^*))$. For each real $q \in \mathbb{R}$ let

$$\begin{aligned} U_{m,q} &= \{x^* \in K : \text{ Re } \langle x^*, x_m\rangle \leq q\}\\ F_{m,q} &= \{x^* \in K : \text{ Re } \langle x^*, x_m\rangle \leq q\}. \end{aligned}$$

CLAIM. If $x \in \mathrm{Ext}(K) \cap U_n$, there is $m, q \in \mathbb{Q}$ with

$$x \in U_{m,q} \subseteq F_{m,q} \subseteq U_n.$$

PROOF. By Hahn-Banach if $y \in K \setminus U_n$ find $m(y), q(y) \in \mathbb{Q}$ with

$$x \in U_{m(y),q(y)} \subseteq F_{m(y),q(y)} \not\ni y.$$

Since $K \setminus U_n$ is compact, choose $y_1, \cdots, y_k \in K \setminus U_n$ with $F_{m(y_1),q(y_1)} \cap \cdots \cap F_{m(y_k),q(y_k)} \subseteq U_n$. Let $K_i = K \setminus U_{m(y_i),q(y_i)}$. As K_i is convex, compact, so is $\text{co}(K_1 \cup \cdots \cup K_k)$. Also $x \notin \text{co}(K_1 \cup \cdots \cup K_k)$, as x is an extreme point of K. So by Hahn-Banach find $m, q \in \mathbb{Q}$ with $x \in U_{m,q}$ and $(K_1 \cup \cdots \cup K_n) \cap F_{m,q} = \emptyset$. Then $x \in U_{m,q} \subseteq F_{m,q} \subseteq U_n$. ⊣

So assuming $U_n \cap \text{Ext}(K) \neq \emptyset$ we can find $m, q \in \mathbb{Q}$ with $\emptyset \neq F_{m,q} \subseteq U_n$.

We will now find effectively a point in $F_{m,q} \cap \text{Ext}(K)$. Notice that $F_{m,q}$ is compact, convex and $K \setminus F_{m,q}$ is convex. As before we can find effectively a point x in $\text{Ext}(F_{m,q})$. If this is in $\text{Ext}(K)$ we are done. Else we can find, effectively from x, $y \neq z \in K$ and $0 < \lambda < 1$ such that $x = \lambda y + (1 - \lambda)z$. Let δ be the line through y, z. Let also y', z' be the end-points of $F_{m,q} \cap \delta$. One of y', z' is in $\text{Ext}(K)$. Since y', z' can be found effectively from y, z we are done. ⊣

So we have

COROLLARY 2.3. *There is a Borel map $D : \Gamma \to (\Delta^{\mathbb{N}})^{\mathbb{N}}$ such that $D(\gamma) = \{d_n(\gamma)\}_{n \in \mathbb{N}}$ is a dense set in $\mathbf{P}(A(\gamma))$.*

Now $\mathbf{P}(A(\gamma))$ is a G_δ set in $\Delta^{\mathbb{N}}$, in fact in a uniform way, i.e., we can find Borel maps $F_n : \Gamma \to K(\Delta^{\mathbb{N}})$, where $K(\Delta^{\mathbb{N}})$ is the compact space of closed subsets of $\Delta^{\mathbb{N}}$ with the Hausdorff metric, such that $\mathbf{P}(A(\gamma)) = \Delta^{\mathbb{N}} \setminus \bigcup_n F_n(\gamma)$. By standard facts then we can define a complete metric on $\mathbf{P}(A(\gamma))$ which gives the topology of $\mathbf{P}(A(\gamma))$. Since the definition of this metric is effective, given a sequence of closed sets whose union is the complement of $\mathbf{P}(A(\gamma))$, we have

COROLLARY 2.4. *There is a family of Borel maps $\rho_{m,n} : \Gamma \to \mathbb{R}$ $(m, n \in \mathbb{N})$ such that if $\rho_\gamma(d_m(\gamma),\, d_n(\gamma)) := \rho_{m,n}(\gamma)$, ρ_γ is the restriction of a complete metric ρ_γ on $\mathbf{P}(A(\gamma))$.*

Thus the completion of $\{d_n(\gamma)\}_{n \in \mathbb{N}}$ under ρ_γ is the space $\mathbf{P}(A(\gamma))$.

§3. The class BD

We have that

$$A \in \text{BD} \Leftrightarrow \text{There is } n \text{ so that every irreducible representation of } A \text{ has dimension } \leq n.$$

In the notation of Dixmier [4], p. 85 this means:

$$\exists n(\hat{A} = {}_n\hat{A}).$$

Now ${}_n\hat{A}$ is closed in $\hat{A}$, so if $\Phi : P(A) \twoheadrightarrow \hat{A}$ is the canonical open and continuous surjection of $P(A)$ onto $\hat{A}$, then $\Phi^{-1}({}_n\hat{A})$ is closed in $P(A)$. So if $\{f_m\}$ is a dense sequence in $P(A)$, then

$$A \in \mathrm{BD} \Leftrightarrow \exists n \forall m (f_m \in \Phi^{-1}({}_n\hat{A})).$$

It follows that

$$\gamma \in \widetilde{\mathrm{BD}} \Leftrightarrow \exists n \forall m [\Phi(h_\gamma(d_m(\gamma))) \text{ has dimension } \leq n].$$

So it is enough to show, in order to complete the proof that BD is Borel, that for each n, m the set

$$D_{n,m} = \{\gamma \in \Gamma : \Phi(h_\gamma(d_m(\gamma))) \text{ has dimension } \leq n\}$$

is Borel. Given $\gamma \in \Gamma$ denote by $f_m(\gamma)$ the pure state $h_\gamma(d_m(\gamma))$. Then the space of the representation $\Phi(f_m(\gamma))$ is $A(\gamma)/N_{m,\gamma}$, where $N_{m,\gamma}$ is the left ideal of $x \in A(\gamma)$ with $f_m(\gamma)(x^*x) = 0$, and with inner product given by $f_m(\gamma)(y^*x)$. As $\{\gamma_k\}$ is dense in $A(\gamma_k)$, $\{\gamma_k/N_{m,\gamma}\}$ is dense in $A(\gamma)/N_{m,\gamma}$. Thus $A(\gamma)/N_{m,\gamma}$ has dimension $\leq n$ iff $\exists y_1 \cdots \exists y_n \in A(\gamma) \forall k \exists \alpha_1 \cdots \alpha_n \in \mathbb{C}(\gamma_k - \sum \alpha_i y_i \in N_{m,\gamma})$. This easily gives an analytic definition of $D_{n,m}$. To give a co-analytic definition of $D_{n,m}$ notice that $A(\gamma)/N_{m,\gamma}$ has dimension $\leq n$ iff $\forall y_1 \cdots y_{n+1} \in A(\gamma) \exists (\alpha_1 \cdots \alpha_{n+1}) \neq \vec{0}$ $(\sum \alpha_i y_i \in N_{m,\gamma})$. Now $\{(\alpha_1, \cdots, \alpha_{n+1}) \in \mathbb{C}^{n+1} : \sum \alpha_i y_i \in N_{m,\gamma}\}$ forms a subspace of $\mathbb{C}^{n+1}$, so given $y_1, \cdots, y_{n+1} \in A(\gamma)$, if $\exists (\alpha_1 \cdots \alpha_{n+1}) \neq \vec{0}$ such that $\sum \alpha_i y_i \in N_{m,\gamma}$, then such $(\alpha_1, \cdots, \alpha_{n+1})$ can be found which is $\Delta^1_1(\gamma, \xi_1, \cdots, \xi_{n+1})$, where $\xi_1, \cdots \xi_{n+1} \in \mathbb{N}^\mathbb{N}$ are such that $\gamma_{\xi_j}(i) \to_{i\to\infty} y_j$. This gives the co-analytic definition that we want.

§4. The class TC

By Dixmier [4], p. 106 to say that A is in TC is equivalent to saying that the (self-adjoint two-sided) ideal of all $x \in A$ such that $\pi \in \hat{A} \mapsto \mathrm{Tr}\, \pi(x)\pi(x)^*$ is finite and continuous on $\hat{A}$ is dense in A. Let $K(A)$ be the Pedersen ideal of A (see Pedersen [14], p. 175). Then

$$A \in \text{ TC iff } \forall x \in K(A)\ (\pi \mapsto \ \mathrm{Tr}\, \pi(x)\pi(x)^* \text{ is finite and continuous}).$$

Now if $\mathcal{C}$ is the class of positive continuous functions on $(0, \infty)$ with compact support, then $K(A)$ is the linear span of all $x \in A_+$ for which there are $x_1, \cdots, x_k \in A_+$ and $f_1, \cdots, f_k \in \mathcal{C}$ with $x \leq f_1(x_1) + \cdots + f_k(x_k)$. So

$$\begin{aligned} A \in \text{ TC } \Leftrightarrow\ & \forall (f_{i,j})_{\substack{1\leq i\leq n \\ 1\leq j\leq n}} \in \mathcal{C} \forall (x_{i,j}) \in A_+ \\ & \forall \alpha_1 \cdots \alpha_n \in \mathbb{C} \forall y_1 \cdots y_m \in A_+ \\ & (y_i \leq \sum_j f_{i,j}(x_{i_j}) \Rightarrow \\ & \pi \in \hat{A} \mapsto \ \mathrm{Tr}\, \pi(\sum \alpha_i y_i)\pi(\sum \overline{\alpha}_i y_i^*) \\ & \text{is finite and continuous}). \end{aligned}$$

Now $\pi \in \hat{A} \mapsto$ Tr $\pi(x)\pi(x^*)$ is finite and continuous iff $f \in P(A) \mapsto$ Tr $\Phi(f)(x)\Phi(f)(x^*)$ is finite and continuous on $P(A)$. This gives, after a few more messy but routine calculations, that TC is co-analytic.

We now want to show that TC is also analytic. This is based on the following facts: (i) the class TC is closed under stable isomorphism; (ii) the stable isomorphism class of $A \in$ TC is determined by its so-called Dixmier-Douady invariant and (iii) by a result of Raeburn-Taylor [16], one can construct explicitly a separable continuous trace C^*-algebra with a given Dixmier-Douady invariant. We will use these facts to outline a sketch of the proof that TC is analytic. The actual details of the proof involve rather horrible but straightforward calculations so we will omit them.

For each Polish locally compact space T, $\{N_i\}_{i\in \mathrm{N}}$ locally finite open cover of T and 2-cocycle $\lambda_{ijk} : \overline{N}_{ijk} \to S^1$ ($N_{ijk} = N_i \cap N_j \cap N_k$), Raeburn-Taylor [16] construct explicitly a separable C^*-algebra

$$A = A(T, \{N_i\}, \ \{\lambda_{ijk}\})$$

whose spectrum is T and its Dixmier-Douady class is represented by

$$(\{N_i\}, \ \{\lambda_{ijk}\}).$$

Let $\mathcal{A}$ be the class of all such $A(T, \{N_i\}, \ \{\lambda_{ijk}\})$. It follows that $\mathcal{A}$ is analytic. Now it is not hard to check that the relation $A \cong_s B$ of stable isomorphism between A, B is analytic (i.e., $\{(\gamma, \delta) \in \Gamma^2 : A(\gamma) \cong_s A(\delta)\}$ is analytic. Since $A \in$ TC $\Leftrightarrow \exists B[B \in \mathcal{A}$ and $B \cong_s A]$ it follows that TC is analytic.

§5. The class N

The argument here is due to Effros. For a unital C^*-algebra A let $M_n(A)$ be the algebra of $n \times n$ matrices on A and let

$$F_n(A) = \ \{M \in M_n(A) : M \geq 0, || \sum M_{ii} || \leq 1\}$$

Let also $M_n(A^*)$ be the set of $n \times n$ matrices on A^* and let

$$S_n(A) = \ \{g \in M_n(A^*) : g \geq 0, g_{ij}(1) = \delta_{ij}\}.$$

Then the approximation characterization of nuclearity (see Choi-Effros [3]) gives the following, for unital separable C^*-algebras,

$$A \in \mathrm{N} \Leftrightarrow \forall q \in N \forall a_1 \cdots, a_p \in A \exists n \exists M \in F_n(A) \exists g \in S_n(A) \forall \ell = 1, \cdots, p$$
$$||a_i - \Phi(M) \circ \Lambda(g)(a)|| < \frac{1}{q+1},$$

where $\Lambda(g)(a) = [g_{ij}(a)]$, $\Phi(M)([a_{ij}]) = \sum a_{ij} M_{ij}$, so that

$$\Phi(M) \circ \Lambda(g)(a) = \sum M_{ij}\ g_{ij}(a).$$

Now since $||\Phi(M)|| \leq 1$ ($\Phi(M) : M_n \to A$) and

$$||\Lambda(g)|| = 1\ (\Lambda(g) : A \to M_n),$$

it is clear we can restrict $a_1, \cdots, a_p$ to any fixed dense subset of A. So if $A(\gamma) = A(\overline{\gamma}) = A$, we have

$$\begin{aligned} A(\gamma) \in \mathrm{N} \Leftrightarrow & \forall q \forall m_1 \cdots m_p \exists n \exists M \in F_n(A) \exists g \in S_n(A) \forall \ell = 1, \cdots, p \\ & ||\overline{\gamma}_{m_\ell} - \sum M_{ij}\ g_{ij}(\overline{\gamma}_{m_\ell})|| < \frac{1}{q+1}. \end{aligned}$$

Now (see Takesaki [18], p. 193), $F_n(A)$ consists of all $M \in M_n(A)$ of the form $[M_{ij}] = [b_i^*\ b_j]$, with $b_1, \cdots, b_n \in A$ and $||\sum b_i^* b_i|| \leq 1$. Again it is clear that we can restrict the M_{ij} above to be of the form $\overline{\gamma}_{k_i} \overline{\gamma}_{k_j}$ with $\sum ||\overline{\gamma}_{k_i}^* \overline{\gamma}_{k_i}|| \leq 1$. So we have

$$\begin{aligned} A(\gamma) \in \mathrm{N} \Leftrightarrow & \forall q \forall m_1 \cdots m_p \exists n \exists k_1 \cdots k_n \exists g \in S_n(A) \\ & (i) \sum ||\overline{\gamma}_{k_i}^* \overline{\gamma}_{k_i}|| \leq 1 \\ & (ii)\ ||\overline{\gamma}_{m_\ell} - \sum_{i,j} \overline{\gamma}_{k_i} \overline{\gamma}_{k_j} g_{ij}(\overline{\gamma}_{m_\ell})|| < \frac{1}{q+1}. \end{aligned}$$

Now $M_n(A^*) = M_n(A)^*$ via the duality $\langle [g_{ij}],\ [a_{ij}] \rangle = \sum g_{ij}(a_{ij})$. So

$$\begin{aligned} [g_{ij}] \in S_n(A) \Leftrightarrow & g_{ij}(1) = \delta_{ij}\ \& \\ & \forall b_1 \cdots b_j \in A \sum g_{ij}(b_i^* b_j) \geq 0 \\ \Leftrightarrow & g_{ij}(1) = \delta_{ij}\ \& \\ & \forall n_1 \cdots n_j \sum g_{ij}(\overline{\gamma}_i^*\ \overline{\gamma}_j) \geq 0. \end{aligned}$$

Let $S_n^M(A) = \{[g_{ij}] \in S_n(A) : ||g_{ij}|| \leq M\}$. This is clearly a compact set in $B_M(A^*)^{n \times n}$ (where $B_M(A^*)$ is the ball of radius M in A^*), in the (product of the) weak*-topology. Now one can replace "$\exists g \in S_n(A)^*$" above by "$\exists M \exists g \in S_n^M(A)$" and the "$g \in S_n^M(A)$" by a countable dense subset of them, which can be found in a Borel way from γ as $S_n^M(A)$ is compact. This shows that N is Borel at least for unital A. If A is not unital and A_1 is the canonical unital C^*-algebra obtained by adding a unit to A then (see Choi-Effros [3]), $A \in \mathrm{N} \Leftrightarrow A_1 \in N$, so we are done in the general case as well.

§6. The class NGCR

We have first that

$$A \in \mathrm{NGCR} \Leftrightarrow \forall I(I \neq \emptyset \text{ closed ideal of } I \Rightarrow I \notin \mathrm{TC}).$$

(This is because any GCR algebra has a non-$\emptyset$ closed TC ideal – Dixmier [4], p. 103). Thus

$$\gamma \in \widetilde{\mathrm{NGCR}} \Leftrightarrow \forall \delta \in \Gamma(A(\delta) \text{ is a closed ideal of } A(\gamma) \Rightarrow \delta \notin \widehat{\mathrm{TC}}).$$

Now

$$\begin{aligned} A(\delta) \text{ is a closed ideal of } A(\gamma) \Leftrightarrow \forall n \forall p \exists q (||\delta_p - \gamma_q|| < \frac{1}{n+1} \ \& \\ \forall n \forall p (\gamma_n \delta_p, \delta_p \gamma_n \in A(\delta)), \end{aligned}$$

so this is a Borel relation of δ, γ, thus NGCR is co-analytic, since $\widehat{\mathrm{TC}}$ is Borel.

To see that NGCR is also analytic we use the following characterization of Glimm ([6], see also Dixmier [4], 9.5.4),

$$\begin{aligned} A \in \mathrm{NGCR} \Leftrightarrow\ & A \text{ is isomorphic to some } A(\gamma) \\ & \text{such that } A''(\gamma) \text{ is a von Neumann algebra of type II.} \end{aligned}$$

It is easy to see now that the map $\gamma \mapsto A''(\gamma)$ from Γ into the space of von Neumann algebras on H with the Effros Borel structure is Borel. Since it is known (see Nielsen [13]), that the class of type II von Neumann algebras is a Borel set in this space, it follows that NGCR is analytic.

§7. The classes FD, CCR and GCR: I

First we will verify that these are all co-analytic.

For FD. We have that

$$A \in \mathrm{FD} \Leftrightarrow \forall \pi \in \hat{A} \exists n (\dim(\pi) \leq n),$$

so that in the notation of §3

$$\gamma \in \widetilde{\mathrm{FD}} \Leftrightarrow \forall x \in \mathbf{P}(A(\gamma)) \exists n \dim(\Phi(h_\gamma(x)) \leq n,$$

so it is enough to show that the set

$$\{(\gamma, x) \in \Gamma \times \Delta^{\mathrm{N}} : x \in \mathbf{P}(A(\gamma)) \ \& \ \dim(\Phi(h_\gamma(x)) \leq n\}$$

is Borel. This is done exactly as in §3.

For CCR. We have

$$A \in \text{ CCR } \Leftrightarrow \forall \pi \in \hat{A} \forall x \in A(\pi(x) \text{ is compact}),$$

so that

$$\begin{aligned} \gamma \in \widetilde{\text{CCR}} &\Leftrightarrow \forall n \forall \pi \in \hat{A}(\gamma)(\pi(\gamma_n) \text{ is compact}) \\ &\Leftrightarrow \forall n \forall x \in \mathbf{P}(A(\gamma))\ (\Phi(h_\gamma(x))\ (\gamma_n) \text{ is compact}). \end{aligned}$$

So it is enough to show that

$$\{(\gamma, x) \in \Gamma \times \Delta^{\mathrm{N}} : x \in \mathbf{P}(A(\gamma))\ \&\ \Phi(h_\gamma(x))\ (\gamma_n) \text{ is compact}\}$$

is co-analytic. Abbreviate, for $x \in \mathbf{P}(A(\gamma))$, $h_\gamma(x) = f_x$ and $\Phi(f_x) = \pi_x$. Then as the space of the representation π_x is $A(\gamma)/N_{f_x}$, where $N_{f_x} = \{y \in A(\gamma) : f_x(y^*y) = 0\}$ with inner product induced by $f_x(y^*z)$, and $\pi_x(y)\ (z/_{N_{f_x}}) = yz/_{N_{f_x}}$, we have (using the weak- norm continuity characterization of compact operators):

$$\begin{aligned} \pi_x(\gamma_n) \text{ is compact } \Leftrightarrow\ & \forall x_1, x_2, \dots \in A(\gamma) \forall y \in A(\gamma)\ (f_x(x_i^* x_i) \leq 1\ \& \\ & f(y^*y) \leq 1\ \& \\ & x_i/_{N_{f_x}} \to^w y/_{N_{f_x}} \Rightarrow \pi_x(\gamma_n)\ (x_i) \to \pi_x(\gamma_n)\ (y)). \end{aligned}$$

Now $x_i/_{N_{f_x}} \to^w y/_{N_{f_x}}$ iff $\forall m(f_x(x_i^* y_m) \to f_x(y^* \gamma_m))$ and $\pi_x(\gamma_n)\ (x_i) \to \pi_x(\gamma_n)\ (y)$ iff $\gamma_n x_i/_{N_{f_x}} \to \gamma_n y/_{N_{f_x}}$ iff $f_x((\gamma_n x_i - \gamma_n y)^*\ (\gamma_n x_i - \gamma_n y)) \to 0$. So clearly the property "$\pi_x(\gamma_n)$ is compact" is co-analytic and we are done.

For GCR. We will use the following characterization of GCR algebras

$$\begin{aligned} A \in \text{ GCR } &\Leftrightarrow A \text{ is of type I} \\ &\Leftrightarrow \text{Every representation of } A \text{ is of type I} \\ &\Leftrightarrow \forall \pi : A \to \mathcal{B}(H)\ (\pi(A)'' \text{ is of type I}). \end{aligned}$$

So

$$\begin{aligned} \gamma \in \widetilde{\text{GCR}} \Leftrightarrow \forall \delta \in \Gamma(&A(\delta) \text{ is a homomorphic} \\ &\text{image of } A(\gamma) \Rightarrow A''(\delta) \text{ is of type I}). \end{aligned}$$

As $\delta \mapsto A''(\delta)$ is Borel from Γ into the space of von Neumann algebras, and the class of type I von Neumann algebras is Borel, it is enough to show that

$$\{(\gamma, \delta) :\ A(\delta) \text{ is a homomorphic image of } A(\gamma)\}$$

is analytic (in $\Gamma \times \Gamma$). Recall from §2 the definition of $\overline{\gamma}$ for each $\gamma \in \Gamma$. Then $A(\delta)$ is a homomorphic image of $A(\gamma) \Leftrightarrow \exists \epsilon \in \Gamma(A(\delta) = A(\epsilon)\ \&\ \exists x \in$

$\mathbb{N}^{\mathbb{N}}$ such that the map $\pi_x(\overline{\gamma}_n) = \overline{\epsilon}_{x(n)}$ is a norm-decreasing "$\mathbb{Q} + i\ \mathbb{Q}$-homomorphism" of $\{\overline{\gamma}_n : n \in \mathbb{N}\}$ onto $\{\overline{\epsilon}_n : n \in \mathbb{N}\}$) so this is clearly analytic.

Next we will show that the above classes are not all Borel, in fact, complete co-analytic and moreover there is no analytic class $\mathcal{A}$ such that $\text{FD} \subseteq \mathcal{A} \subseteq \text{GCR}$. This will be done by finding a complete co-analytic set P (in some Polish space X) and a Borel map $H : X \to \Gamma$ such that

$$x \in P \Rightarrow H(x) \in \widetilde{\text{FD}}$$
$$x \notin P \Rightarrow H(x) \notin \widetilde{\text{GCR}}.$$

This will be done by using a certain class of homeomorphisms of $2^{\mathbb{N}}$ and their transformation group C^*-algebras, i.e., crossed products.

§8. Classifying classes of homeomorphisms of $2^{\mathbb{N}}$

Let $H = H(2^{\mathbb{N}})$ be the Polish space of homeomorphisms with the metric

$$d(g,h) = \max_x \delta(g(x), h(x)) + \max_x \delta(g^{-1}(x), h^{-1}(x)),$$

where δ is the metric on $2^{\mathbb{N}}$.

A homeomorphism $f \in H(2^{\mathbb{N}})$ will be called **quasi-periodic** if every orbit of f is finite. It will be called **smooth** if the equivalence relation induced by f, i.e.,

$$xEy \Leftrightarrow \exists n (f^n(x) = y)$$

is smooth, i.e., has a Borel selector.

THEOREM 8.1. *Let* QP,S *be the classes of quasi-periodic, smooth homeomorphisms of* $2^{\mathbb{N}}$ *resp. Then* QP,S *are complete co-analytic sets in* $H(2^{\mathbb{N}})$*. In fact there is complete co-analytic set* P *in some Polish space* X *and Borel* $F : X \to H(2^{\mathbb{N}})$ *such that*

$$x \in P \Rightarrow F(x) \in \text{QP},$$
$$x \notin P \Rightarrow F(x) \notin \text{S}.$$

(In particular, there is no analytic set R *with* $\text{QP} \subseteq R \subseteq \text{S}$*).*

To prove 8.1 we will make use of a specific class of homeomorphisms of $2^{\mathbb{N}}$, the so-called **Lipschitz** ones. Denote by 2^n the set of n-tuples of 0's and 1's. Given permutations π of 2^n and ρ of 2^m with $n \leq m$ we write $\pi \subseteq \rho$ if

$$\rho(x_1, \cdots, x_m)|n = \pi(x_1, \cdots, x_n).$$

If π_n are permutations of 2^n and $\pi_1 \subseteq \pi_2 \subseteq \pi_3 \subseteq \cdots$, then $f : 2^{\mathrm{N}} \to 2^{\mathrm{N}}$ defined by

$$f(x_1, x_2, \cdots) = \bigcup_n \pi_n(x_1, \cdots, x_n)$$

is a homeomorphism of 2^{N}. These are the **Lipschitz homeomorphisms** if 2^{N}. Clearly (π_n) is uniquely determined by f.

Let $L \subseteq \prod_n (2^n)!$ be the set of all (π_n) with $\pi_1 \subseteq \pi_2 \subseteq \cdots$. Clearly, L is a closed set in $\prod_n (2^n)!$, so a compact metric space. The map $(\pi_n) \mapsto f$ is clearly Borel. It will be thus enough to show

THEOREM 8.2. *(i) For a Lipschitz homeomorphism f,*

$$f \text{ is smooth } \Leftrightarrow f \text{ is quasi} - \text{periodic.}$$

(ii) The set of quasi-periodic ($\equiv$ smooth) $(\pi_n) \in L$ is a complete co-analytic set in L.

(Thus in 8.1. we can take $X = L$, $P = \{(\pi_n) : (\pi_n)$ is quasi-periodic $\}, F((\pi_n)) =$ the corresponding f.)

PROOF OF 8.2. First we show that if f is smooth, then f is quasi-periodic, the converse being obvious.

Call a point $x \in 2^{\mathrm{N}}$ **strictly recurrent** for f if for each nbhd U of x, $f^n(x) \in U \setminus \{x\}$ for some n. By Katznelson-Weiss [10] if f is smooth then no point $x \in 2^{\mathrm{N}}$ is strictly recurrent. (This is actually a special case of Glimm [7]). We will now show that this implies that every orbit of f is finite. For that we will use a "tree picture" of Lipschitz homeomorphisms, see Jackson-Dougherty-Kechris [9].

We define the **orbit tree** T_f of f as follows: Given n and an orbit $\mathcal{O}$ of π_n (on 2^n), exactly one of the following happens: when we look at π_{n+1} either $\mathcal{O}$ extends to one orbit or two orbits. (In particular $|\mathcal{O}| = 2^m$ for some m). So we can form a binary tree as follows: The nth level nodes of the tree are the orbits of π_n on 2^n. Every nth level node has one or two $(n+1)$th level nodes extending it according to the above cases.

For every $x \in 2^{\mathrm{N}}$ there is a unique infinite path $\alpha_x \in [T_f]$ (= the set of infinite paths through T_f) such that $x|n \in \alpha_x(n)$. Now if α_x has the property that for infinitely many n, $\alpha_x(n)$ doubles in the next level $\alpha_x(n+1)$, then x is strictly recurrent, a contradiction. This can be seen as follows: Fix a nbhd U of x. We can assume $U = \{y : y|n = s\}$, where s is a sequence of length n and n is such that $\alpha_x(n)$ doubles in the next level. Say the orbit $\alpha_x(n)$ has 2^m elements. Let $s\hat{\ }i, s\hat{\ }j$ be the level $n+1$ extensions of s, with $s\hat{\ }i = x|(n+1)$. Then $\pi_{n+1}^{2^m+1}(s\hat{\ }i) = s\hat{\ }j$, so $f^{2^m+1}(x) = x'$ where $x'|(n+1) = s\hat{\ }j$. Thus $f^{2^m+1}(x) \in U$, but $f^{2^m+1}(x) \neq x$.

So we have shown that for every $x \in 2^{\mathrm{N}}$ there is some n_0 such that $\alpha_x(n)$ splits in two orbits for all $n \geq n_0$. Say the orbit $\alpha_x(n_0)$ has 2^{m_0}

elements. Then clearly the orbit of x has 2^{m_0} elements too and we are done.

Finally, the proof that the set of quasi-periodic $(\pi_n) \in L$ is a complete co-analytic set is given in Kechris [11]. $\dashv$

§9. The clases FD, CCR, GCR: II

We will now prove the result stated at the end of §7.

THEOREM 9.1. *There is a complete co-analytic set P, in some Polish space X, and a Borel map $G : X \to \Gamma$ such that*

$$x \in P \Rightarrow G(x) \in \widetilde{\mathrm{FD}}$$
$$x \notin P \Rightarrow G(x) \notin \widetilde{\mathrm{GCR}}.$$

PROOF. Take $X = L$ the space of Lipschitz homeomorphisms of 2^{N} as in §8.

Each $\pi \in L$ defines an obvious action, also denoted by π, of $\mathbb{Z}$ into $C(2^{\mathrm{N}})$. Denote by $C(2^{\mathrm{N}}) \ltimes_{\pi} \mathbb{Z} = C^*(\pi)$ the corresponding C^*-crossed product of $C(2^{\mathrm{N}})$ by $\mathbb{Z}$ with respect to π (see Tomiyama [19], p. 68–69). Then by 4.1.9 in Tomiyama [19],

$$\pi \text{ in quasi} - \text{periodic } \Rightarrow C^*(\pi) \in \text{ FD}.$$

Also by Gootman [8]

$$\pi \text{ is smooth } \Leftrightarrow C^*(\pi) \in \text{ GCR},$$

in particular

$$\pi \text{ is not quasi} - \text{periodic } \Rightarrow c^*(\pi) \notin \text{ GCR}.$$

We will show that there is a Borel map $G : L \to \Gamma$ such that $A(H(\pi)) \cong C^*(\pi)$. It follows by taking $P = \{\pi \in L : \pi \text{ is not quasi-periodic }\}$ that

$$\pi \in P \Rightarrow G(\pi) \in \widetilde{\mathrm{FD}}$$
$$\pi \notin P \Rightarrow G(\pi) \notin \widetilde{\mathrm{GCR}}$$

and our proof is complete.

To construct G : First since $\mathbb{Z}$ is amenable we can replace $C^*(\pi)$ by the reduced product $C(2^{\mathrm{N}}) \ltimes_{\pi r} \mathbb{Z}$ (see Tomiyama [19], 3.2.3). An isomorphic copy of $C(2^{\mathrm{N}}) \ltimes_{\pi r} \mathbb{Z}$ can be constructed as follows:

We have that $\mathbb{Z}$, via π, acts on $C(2^{\mathbb{N}})$ as follows: $n \cdot_\pi f = f \circ \pi^n$. Fix a canonical faithful representation λ of $C(2^{\mathbb{N}})$ into $\mathcal{B}(H)$ so that if $\{f_n\}$ is some fixed dense sequence in $C(2^{\mathbb{N}})$ and $\lambda(f_n) = c_n$, (so that $c = (c_n) \in \Gamma$), then the maps $\pi \mapsto (n \cdot_\pi c_m)_{n,m}$, from L into Γ, where $n \cdot_\pi \lambda(f) = \lambda(n \cdot_\pi f)$ is the isomorphic action of $\mathbb{Z}$ via π in $\lambda(C(2^{\mathbb{N}})) = A(c)$, are Borel.

Now let $H' = \ell^2(\mathbb{Z}, H)$. For each $n \in \mathbb{Z}$ let u_n be the unitary operator on H' given by

$$u_n \xi(m) = \xi(m-n),\ \xi \in \ell^2(\mathbb{Z}, H).$$

Also consider the representation $\rho_\pi : A(c) \to \mathcal{B}(H')$ given by

$$(\rho_\pi(a)\xi)(n) = (-n) \cdot_\pi a\xi(n),\ \xi \in \ell^2(\mathbb{Z}, H), n \in \mathbb{Z}$$

(see Tomiyama [19], p. 71). Then $C(2^{\mathbb{N}}) \ltimes_{\pi r} \mathbb{Z}$ is isomorphic to the C^*-subalgebra of $\mathcal{B}(H')$ generated by $\{\rho_\pi(a) : a \in A(c)\} \cup \{u_n : n \in \mathbb{Z}\}$, therefore by $\{\rho_\pi(c_m) : m \in \mathbb{Z}\} \cup \{u_n : n \in \mathbb{Z}\} =^{def} \{\delta^\pi_n : n \in \mathbb{Z}\}$, where clearly $\pi \mapsto \delta^\pi = (\delta^\pi_n)$ is Borel from L into $\Gamma' = \Gamma(H')$. Fixing an isomorphism of H and H', we finally obtain a Borel map $G : L \to \Gamma$ such that $A(G(\pi)) \cong C(2^{\mathbb{N}}) \ltimes_{\pi r} \mathbb{Z}$ and the proof is complete.

REMARK. One can also see that GCR is complete co-analytic using a construction of Behncke, Krauss and Leptin [1], which in turn generalizes a construction of Dixmier [4], 4.7.17: To each countable linear order $\mathcal{L}$ one can assign, in a Borel way, a separable C^*-algebra $A(\mathcal{L})$ such that

$$\mathcal{L} \text{ is wellordered } \Leftrightarrow A(\mathcal{L}) \text{ in GCR}.$$

Since the set of linear orders on $\mathbb{N}$, which are wellordered is complete co-analytic, the result follows.

REMARK. Since the crossed products we are using are unital, all the results here hold for unital algebras as well.

REMARK. The algebras $A(\mathcal{L})$ are in AF, so the result about GCR holds within the class of AF algebras as well.

§10. A co-analytic rank on FD

Given a co-analytic set $P \subseteq X$, in some Polish space X, a **co-analytic rank** on P is a map $\varphi : P \to \omega_1$ (= the first uncountable ordinal), such that for each countable ordinal $\alpha < \omega_1$ the set $P_\alpha = \{x \in P : \varphi(P) \leq \alpha\}$ is Borel "uniformly" in α. More precisely, this means that there are analytic and co-analytic, resp., relations $R, S \subseteq X^2$ such that

$$x \in P \Rightarrow [\varphi(y) \leq \varphi(x) \Leftrightarrow R(x,y) \Leftrightarrow S(x,y)].$$

One of the basic properties of co-analytic ranks is the **boundedness theorem**: If $S \subseteq P$ is analytic, there is $\alpha < \omega_1$ with $\sup\{\varphi(x) : x \in S\} < \alpha$.

We will now define a rank φ on FD and verify that it is a co-analytic rank (i.e., $\tilde{\varphi} : \widetilde{\mathrm{FD}} \to \omega_1$ given by $\tilde{\varphi}(\gamma) = \varphi(A(\gamma))$ is co-analytic).

Fix a separable C^*-algebra A. Consider $\hat{A}$ and define a transfinite sequence $U_0 \subseteq U_1 \subseteq \cdots \subseteq U_\xi \subseteq \cdots$ of open subsets of $\hat{A}$ as follows:

$$\begin{aligned} U_0 &= \emptyset, \\ U_{\xi+1} &= \{\pi \in \hat{A} : \text{There is open nbhd } V \text{ of } \pi \text{ and positive integer } k \\ &\qquad \text{such that } \forall \pi' \in V(\pi' \in U_\xi \text{ or } \dim(\pi') \leq k)\}, \\ U_\lambda &= \bigcup_{\xi<\lambda} U_\xi, \text{ if } \lambda \text{ is limit.} \end{aligned}$$

Let I_ξ be the closed ideal of A such that $\widehat{I_\xi} = U_\xi$. Since these notions depend on A we will write $U_\xi(A)$, $I_\xi(A)$ if necessary. Notice that the I_ξ can be defined by the following induction:

$$\begin{aligned} I_0 &= \emptyset, \\ I_{\xi+1}/I_\xi &= I_1(A/I_\xi), \\ I_\lambda &= \overline{\bigcup_{\xi<\lambda} I_\xi}, \lambda \text{ limit.} \end{aligned}$$

Note also that

$$\begin{aligned} \widehat{I_1}(A) = U_1(A) &= \bigcup_k \operatorname{Int}\{\pi \in \hat{A} : \dim(\pi) \leq k\} \\ &= \bigcup_k \operatorname{Int}({}_k\hat{A}). \end{aligned}$$

CLAIM.

$$\begin{aligned} A \in \mathrm{FD} &\Leftrightarrow \exists \xi < \omega_1 (U_\xi = \hat{A}) \\ &(\Leftrightarrow \exists \xi < \omega_1 (I_\xi = \hat{A})). \end{aligned}$$

PROOF. $\Rightarrow$: Since $A \in \mathrm{FD}$, $\hat{A} =_k \hat{A}$. Since ${}_k\hat{A}$ is closed (Dixmier [4], p. 85) and $\hat{A}$ is a Baire space (Dixmier [4], p. 79), $\widehat{I_1}(A) \neq \emptyset$, by the Baire Category Theorem. Now notice that $A \in \mathrm{FD} \Leftrightarrow \forall$ closed ideal $I(A/I \in \mathrm{FD})$, so if ξ is the least (countable) ordinal such that $I_\xi = I_{\xi+1}$ we must have $I_\xi = A$, so $U_\xi = \hat{A}$.

$\Leftarrow$: Say $\pi \in U_{\xi+1} - U_\xi$. Then for some $k, \dim(\pi) \leq k$, so $A \in \mathrm{FD}$. $\dashv$

Now define for $A \in \mathrm{FD}$,

$$\varphi(A) = \text{ least } \xi \text{ such that } U_{\xi+1} = \hat{A}$$

(we use this instead of the least ξ for which $U_\xi = \hat{A}$ for technical reasons). Then $\varphi : \mathrm{FD} \to \omega_1$ is a rank. We want to verify that it is a co-analytic rank.

Let $\Phi : P(A) \twoheadrightarrow \hat{A}$ be the canonical open, continuous surjection. Put

$$\mathcal{O}_\xi = \Phi^{-1}(U_\xi).$$

Then we can define these inductively by

$$\begin{aligned} \mathcal{O}_0 &= \emptyset, \\ f \in \mathcal{O}_{\xi+1} &\Leftrightarrow \exists \text{ open nbhd } V \text{ of } f \text{ and } k > 0 \text{ such that} \\ &\qquad \forall f \in V (f \in \mathcal{O}_\xi \text{ or } \dim(\Phi(f)) \leq k)), \\ \mathcal{O}_\lambda &= \bigcup_{\xi<\lambda} \mathcal{O}_\xi, \ \lambda \text{ limit.} \end{aligned}$$

Now fix any sequence of basic open sets $\{V_n\}$ in $P(A)$, where some V_n maybe $\emptyset$, and define the monotone inductive definition $\Psi_A(n,S) = \Psi(n,S)$ on $\mathbb{N}$ ($n \in \mathbb{N}, S \subseteq \mathbb{N}$) by

$$\Psi(n,S) \Leftrightarrow \exists k \forall f \in V_n [f \in \bigcup_{m \in S} V_m \vee \dim(\pi_f) \leq k].$$

Denote by Ψ^ξ the ξth iterate of Ψ, i.e.,

$$\begin{aligned} \Psi^0 &= \emptyset, \\ \Psi_{\xi+1} &= \{n : \Psi(n, \Psi^\xi)\}, \\ \Psi^\lambda &= \bigcup_{\xi<\lambda} \Psi^\xi, \lambda \text{ limit.} \end{aligned}$$

CLAIM. $\mathcal{O}_\xi = \bigcup \{V_n : n \in \Psi_\xi\}$.

This is easy to check by induction. From this it is easy to see that $\mathcal{O}_{\xi+1} = \hat{A}$ iff $\Psi^{\xi+1} = \mathbb{N}$, so

$$A \in \mathrm{FD} \Leftrightarrow \exists \xi < \omega_1 \ (\Psi^{\xi+1} = \mathbb{N})$$

and

$$\varphi(A) = \text{ least } \xi \text{ such that } \Psi^{\xi+1} = \mathbb{N}.$$

We will now verify two things:

i) For $\gamma \in \widetilde{\mathrm{FD}}$, $\varphi(A(\gamma)) < \omega_1^{CK}(\gamma)$.

ii) If $\mathrm{WO} \subseteq \mathbb{N}^{\mathbb{N}}$ is the set of codes of countable ordinals, there is an analytic relation S and a co-analytic relation R such that for $w \in \mathrm{WO}$:

$$\varphi(A(\gamma)) \leq |w| \Leftrightarrow R(\gamma, w) \Leftrightarrow S(\gamma, w),$$

where $|w|$ is the ordinal coded by w.

From this we get for any $\gamma \in \widetilde{\mathrm{FD}}$:

$$\begin{aligned}\varphi(A(\delta)) \leq \varphi(A(\gamma)) \Leftrightarrow \forall w \in \Delta_1^1(\gamma)(\ & w \in \mathrm{WO}\ \& \\ & \varphi(A(\gamma)) \leq |w| \Rightarrow \varphi(A(\delta)) \leq |w|),\end{aligned}$$

which shows that this relation is analytic, and also

$$\begin{aligned}\varphi(A(\delta)) \leq \varphi(A(\gamma)) \Leftrightarrow \exists w \in \Delta_1^1(\gamma)\{& w \in \mathrm{WO}\ \&\ \varphi(A(\gamma)) \leq |w|\ \& \\ & \forall v \in \mathrm{WO}(|v| \leq |w| \Rightarrow \\ & (\varphi(A(\gamma)) \leq |v| \Rightarrow \varphi(A(\delta)) \leq |v|)),\end{aligned}$$

which shows that this relation is co-analytic and we are done.

Proof of i) and half of ii): Fix a sequence of basic open sets $\{G_n\}$ in $\Delta^{\mathbb{N}}$. Define then the inductive definition $\Psi^*(n, S; \gamma)$, with parameter $\gamma \in \Gamma$, as follows: Put $h_\gamma(G_n) = \tilde{V}_n^\gamma$ and $V_n^\gamma = P(A) \cap \tilde{V}_n^\gamma$. So $\{V_n^\gamma\}$ is an open basis in $P(A(\gamma))$. Then let

$$\begin{aligned}\Psi^*(n, S; \gamma) &\Leftrightarrow \exists k \forall f \in V_n^\gamma [f \in \bigcup_{m \in S} V_n^\gamma \vee \dim(\pi_f) \leq k] \\ &\Leftrightarrow \exists k \forall x \in \Delta^{\mathbb{N}} [x \in \mathbf{P}(A(\gamma)) \cap G_n \\ &\qquad \Rightarrow x \in \bigcup_{m \in S} G_m \vee \dim(\Phi(h_\gamma(x)) \leq k)].\end{aligned}$$

Thus Ψ^* is a co-analytic monotone inductive definition. Moreover

$$\gamma \in \widetilde{\mathrm{FD}} \Leftrightarrow \exists \xi < \omega_1(\{n : (\Psi^*)^{\xi+1}(n; \gamma)\} = \mathbb{N}),$$

where

$$\begin{aligned}(\Psi^*)^0 &= \emptyset, \\ (\Psi^*)^{\xi+1}(n; \gamma) &\Leftrightarrow \Psi^*(n, \{n : (\Psi^*)^\xi(n; \gamma)\}; \gamma) \\ (\Psi^*)^\lambda &= \bigcup_{\xi < \lambda} (\Psi^*)^\xi, \lambda \text{ limit},\end{aligned}$$

and

$$\varphi(A(\gamma)) = \text{ least } \xi(\{n : (\Psi^*)^{\xi+1}(n;\gamma)\} = \mathbb{N}).$$

So, by standard facts on co-analytic monotone inductive definitions,

$$\varphi(A(\gamma)) < \omega_1^{CK}(\gamma)$$

and there is co-analytic relation R as in ii).

Proof of the rest of ii): We now have to show that the relation $\varphi(A(\gamma)) \leq \xi$ is analytic uniformly in ξ. As usual this will be done by effective transfinite induction on ξ.

$\boldsymbol{\xi = 0}$:

$$\varphi(A(\gamma)) \leq 0 \Leftrightarrow I_1(A(\gamma)) = \hat{A}(\gamma).$$

Now $\widehat{I_1} = U_1$ is the union of the interiors of ${}_k\hat{A}$. Say $\text{Int}({}_k\hat{A}) = (U_1)_k$. So $U_1 = \bigcup_k (U_1)_k$. Let $(I_1)_k$ be the closed ideal such that $(\widehat{I_1})_k = (U_1)_k$. Then $(I_1)_1 \subseteq (I_1)_2 \subseteq \cdots$ and $I_1 = \overline{\bigcup_k (I_1)_k}$, so $I_1 = \hat{A}$ iff $\bigcup_k (I_1)_k$ is dense in $\hat{A}$. For each $y \in A$ denote by $I(y)$ the closed ideal generated by y. Then the above imply the following:

CLAIM.

$$\varphi(A(\gamma)) \leq 0 \Leftrightarrow \forall n \forall m \exists y \in A(\gamma) \exists k (||y - \gamma_n|| < \frac{1}{m+1} \;\&$$

$$I^*(y) \text{ has only } \leq k - \text{dimensional}$$

$$\text{irreducible representations}).$$

PROOF. Assume $\varphi(A(\gamma)) \leq 0$. Then $\bigcup_k (I_1)_k$ is dense in A $(= A(\gamma))$. Given γ_n and m there is then k and $y \in (I_1)_k$ with $||y - \gamma_n|| < \frac{1}{m+1}$. Now $I^*(y) \subseteq (I_1)_k$, so $\widehat{I^*(y)} \subseteq \widehat{(I_1)_k} \subseteq_k \hat{A}$, so $I^*(y)$ has only $\leq k$-dimensional irreducible representations.

Assume the expression on the right above is true. We have to show that $\bigcup_k (I_1)_k$ is dense in A. Fix γ_n, m. Then find y, k such that $||y - \gamma_n|| < \frac{1}{m+1}$, and $I^*(y)$ has only $\leq k-$dimensional irreducible representations, i.e., $\widehat{I^*(y)} \subseteq_k \hat{A}$. Then $\widehat{I^*(y)} \subseteq (U_1)_k = \text{Int}({}_k\hat{A})$, so $\widehat{I^*(y)} \subseteq \widehat{(I_1)}_k$, i.e., $I^*(y) \subseteq (I_1)_k$, so $y \in (I_1)_k$ and we are done.

Now given $y \in A(\gamma)$ (represented as a limit of a subsequence of γ) we can find in a Borel way from y, γ a $\delta \in \Gamma$ with $A(\delta) = I^*(y)$, which easily implies, using also §3, that $\varphi(A(\gamma)) \leq 0$ is an analytic property. This finishes the case $\xi = 0$.

To do the general case we will need the following two facts:

LEMMA 10.1. *$I_{\xi+1}$ is the largest closed ideal of A with $\varphi(I) \leq \xi$.*

LEMMA 10.2. *If $I \subseteq I'$ are closed ideals and $\varphi(A/I) \leq 0$, then $\varphi(A/I') \leq 0$.*

Granting these lemmas we complete the proof as follows:

$\xi \to \xi+1$: Assume $\varphi(A(\gamma)) \leq \xi$ is analytic. Then (by the lemmas above)

$$\begin{aligned}\varphi(A(\gamma)) \leq \xi + 1 \Leftrightarrow\ & \exists I (I \text{ is a closed ideal} \\ & \text{and } \varphi(I) \leq \xi \ \& \\ & \varphi(A(\gamma)/I) \leq 0) \\ \Leftrightarrow\ & \exists \delta \in \Gamma (A(\delta) \text{ is a closed} \\ & \text{ideal of } A(\gamma) \ \& \ \varphi(A(\delta)) \leq \xi \\ & \& \ \varphi(A(\gamma)/A(\delta)) \leq 0).\end{aligned}$$

Now the condition

$$A(\delta) \text{ is a closed ideal of } A(\gamma)$$

is clearly analytic, and so is the condition

$$A(\delta) \text{ is a closed ideal of } A(\gamma) \ \& \ A(\epsilon) \cong A(\gamma)/A(\delta).$$

This shows that $\varphi(A(\gamma)) \leq \xi + 1$ is analytic too.

$\xi < \lambda \to \lambda$ (λ limit ordinal): Assume $\varphi(A((\gamma)) \leq \xi$ is analytic for all $\xi < \lambda$. Then

$$\begin{aligned}\varphi(A(\gamma)) \leq \lambda \Leftrightarrow\ & \exists \text{ increasing sequence of closed ideals of } A(\gamma), \\ & J_1 \subseteq J_2 \subseteq \cdots \subseteq J_\xi \subseteq \cdots, \\ & \text{such that } \varphi(J_\xi) \leq \xi \ \& \ \varphi(A(\gamma)/\overline{\bigcup_{\xi<\lambda} J_\xi}) \leq 0 \\ \Leftrightarrow\ & \exists \gamma_1, \gamma_2, \cdots, \gamma_\xi, \cdots \in \Gamma \ (\xi < \lambda) \\ & (A(\gamma_1) \subseteq A(\gamma_2) \subseteq \cdots \subseteq A(\gamma_\xi) \subseteq \cdots \\ & \text{and each } A(\gamma_\xi) \text{ is a closed ideal} \\ & \text{of } A(\gamma) \text{ and } \varphi(A(\gamma_\xi)) \leq \xi \text{ and} \\ & \varphi(A(\gamma)/A(\delta(\gamma_1, \gamma_2, \cdots))) \leq 0),\end{aligned}$$

where $\delta(\gamma_1, \gamma_2, \cdots) \in \Gamma$ is an enumeration of $(\gamma_1)_n, \cdots, (\gamma_\xi)_n, \cdots (n \in \mathbb{N}, \xi < \lambda)$. This shows that $\varphi(A(\gamma)) \leq \lambda$ is analytic as well.

PROOF OF 10.1. First one can easily show by induction on ξ that

$$U_\xi(\hat{I}) = U_\xi \cap \hat{I}.$$

Thus

$$\begin{aligned}\varphi(\hat{I}) \leq \xi &\Leftrightarrow U_{\xi+1}(\hat{I}) = \hat{I}\\ &\Leftrightarrow U_{\xi+1} \cap \hat{I} = \hat{I}\\ &\Leftrightarrow \hat{I} \subseteq U_{\xi+1}\\ &\Leftrightarrow I \subseteq I_{\xi+1}.\end{aligned}$$ ⊣

PROOF OF 10.2. It is enough to assume $I = \emptyset$. So let A be such that $\varphi(A) \leq 0$ and let I be closed ideal in A. We will show that $\varphi(A/I) \leq 0$. Now $\widehat{A/I} = \hat{A} \setminus \hat{I}$, so since $U_1(A) = \hat{A}$, it is clear that $U_1(\hat{A} \setminus \hat{I}) = \hat{A} \setminus \hat{I}$, i.e., $\varphi(A/I) \leq 0$. ⊣

For each countable ordinal ξ denote by FD_ξ the class of all $A \in \mathrm{FD}$ with $\varphi(A) \leq \xi$. So FD_ξ is Borel. Then $\mathrm{FD}_0 \subseteq \mathrm{FD}_1 \subseteq \cdots \subseteq \mathrm{FD}_\xi \subseteq \cdots$, $\mathrm{FD} = \bigcup_\xi \mathrm{FD}_\xi$ and for each ξ there is $\eta > \xi$ with $\mathrm{FD}_{\eta+1} \setminus \mathrm{FD}_\eta \neq \emptyset$, i.e., this is a hierarchy that doesn't stop at any countable ordinal. Also BD$\subseteq \mathrm{FD}_0$. The class FD_0 consists of all A with the following property: For any $\pi \in \hat{A}$ there is k and closed ideal $I \subseteq A$ with $\pi(I) \neq \emptyset$ and I has at most k-dimensional irreducible representations. Equivalently this can be expressed as follows: For each A there is a largest closed ideal $(I_1)_k \subseteq A$ having only at most $\leq k$-dimensional irreducible representations, i.e., $\widehat{(I_1)}_k = \mathrm{Int}({}_k\hat{A})$. Then $A \in \mathrm{FD}_0 \Leftrightarrow$ the ideal $\bigcup_k (I_1)_k$ is dense in A.

REMARK. For each class of separable C^*-algebras $\mathcal{A}$ let $\mathcal{A}^s$ denote its closure under stable isomorphisms, i.e., $A \in \mathcal{A}^s \Leftrightarrow \exists B \in \mathcal{A}(B$ is stably isomorphic to $A)$. Now going back to §4 one can easily check that the class $\mathcal{A}$ discussed there is contained in FD_0. (In the notation of Theorem 1 in Raeburn-Taylor [16], for each $A \in \mathcal{A}$ and $t \in \hat{A}$, t is finite dimensional with dimension n_t, where $n_t = |\{i \in I\colon t \in N_i\}|$. As $\{N_i\}$ is a locally finite cover, given $t \in \hat{A}$ there is an open set $V \subseteq \hat{A}$ with $t \in V$ and j_0 such that $V \cap N_j = \emptyset$, if $j > j_0$. So $n_t \leq j_0$, if $t \in V$, i.e., $V \subseteq {}_{j_0}\hat{A}$, so $U_1(A) = \bigcup_k \mathrm{Int}({}_k\hat{A}) = \hat{A}$, i.e., $\varphi(A) \leq 0$).

So we have

$$\mathrm{FD}_0^s = \mathrm{TC}.$$

In Brown [2], p. 342 it is mentioned that P. Green has shown that

$$\mathrm{FD}^s = \mathrm{GTC}.$$

Since GTC is not analytic and FD_ξ^s is, it follows that for each $\xi < \omega_1$

$$\mathrm{FD}_\xi^s \subsetneqq \mathrm{GTC}.$$

§11. A co-analytic rank on CCR

Fix a separable Hilbert space $\tilde{H}$. Denote by $F_n(\tilde{H})$ the class of operators in $\mathcal{B}(\tilde{H})$ of rank $\leq n$ $(n = 1, 2, \cdots)$. We have the following closure properties of $F_n(\tilde{H})$:

a) $F_n(\tilde{H})$ is closed in the weak operator topology.

Proof. Let $S_n \in F_n(\tilde{H}), S_m \to^w S$. If $S \notin F_n(\tilde{H})$ then for some $\xi_1, \cdots, \xi_{n+1} \in \tilde{H}$, $S(\xi_1), \cdots, S(\xi_{n+1})$ are linearly independent. On the other hand, $S_m(\xi_1), \cdots, S_m(\xi_{n+1})$ are linearly dependent, so $\sum \alpha_{m,j} S_m(\xi_j) = 0$ for some $\alpha_{m,j}$, not all 0. By going to subsequences and renumbering $\xi_1, \cdots, \xi_{n+1}$, we can assume that for some $J \subseteq \{1, \cdots, n+1\}$, $\alpha_{m,j} = 0$, if $j \notin J$, and $\alpha_{m,j} \neq 0$, if $j \in J$, $\max\limits_{j \in J} |\alpha_{m,j}| = |\alpha_{m,1}| = 1$, and finally that $\alpha_{m,j} \to_m \alpha_j$. Fix $\eta \in \tilde{H}$. Then $\sum\limits_{j \in J} \alpha_{m,j} \langle S_m(\xi_j), \eta \rangle = 0$, so as $S_m \to^w S$, $\sum\limits_{j \in J} \alpha_j \langle S(\xi_j), \eta \rangle = 0$. Since η is arbitrary, $\sum\limits_{j \in J} \alpha_j S(\xi_j) = 0$, thus $\alpha_j = 0$, for all j, contradicting that $|\alpha_1| = 1$.

b) $F_n(\tilde{H})$ is hereditary, i.e., if $0 \leq S \leq T$, $T \in F_n(\tilde{H})$, then $S \in F_n(\tilde{H})$.

Proof. Let $X = T\tilde{H}$, so that $\dim(X) \leq n$. If $x \in X^\perp$, then

$$0 \leq \langle Sx, x \rangle \leq \langle Tx, x \rangle = 0,$$

so $\langle Sx, x \rangle = \langle S^{1/2}x, S^{1/2}x \rangle = 0$, i.e, $S^{1/2}X^\perp = \{0\}$, so $S^{1/2}\tilde{H} = S^{1/2}X$, i.e., $S^{1/2} \in F_n(\tilde{H})$ and thus $S = S^{1/2}S^{1/2} \in F_n(\tilde{H})$.

c) $F_n(\tilde{H})^* = F_n(\tilde{H})$ and $TF_n(\tilde{H})S \subseteq F_n(\tilde{H})$ for any $T, S \in F_n(\tilde{H})$.

Now fix a separable C^*-algebra A. Consider the following subset $R = R^A \subseteq A$ defined by

$$x \in R \Leftrightarrow \forall \pi \in \hat{A} \exists n (\pi(x) \in F_n(H_\pi)),$$

where H_π is the space of the representation π. Note that this definition makes sense since if π, π' are equivalent representations and $\pi(x) \in F_n(H_\pi)$, then $\pi'(x) \in F_n(H_{\pi'})$.

Denote by $K_n^x = K_n^{A,x} \subseteq \hat{A}$ the set

$$K_n^x = \{\pi \in \hat{A} : \pi(x) \in F_n(H_\pi)\}.$$

LEMMA. *K_n^x is a closed subset of $\hat{A}$.*

PROOF. This is a uniform version of the preceding argument for $F_n(\tilde{H})$.

Let $f \mapsto [\pi_f]$ be the canonical surjection of $P(A)$ onto $\hat{A}$. We have to show that

$$\tilde{K}^x_n = \{f \in P(A) : \pi_f(x) \in F_n(H_{\pi_f})\}$$

is closed in $P(A)$. Recall that H_{π_f} is A/N_f, where $N_f = \{x \in A : f(x^*x) = 0\}$ with inner product $f(y^*x)$. Also $\pi_f(x)(y/N_f) = xy/N_f$. So let $f_i \in \tilde{K}^x_n, f_i \to^{w^*} f \in P(A)$. We will show that $f \in \tilde{K}^x_n$. Otherwise there are $x_1, \cdots, x_{n+1} \in A$ with $\pi_f(x)(x_1/N_f), \cdots, \pi_f(x)(x_{n+1}/N_f)$ linearly independent. On the other hand, $\pi_{f_i}(x)(x_1/N_{f_i}), \cdots, \pi_{f_i}(x)(x_{n+1}/N_{f_i})$ are linearly dependent, say

$$\sum \alpha_{i,j}\pi_{f_i}(x)(x_j/N_{f_i}) = 0,$$

where at least one of the $\alpha_{i,j}$ is $\neq 0$ for each fixed i. Without loss of generality we can assume that there is $J \subseteq \{1, \cdots, n+1\}$ such that $\alpha_{i,j} = 0$ if $j \notin J$ and $\alpha_{i,j} \neq 0$ if $j \in J$, for each i. Also we can assume that $\max_{j \in J} |\alpha_{i,j}| = |\alpha_{i,1}| = 1$ and that $\alpha_{i,j} \to_{i\to\infty} \alpha_j$ for $j \in J$. Since

$$\sum_{j\in J} \alpha_{i,j}\langle \pi_{f_i}(x)(x_j/N_{f_i}), y/N_{f_i}\rangle = 0$$

for all i and $y \in A$, i.e.,

$$\sum_{j\in J} \alpha_{i,j} f_i(y^*xx_j) = 0, \ \forall i \forall y \in A,$$

it follows that

$$\sum_{j\in J} \alpha_j f(y^*xx_j) = 0$$

or

$$\sum_{j\in J} \alpha_j\langle \pi_f(x)\ (x_j/N_f),\ y/N_f\rangle = 0,$$

thus $\alpha_j = 0$ contradicting that $|\alpha_1| = 1$. ⊣

So we have for $x \in A$:

$$x \in R \Leftrightarrow \forall \pi \in \hat{A} \exists n(\pi \in K^x_n),$$

where K^x_n is closed in $\hat{A}$. Now define by transfinite induction open sets $U^x_\xi \subseteq \hat{A}$ as follows

$$U^x_1 = \{\pi \in \hat{A} : \exists n \exists \text{ open nbhd } V \ni \pi \forall \pi' \in V(\pi' \in K^x_n)\} = \bigcup_n \operatorname{Int}(K^x_n),$$

$$U^x_\lambda = \bigcup_{\xi<\lambda} U^x_\xi, \ \lambda \text{ limit},$$

$$U^x_{\xi+1} = \{\pi \in \hat{A} : \exists n \exists \text{ open nbhd } V \ni \pi \forall \pi' \in V(\pi' \in U^x_\xi \vee \pi' \in K^x_n)\}.$$

Then it is easy to check, using the Baire Category Theorem in $\hat{A}$, that

$$x \in R \Leftrightarrow \exists \xi < \omega_1 (U^x_{\xi+1} = \hat{A}).$$

Thus define for $x \in R$, the rank

$$||x||_R = \text{ least } \xi(U^x_{\xi+1} = \hat{A}),$$

and put

$$\begin{aligned} R_\xi &= \{x \in R : \ ||x||_R \leq \xi\} \\ &= \{x \in R : \ U^x_{\xi+1} = \hat{A}\}. \end{aligned}$$

LEMMA. *$A \in$ CCR $\Leftrightarrow \exists \xi < \omega_1[(R_\xi)_+$ is dense in $A_+]$. (Here $B_+ =$ the positive elements of B.)*

PROOF. $\Leftarrow$: Assume $(R_\xi)_+$ is dense in A_+ for some $\xi < \omega_1$. If $x \in (R_\xi)_+$, then $\forall \pi \in \hat{A} \exists n(\pi \in K^x_n)$, so in particular $\forall \pi \in \hat{A}(\pi(x)$ is compact). So there is a dense subset of A, i.e., the linear span of $(R_\xi)_+$, which is mapped into compact operators under any $\pi \in \hat{A}$, so $A \in$ CCR.

$\Rightarrow$: Fix a countable dense set S in A_+. Let $f_n : [0,1] \to [0,1]$, $f_n(t) \leq t$ be nonnegative, continuous, vanishing in a nbhd of 0, such that $f_n(t) \to t$ uniformly as $n \to \infty$. So $f_n(s) \to s$ for $s \in S$ and $f_n(s) \in A_+$ as well. Since $\forall \pi \in \hat{A}(\pi(s)$ is compact), $\forall n \forall \pi \in \hat{A}(\pi(f_n(s)) = f_n(\pi(s))$ has finite rank), i.e., $\forall n \forall s \in S(f_n(s) \in R)$. So let $\xi = \sup_{n,s \in S} ||f_n(s)||_R < \omega_1$. As $\{f_n(s) : \ n \in \mathbb{N}, s \in S\}$ is dense in A_+, it follows that $(R_\xi)_+$ is dense in A_+. ⊣

We can now define the following rank on CCR:

$$\psi(A) = \text{ least } \xi \text{ such that } (R_\xi)_+ \text{ is dense in } A_+.$$

We will show that this is a co-analytic rank on CCR.

Suppose $A = A(\gamma), \gamma \in \Gamma$. It is easy to define a Borel function $\gamma \mapsto \hat{\gamma}$ from Γ into Γ such that $\hat{\gamma}_n \in A_+$ and $\{\hat{\gamma}_n : n \in \mathbb{N}\}$ is dense in A_+. (For instance we can take $\hat{\gamma}_n = \gamma_n \gamma^*_n$)

Now let $f_m : [0,1] \to [0,1]$, $f_m(t) \leq t$ be continuous, vanishing in a nbhd of 0, and $f_m(t) \to t$ uniformly as $m \to \infty$. So $f_m(\hat{\gamma}_n) \in A_+$ and $f_m(\hat{\gamma}_n) \to \hat{\gamma}_n$. So if $\tilde{\gamma}_{m,n} = f_m(\hat{\gamma}_n) \in R_\xi$, then $\psi(A) \leq \xi$. We will now prove the converse.

LEMMA. *Let $A = A(\gamma), \gamma \in \Gamma$. If $||A|| \leq \xi$, i.e., $(R_\xi)_+$ is dense in A_+, then $\tilde{\gamma}_{m,n} \in (R_\xi)_+, \forall m, n$.*

PROOF. We establish first some properties of R_ξ.

1) R_ξ is hereditary: For that it is enough to show that if $0 \leq x \leq y$, $y \in A_+$ then for all $\xi, U_\xi^y \subseteq U_\xi^x$. This is proved by induction on ξ using the fact that $F_n(\tilde{H})$ is hereditary.

2) For $z \in A$, $zR_\xi z \subseteq R_\xi$: For that again it is enough to show that for $y \in A, U_\xi^y \subseteq U_\xi^{zyz}$, which is again proved by induction on ξ using that for $T \in \mathcal{B}(\tilde{H})$, $TF_n(\tilde{H})T \subseteq F_n(\tilde{H})$.

We now use an argument as in Pedersen [14], p.175: Fix $\tilde{\gamma}_{m,n} = f_m(\hat{\gamma}_n)$. Let $g : [0,1] \to [0,1]$ be equal to 1 on the support of f_m and $g(0) = 0$, $0 \leq g \leq 1$. Consider $x = g(\tilde{\gamma}_{m,n})$. As $f_m g = g$, we have $\tilde{\gamma}_{m,n}x = \tilde{\gamma}_{m,n}$. Also $x \in A_+$. As $(R_\xi)_+$ is dense in A_+, there is $y \in (R_\xi)_+$ with $||x-y|| < \frac{1}{2}$. Then $\tilde{\gamma}_{m,n} = \tilde{\gamma}_{m,n}^{1/2} x \tilde{\gamma}_{m,n}^{1/2} = \tilde{\gamma}_{m,n}^{1/2}(x-y)\tilde{\gamma}_{m,n}^{1/2} + \tilde{\gamma}_{m,n}^{1/2} y \tilde{\gamma}_{m,n}^{1/2} \leq \frac{1}{2}\tilde{\gamma}_{m,n} + \tilde{\gamma}_{m,n}^{1/2} y \tilde{\gamma}_{m,n}^{1/2}$, thus $\tilde{\gamma}_{m,n} \leq \frac{1}{2}\tilde{\gamma}_{m,n}^{1/2} y \tilde{\gamma}_{m,n}^{1/2}$, so $\tilde{\gamma}_{m,n} \leq zR_\xi z$, where $z = \frac{1}{4}\tilde{\gamma}_{m,n}^{1/2}$, therefore $\tilde{\gamma}_{m,n} \in R_\xi$. ⊣

So we have

$$\begin{aligned} \psi(A(\gamma)) &= \text{least } \xi \text{ such that } (R_\xi)_+ \text{ is dense in } A(\gamma)_+ \\ &= \text{least } \xi \text{ such that } \tilde{\gamma}_{m,n} \in R_\xi \text{ for all } m, n, \end{aligned}$$

i.e.,

$$\psi(A(\gamma)) = \sup_{m,n} ||\tilde{\gamma}_{m,n}||_{R^{A(\gamma)}}.$$

¿From this it immediately follows that in order to show that ψ is a co-analytic norm on CCR, it is enough to show that for each fixed m, n the function

$$\theta(\gamma) = ||\tilde{\gamma}_{m,n}||_{R^{A(\gamma)}}$$

is a co-analytic norm on the co-analytic set

$$R_{m,n}(\gamma) \Leftrightarrow \gamma \in \Gamma \wedge \tilde{\gamma}_{m,n} \in R^{A(\gamma)}$$

$(R_{m,n} \subseteq \Gamma)$. To see this, first define the monotone inductive definition;

$$\begin{aligned} \Psi_{m,n}(p, S; \gamma) &\Leftrightarrow \exists k \forall f \in V_p^\gamma [f \in \bigcup_{m \in S} V_m^\gamma \vee \pi_f(\tilde{\gamma}_{m,n}) \in F_k(H_{\pi_f})] \\ &\Leftrightarrow \exists k \forall x \in \Delta^{\mathrm{N}} [x \in \mathbf{P}(A(\gamma)) \cap G_n \Rightarrow \\ &\qquad x \in \bigcup_{m \in S} G_m \vee \\ &\qquad \Phi(h_\gamma(x)(\tilde{\gamma}_{m,n})) \in F_k(H_{\Phi(h_\gamma(x))})]. \end{aligned}$$

It is easy to check that this is a co-analytic monotone inductive definition. Since now

$$R_{m,n}(\gamma) \Leftrightarrow \exists\xi(\{p : \Psi^{\xi+1}(p;\gamma) = \mathbb{N}\})$$

and

$$||\tilde{\gamma}_{m,n}||_{R^{A(\gamma)}} = \text{ least } \xi(\{p : \Psi^{\xi+1}(p;\gamma)\} = \mathbb{N}\})$$

(this is proved as in §10), it follows that for γ with $R_{m,n}(\gamma)$ we have $\theta(\gamma) < \omega_1^{CK}(\gamma)$, while for some co-analytic relation P and $w \in \mathrm{WO}$.

$$\theta(\gamma) \leq |w| \Leftrightarrow P(\gamma, w).$$

It remains to find an analytic relation S with

$$\theta(\gamma) \leq |w| \Leftrightarrow S(\gamma, w),$$

for $w \in \mathrm{WO}$.

Let us go back temporarily to an arbitrary separable A. Given $x \in A$ and a closed ideal $I \subseteq A$, define a relativized $U_\xi^x(I)$ as follows:

$$\begin{aligned} U_1^x(I) &= \{\pi \in \hat{I} : \exists n \exists \text{ open } V \subseteq \hat{I}, V \ni \pi \forall \pi' \in V(\pi' \in K_n^x)\}, \\ U_{\xi+1}^x(I) &= \{\pi \in \hat{I} : \exists \text{ open } V \subseteq \hat{I}, V \ni \pi \forall \pi' \in V[\pi' \in U_\xi^x(I) \vee \pi' \in K_n^x]\}, \\ U_\lambda^x(I) &= \bigcup_{\xi<\lambda} U_\xi^x. \end{aligned}$$

CLAIM. $U_\xi^x(I) = \hat{I} \cap U_\xi^x$ (where $U_\xi^x = U_\xi^x(A)$). (Easy by induction).

For $x \in R^I$, let

$$||x, I|| = \text{ least } \xi(U_{\xi+1}^x(I) = \hat{I}).$$

Since U_ξ^x is open in $\hat{A}$, let I_ξ^x be the closed ideal such that $\hat{I}_\xi^x = U_\xi^x$.

CLAIM. $I_{\xi+1}^x$ is the largest ideal I with $||x, I|| \leq \xi$.

PROOF.

$$\begin{aligned} ||x, I|| \leq \xi &\Leftrightarrow U_{\xi+1}^x(I) = \hat{I} \\ &\Leftrightarrow U_{\xi+1}^x \cap \hat{I} = \hat{I} \\ &\Leftrightarrow \hat{I} \subseteq U_{\xi+1}^x \\ &\Leftrightarrow I \subseteq I_{\xi+1}^x. \end{aligned}$$

We will now show that the relation

$$||x, I|| \leq \xi$$

is uniformly analytic in x, I, ξ, which means that there is analytic T such that for $\gamma \in \Gamma, \delta \in \Gamma$ with $A(\delta) = I$ a closed ideal in $A(\gamma) = A$, $\epsilon \in \mathbb{N}^{\mathbb{N}}$ which represents a subsequence of $\{\gamma_n\}$ converging to x, and $w \in \text{WO}$, with $|w| = \xi$ we have

$$||x, I|| \leq \xi \Leftrightarrow T(\gamma, \delta, \epsilon, w).$$

This is done as usual by effective transfinite induction on ξ:

$\boldsymbol{\xi = 0}$:

$$\begin{aligned} ||x, I|| \leq 0 &\Leftrightarrow U_1^x(I) = \hat{I} \\ &\Leftrightarrow \hat{I} \subseteq U_1^x \\ &\Leftrightarrow \bigcup_k \text{Int}\,(K_k^x) \supseteq \hat{I} \\ &\Leftrightarrow I \subseteq \overline{\bigcup_k I_k^x}, \text{ where } \widehat{I_k^x} = \text{Int}(K_k^x), I_k^x \text{ closed ideal in } A \\ &\Leftrightarrow \forall n \forall m \exists y \exists k (||y - \delta_n|| < \frac{1}{m+1} \ \&\ I(y) \subseteq I_k^x), \end{aligned}$$

where $I(y)$ is the closed ideal generated by y. Now "$I(y) \subseteq I_k^x$" iff "$\widehat{I(y)} \subseteq K_k^x$" iff "$\forall \pi \in \hat{I}(y)\ \pi(x) \in F_k(H_\pi)$". Since it is enough to restrict "$\forall \pi \in \hat{I}(y)$" to a dense subset of $\hat{I}(y)$, this is clearly a Borel condition, so $||x, I|| \leq 0$ is analytic.

$\boldsymbol{\xi \to \xi + 1}$:

$$\begin{aligned} ||x, I|| \leq \xi + 1 &\Leftrightarrow U_x^{\xi+2}(I) = \hat{I} \\ &\Leftrightarrow \hat{I} \subseteq I_{\xi+2}^x. \end{aligned}$$

For $J \subseteq I$ closed ideals, put

$$\begin{aligned} Q(x, J, I) \Leftrightarrow \forall \pi \in (\widehat{I/J}) \exists V \text{ open in } (\widehat{I/J}) \\ \exists n \forall \pi' \in V(\pi'(x) \in F_n(H_{\pi'})). \end{aligned}$$

Then we have

$$\begin{aligned} ||x, I|| \leq \xi + 1 \Leftrightarrow \exists J (J \subseteq I \text{ closed ideal } \& \\ Q(x, J, I) \ \&\ ||x, J|| \leq \xi). \end{aligned}$$

(For this notice that

$$Q(x, J, I) \ \&\ J \subseteq J' \subseteq I \Rightarrow Q(x, J', I).)$$

As the relation Q is also analytic (which can be seen as in the argument for the case $\xi = 0$), this shows that $||x, I|| \leq \xi + 1$ is analytic as well.

$\boldsymbol{\xi < \lambda \to \lambda}$: Again we have

$$\begin{aligned}||x, I|| \leq \lambda \Leftrightarrow \exists J_1, \cdots, J_\xi, \cdots (\xi < \lambda) \\ (J_1 \subseteq J_2 \subseteq \cdots \subseteq J_\xi \subseteq \cdots \,\& \\ ||x, J_\xi|| \leq \xi \,\&\, Q(x, \overline{\bigcup_{\xi<\lambda} J_\xi}, I),\end{aligned}$$

so we are done as before. ⊣

§12. The problem of co-analytic ranks on GCR

The natural way to define a co-analytic rank on GCR is to use one of the canonical composition series for GCR algebras. (For an exposition of this see Dixmier [4] or Pedersen [14]). For example, one can take for each $A \in$ GCR the canonical composition series J_α such that $J_{\alpha+1}/J_\alpha = \text{CCR}(A/J_\alpha) =$ the largest CCR ideal of A/J_α, and define the rank $\rho(A) =$ least ξ such that $(J_\xi = A)$. Unfortunately, this rank is not a co-analytic rank, since the set of A's with $\rho(A) = 1$ is the class CCR, which is not Borel.

Another possibility would be to use the composition series I_α where $I_{\alpha+1}/I_\alpha =$ the largest ideal of A/I_α of type I_0 (see Pedersen [14]). If $\rho'(A) =$ least ξ such that $(I_\xi = A)$, then the set of A's with $\rho'(A) = 0$ is the set of A's in I_0, which is easy to see that is an analytic set, but we don't know if it is Borel or not. It is open if ρ' is a co-analytic rank or not.

Denoting by $I_0(A)$, CCR(A), resp., the largest closed ideals of A of type I_0, in CCR, resp. and assuming we have a uniform way of assigning to each A a closed ideal $K(A)$ with $I_0(A) \subseteq K(A) \subseteq \text{CCR}(A)$ we define a composition series K_α with $K_{\alpha+1}/K_\alpha = K(A/K_\alpha)$. This could be used to define a rank

$$\rho_K(A) = \text{ least } \xi \text{ such that } (K_\xi = A).$$

It is conceivable that with the appropriate choice of $K(A)$ this would give a co-analytic rank. For example, if

$$K_0(A) = \overline{L(A)},$$

where

$$\begin{aligned}L(A) = \{x \in A : \forall \pi \in \hat{A} \exists n \exists \text{ open nbhd } V \ni \pi \\ \forall \pi' \in V(\pi'(x) \in F_n(H_{\pi'}))\},\end{aligned}$$

then one can easily check that $L(A)$ is an ideal of A, so $K_0(A)$ is a closed ideal of A. Also $I_0(A) \subseteq K_0(A) \subseteq \text{CCR}(A)$. If ρ_{K_0} is the corresponding rank, then $\{A : \rho_{K_0}(A) = 0\} = \{A : \psi(A) = 0\}$, where ψ is the co-analytic rank on CCR constructed in §11. So $\{A : \rho_{K_0}(A) = 0\}$ is Borel in this case. We don't know however if this ρ_{K_0} is a co-analytic rank on GCR.

REMARK. Given a closed ideal $K(A)$ as before, notice that the set

$$\{A : K(A) \neq \emptyset\}$$

is Borel. This is because

$$K(A) \neq \emptyset \Leftrightarrow A \notin \text{NGCR},$$

$(K(A) \neq \emptyset \Rightarrow \text{CCR}(A) \neq \emptyset \Rightarrow A \notin \text{NGCR} \Rightarrow I_0(A) \neq \emptyset \Rightarrow K(A) \neq \emptyset)$. So it is an interesting question to see if there is a Borel function that picks, given a code for $A = A(\gamma) \notin$ NGCR, an element in $K(A)$. For $K(A) = I_0(A)$ this amounts to picking in a Borel way from γ an abelian element (see Pedersen [14]) of $A(\gamma) = A$ for $A \notin$ NGCR. We do not know the answer here. Similarly, one can ask if one can pick in a Borel way a dense sequence in $K(A)$. This is not possible for $K(A) =$ CCR(A), otherwise CCR would be Borel. Again, we do not know the answer to any of these questions for $I_0(A)$ or $K_0(A)$.

§13. Classifying classes of (second countable) locally compact groups

We refer here to Sutherland [17] for a Borel parametrization of Polish groups. (For another parametrization, see 12.E, 2) of Kechris [11].) In particular, this gives a Borel parametrization of second countable locally compact groups. Various standard classes of such groups have been verified in Sutherland [17] to be Borel. To these we can add the class of amenable groups (use, e.g., to characterization 18.13 of Pier [15]), Lie groups (use, e.g., the characterization of Lie groups as those containing no small subgroups – a strong version of the Hilbert 5th problem), nilpotent groups, connected groups, etc. (Also note that the operation $G \to \hat{G}$ on abelian groups is Borel using constructive proofs of the existence of Haar measure.) However, we conjecture that the class of type I (postliminal) groups is not Borel. (It is easily co-analytic.) An approach to such proof would be as follows: First one would find for each countable ordinal ξ a group G such that $\rho(C^*(G)) > \xi$ (see §12 for the definition of $\rho(A)$). If one can then solve the problem (see §12) of constructing a co-analytic rank ρ^* on GCR, which moreover has the property that $\rho^* \geq \rho$ (which any reasonable approach for constructing ρ^* would surely satisfy), then by the boundedness theorem for co-analytic ranks we would be done. Similar approaches might work for showing that the liminal (CCR) groups do not form a Borel class (here one would only have to construct groups with arbitrarily large $\psi(C^*(G))$).

REFERENCES

[1] H. Behncke, F. Krauss & H. Leptin, *C^*-algebren mit geordneten ideal folgen*, J. Funct. Anal. **10**, 1972, 204–211.

[2] L. Brown, *Stable isomorphism of hereditary subalgebras of C^*-algebras*, Pac. J. Math. **71 (2)**, 1977, 335–348.

[3] M-D. Choi & E. Effros, *Nuclear C^*-algebras and the approximation property*, Amer. J. Math. **100 (1)**, 1978, 61–79.

[4] J. Dixmier, *C^*-algebras*, North Holland, 1977.

[5] P. A. Fillmore, *A User's Guide to Operator Algebras,* Wiley, 1996.

[6] J. Glimm, *Type I C^*-algebras*, Ann. of Math. **73** (1961), 572–612.

[7] J. Glimm, *Locally compact transformation groups*, Trans. Amer. Math. Soc. **101** (1961), 124–138.

[8] E. Gootman, *The type of some C^*- and W^*-algebras associated with transformation groups*, Pac. J. Math. **48 (11)**, 1973, 93–106.

[9] S. Jackson, R. Dougherty, A. S. Kechris, *The structure of hyperfinite Borel equivalence relations*, Trans. Amer. Math. Soc. **341 (1),** 1994, 193–225.

[10] Y. Katznelson & B. Weiss, *The construction of quasi-invariant measures*, Israel J. Math. **12** (1972), 1–4.

[11] A. S. Kechris, *Classical Descriptive Set Theory*, Springer Verlag, 1995.

[12] Y. N. Moschovakis, *Descriptive Set Theory,* North Holland, 1980.

[13] O. Nielsen, *Direct Integral Theory*, Marcel Dekker, 1980.

[14] G. Pedersen, *C^*-algebras and their Automorphism Groups*, Academic Press, 1979.

[15] J-P. Pier, *Amenable Locally Compact Groups*, Wiley, 1984.

[16] I. Raeburn and J. L. Taylor, *Continuous trace C^*-algebras with given Dixmier-Douady class*, J. Austral. Math. Soc. (Series A) **38** (1985), 394–407.

[17] C. Sutherland, *A Borel parametrization of Polish groups*, Publ. RIMS, Kyoto Univ., **21** (1985), 1067–1086.

[18] M. Takesaki, *Theory of Operator Algebras I*, Springer-Verlag, 1979.

[19] J. Tomiyama, *Invitation to C^*-algebras and Topological Dynamics*, World Scientific, 1987.

DECIDABILITY AND UNDECIDABILITY IN THE ENUMERABLE TURING DEGREES

STEFFEN LEMPP

Department of Mathematics
University of Wisconsin
Madison, WI 53706–1388, USA

ABSTRACT. We survey some recent work on the (recursively) enumerable Turing degrees, with particular emphasis on work relating to decidability and undecidability.

Two fundamental notions of mathematics are those of a computable set and of an enumerable set. A set S is called *computable* (or *recursive*) if there is an effective algorithm which for any input x can compute whether x is an element of S. A set S is called *(recursively) enumerable* if there is an effective algorithm listing all elements of S.

Clearly, these notions only make sense for countable sets. However, all "basic" countable sets (such as the set of all k-tuples of natural numbers for some fixed k, the set of all finite strings of 0's and 1's, the set of words over some finite alphabet, or the set of first-order formulas over some finite language) are easily seen to be in effective 1–1 correspondence with the set of natural numbers. (The "code number" for an element of such a countable set is often called its *Gödel number*.) Thus, for investigating computability, one can restrict oneself to studying computability over the natural numbers.

The above intuitive notions of a computable set and an enumerable set were made precise in various equivalent ways in the mid-1930's and later. For example, Turing [Tu36] defined effective algorithm to mean what is now called a *Turing machine*, i.e., a very simple-minded computer with no limitations on run time or memory space. This (and its many equivalent definitions) is nowadays generally accepted as *the* way to define computability (see also Soare [Sota]).

1980 *Mathematics Subject Classification* (1985 *Revision*). 03D25.

Key words and phrases. (recursively) enumerable Turing degrees, decidability.

This research was partially supported by NSF grant DMS-9504474 and a grant of the British Engineering and Physical Sciences Research Council. The author would like to thank the Leeds logicians for their hospitality during his half-year stay in Leeds, and the Chinese logicians for their hospitality during his two-week visit to China. .

There is a close connection between computability and definability in arithmetic. By Post's Theorem [Po48] (see [Kl52, p. 293]), a set is computable iff there are first-order formulas φ and ψ (in the language of arithmetic) in which all quantifiers are bounded such that for all x,

$$x \in S \quad \text{iff} \quad \exists y \varphi(x, y) \quad \text{iff} \quad \forall y \psi(x, y);$$

and a set is enumerable iff there is a first-order formula φ (in the language of arithmetic) in which all quantifiers are bounded such that for all x,

$$x \in S \quad \text{iff} \quad \exists y \varphi(x, y).$$

More obviously, and also historically, there is a close connection between computability/enumerability and decidability/axiomatizability. These connections often led to deep theorems, such as Matiyasevich's solution [Ma70, Ma93] to Hilbert's 10th Problem (based on previous work by Davis, J. Robinson, and others). He showed that a set S is enumerable iff there is a polynomial $p(\overline{x}, \overline{y}) \in \mathbf{Z}$ such that

$$S = \{\overline{x} \mid \exists \overline{y} p(\overline{x}, \overline{y}) = 0\}.$$

Given that most sets are noncomputable, the question arises as to how to compare them in terms of their information content, i.e., how to measure noncomputable information. The most general "effective" *reducibility* was first defined by Turing [Tu39]: A set S is *Turing reducible* to a set T (denoted by $S \leq_{\mathrm{T}} T$) if there is an *oracle Turing machine* computing S (with *oracle* T, i.e., such that the Turing machine can query membership information about the set T). This reducibility gives a prepartial ordering (i.e., a reflexive and transitive relation) on the power set of $\mathbf{N}$. We can now define two sets S and T to be *Turing equivalent* (denoted by $S \equiv_{\mathrm{T}} T$) if they are Turing reducible to each other. This gives an equivalence relation on the power set of $\mathbf{N}$. The equivalence class of a set S is called its *Turing degree* (denoted by $\deg_{\mathrm{T}} S$ or simply $\deg S$) and intuitively denotes the "information content" of the set S while stripping away all the facts about S inessential from a computational point of view, such as whether a particular number is an element of S.

These Turing degrees then form a quotient structure of the power set of $\mathbf{N}$, partially ordered by the relation induced on it by Turing reducibility. We denote this structure by $\mathbf{D}$, and the substructure of the Turing degrees of the enumerable sets by $\mathbf{E}$. Note that the latter structure can also be defined as the set of Turing degrees of solution sets of diophantine equations (by the above-mentioned result of Matiyasevich [Ma70, Ma93]) or as the set of Turing degrees of word problems of finitely presented groups (by Boone [Bo66], Clapham [Cl64], and Fridman [Fr67]).

There is one more operation on the Turing degrees that will be used later, namely, the so-called *Turing jump*. Given a set S, its jump is defined as the set $S' = \{e \mid e$th oracle Turing machine with oracle S halts on input $e\}$. This operation can be iterated: $S^{(0)} = S$, and $S^{(n+1)} = (S^{(n)})'$, and we have $S <_{\mathrm{T}} S'$, and that $S \leq_{\mathrm{T}} T$ implies $S' \leq_{\mathrm{T}} T'$. So this jump operation induces a well-defined operation on the Turing degrees: The jump of a degree $\mathbf{a}$ is A' for any set $A \in \mathbf{a}$. (A more intuitive definition of the nth jump of a set S is that it is the set of all Σ_n-sentences (in the language of arithmetic with a unary predicate for S) true in $\mathbf{N}$.)

1. Studying the enumerable Turing degrees. This partially ordered set of the enumerable Turing degrees has received a considerable amount of attention from researchers over the past fifty years, starting with the seminal paper of Post [Po44]. There are, loosely speaking, the following aspects under which one can consider this structure:

(1) *First-order aspects:* Typical questions here include the algebraic structure of $\mathbf{E}$ as a partially ordered set, especially the investigation of its finite substructures. Many of these were resolved in the 1960's and 1970's but some key problems remain open.
(2) *Logical aspects:* The main question here is the decidability of the first-order theory and of its fragments. This was the main topic of research in the 1980's; the most interesting remaining open question is the decidability of the Π_2-theory.
(3) *Model-theoretic aspects:* This aspect concerns mainly questions of definability and of the type structure. There are some exciting recent results but much remains to be done.
(4) *Second-order aspects:* The main topic here is that of automorphisms of the structure. Once Cooper's result [Cota] about a nontrivial automorphism is fully understood, this is likely to be the most active area in the near future.

2. First-order aspects. Very early work showed that the enumerable degrees form a countable *bounded upper semilattice* (i.e., a partially ordered set with least and greatest element in which any two elements have a least upper bound). The degree $\mathbf{0}$ of the computable sets and the degree $\mathbf{0}'$ of the halting problem are the least and the greatest element, respectively; the least upper bound of the degrees of S and T is $\deg(S \oplus T) = \deg(\{2x \mid x \in S\} \cup \{2x+1 \mid x \in T\})$.

The first nontrivial result was the theorem of Friedberg [Fr57] and Muchnik [Mu56] that the structure contains incomparable elements. This early period of algebraic investigations climaxed with the Sacks Density Theorem [Sa64] that the enumerable degrees are densely ordered. This theorem in particular led Shoenfield to his famous conjecture which can be phrased as follows:

Shoenfield Conjecture [Sh65]. *Given any two finite bounded upper semilattices $U \subset V$, any embedding of U into $\mathbf{E}$ (as an upper semilattice) can* *be extended to an embedding of V into* **E**.

Since this conjecture would have allowed back-and-forth constructions, it would have implied a number of "nice" results about **E**, such as the saturatedness of **E** and the $\aleph_0$-categoricity and decidability of its first-order theory. However, Shoenfield's Conjecture was almost immediately shown to fail quite dramatically. We mention here three

Refutations of Shoenfield's Conjecture.

(1) (Lachlan [La66] and Yates [Ya66]) *There is a minimal pair in* **E**, *i.e., there are nonzero degrees with infimum 0. (This precludes that any embedding of the diamond lattice as an upper semilattice can be extended to an embedding of the five-element lattice obtained by adding a new element below the two atoms of the diamond lattice.)*

(2) (Yates (unpublished) and Cooper [Co74]) *There is a noncuppable degree in* **E**, *i.e., there is a nonzero degree* **a** *such that no incomplete degree* **b** *joins* **a** *to* $\mathbf{0}'$. *(This precludes that any embedding of the three-element linear order can be extended to an embedding of the diamond lattice.)*

(3) (Cooper, Sui, Yi [CSYta]) *There is a superminimal pair in* **E**, *i.e., there is a minimal pair* $\mathbf{a}_0$ *and* $\mathbf{a}_1$ *such that for any* $i \leq 1$, *every degree* $\mathbf{x} \in (\mathbf{0}, \mathbf{a}_i]$ *joins* $\mathbf{a}_{1-i}$ *to* $\mathbf{a}_0 \cup \mathbf{a}_1$. *(This precludes that any embedding of the diamond lattice can be extended to an embedding of the six-element lattice obtained by inserting two new elements, one below one old atom, and the other as the join of the first new element and the other old atom.)*

3. Logical aspects. Shoenfield's Conjecture, even though it failed, was crucial in that it stimulated the next generation of algebraic investigations which revealed the "not so nice" structure of **E** and culminated nearly two decades later in the proof of the undecidability of its theory by Harrington and Shelah [HS82]. Soon afterwards, Harrington and Slaman (unpublished, see [SWta]) succeeded in showing that in fact the first-order theory of **E** is as complicated as possible, namely, as complicated as first-order arithmetic.

Since the proof of this result is not readily available in the literature, we will briefly sketch it here, with some later simplifications due to Slaman and Woodin [SWta] as well as Nies, Shore, and Slaman [NSSta]. Clearly, the first-order theory of the enumerable degrees can be interpreted in first-order arithmetic in the usual way. The proof in the other direction proceeds in several steps:

Step 1: We code the natural numbers with addition and multiplication by a computable partial ordering $(P, \leq)$ such that the natural numbers n are coded by the minimal elements p_n of P and such that the operations are coded, e.g., by

$$m + n = k \text{ iff} \quad \exists p, q \in P \; (p \text{ is minimal over } p_m \text{ and } p_n, \text{ and } q \text{ is maximal in } P \text{ and minimal over } p \text{ and over } p_k),$$

and similarly for multiplication.

Step 2: Code this partial order $(P, \leq)$ into $\mathbf{E}$ with parameters (i.e., degrees) $\mathbf{a}$, $\mathbf{b}$, $\mathbf{c}$, and $\mathbf{d}$ such that the elements p of P are coded by the minimal degrees $\mathbf{x} = \mathbf{x}_p \leq \mathbf{a}$ with the property that $\mathbf{c} \leq \mathbf{x} \cup \mathbf{b}$, and such that $p \leq q$ in P iff $\mathbf{x}_p \cup \mathbf{d} \leq \mathbf{x}_q \cup \mathbf{d}$. We will call such a quadruple of parameters $\mathbf{a}$, $\mathbf{b}$, $\mathbf{c}$, and $\mathbf{d}$ a *coded standard model of arithmetic.*

The remaining steps are now needed to sort out the coded standard models of arithmetic from other coded models of (a finite fragment of) Peano arithmetic (PA).

Step 3: For any nonzero degree $\mathbf{a} \in \mathbf{E}$, there are *low* parameters below $\mathbf{a}$ coding a standard model of arithmetic. (A degree $\mathbf{a}$ is low if $\mathbf{a}' = \mathbf{0}'$.)

Step 4: Given two coded models M_0 and M_1 of (a finite fragment of) Peano arithmetic, we want to code an *embedding* (i.e., an order-preserving injection) f from M_0 into M_1 in a similar fashion. If M_0 is a standard model of arithmetic coded by low parameters, then, for any coded model M_1 of (a finite fragment of) Peano arithmetic, such a coded map f always exists.

Step 5: We can interpret first-order arithmetic in the first-order theory of $\mathbf{E}$ as follows: Fix a sentence φ in the language of arithmetic. Then φ is true in $\mathbf{N}$ iff in $\mathbf{E}$, the following sentence holds:

$$\begin{aligned} &\exists \mathbf{a} > \mathbf{0} \; \exists M_0 \models \text{ (finite fragment of) PA coded below } \mathbf{a} \\ &\quad ((\forall M_1 \models \text{ (finite fragment of) PA coded below } \mathbf{a} \\ &\quad\quad \exists \text{ embedding } f : M_0 \to M_1 \text{ coded below } \mathbf{a}) \text{ and } M_0 \models \varphi). \end{aligned}$$

Note that the clause $(\forall M_1 \exists f \ldots)$ here ensures that M_0 is a standard model of arithmetic since by Step 3, some standard model must be coded below $\mathbf{a}$ by low parameters, and by Step 4, M_0 is "more standard" than any other model of (a finite fragment of) Peano arithmetic coded below $\mathbf{a}$.

Once undecidability of a theory has been established, the immediate next question is at what quantifier-level undecidability first occurs since mathematicians are usually only interested in statements with a small number of alternations of quantifiers.

By an old observation of Sacks [Sa63], the Π_1-theory of the enumerable degrees is decidable since any finite partial order can be embedded into $\mathbf{E}$. By a recent result of Lempp, Nies, and Slaman [LNSta], the Π_3-theory is undecidable. This result is shown by coding finite bipartite graphs (in a language without equality) into $\mathbf{E}$ using only Σ_1-formulas with parameters, and then applying Nies's Transfer Lemma [Ni96] to transfer the hereditary undecidability of the Π_3-theory of bipartite graphs without equality to that of $\mathbf{E}$. The gap remaining in this line of research thus is at the Π_2-theory, which we will discuss in more detail in the last section.

4. Model-theoretic aspects. The above research focusing on undecidability also led to some results concerning the type structure and questions of definability in $\mathbf{E}$.

Lerman, Shore, and Soare [LSS84] exhibited infinitely many 3-types realized in $\mathbf{E}$ by embedding an infinite number of lattices into $\mathbf{E}$, all generated under meet and join by three elements. This showed the non-$\aleph_0$-categoricity of the first-order theory of $\mathbf{E}$, thus disproving another consequence of Shoenfield's Conjecture. Later, Ambos-Spies and Soare [AS89] found infinitely many 1-types realized in $\mathbf{E}$ (namely, degrees bounding n but not $n+1$ many degrees forming pairwise minimal pairs). And Ambos-Spies and Shore [AS93] showed that continuum many 1-types are consistent with the first-order theory of $\mathbf{E}$ (namely, that given any subset $S \subseteq \omega$ with at least three elements, there is a degree coding a partial ordering with maximal chain of length $k+1$ (incomparable to all other elements of the partial ordering) iff $k \in S$).

Clearly, the types of the least and the greatest element of $\mathbf{E}$ are isolated. But it is open whether there are any other isolated 1-types, and whether in fact all 1-types are isolated, i.e., whether $\mathbf{E}$ is a prime model of its theory.

This naturally leads to questions of definability. A fair number of results were shown in this respect over the years. The most exciting is probably the following recent

Theorem (Nies, Shore, Slaman [NSSta]). *If $\mathbf{S} \subseteq \mathbf{E}$ is definable in first-order arithmetic, and closed under double jump (i.e., for any degrees $\mathbf{a}$ and $\mathbf{b}$, $\mathbf{a}'' = \mathbf{b}''$ and $\mathbf{a} \in \mathbf{S}$ implies $\mathbf{b} \in \mathbf{S}$), then $\mathbf{S}$ is definable in $\mathbf{E}$ (in the language of partial ordering).*

This result has a number of interesting consequences; e.g., it shows the definability of the classes of the high_n and low_n enumerable degrees (for $n \geq 2$) in terms of the partial ordering only. By a small trick, they also obtained the definability of the class of the high (i.e., high_1) enumerable degrees. (They show that $\mathbf{a}$ is high iff for any $\mathbf{b}$ there is $\mathbf{c} \leq \mathbf{a}$ with $\mathbf{b}'' = \mathbf{c}''$. Here, an enumerable degree $\mathbf{a}$ is *low_n* if $\mathbf{a}^{(n)} = \mathbf{0}^{(n)}$, and *$high_n$* if $\mathbf{a}^{(n)} = \mathbf{0}^{(n+1)}$.)

Obviously, the least and the greatest element of $\mathbf{E}$ are definable. It is open whether any other enumerable degrees are definable in $\mathbf{E}$ (in the language of partial ordering without parameters). The following question addresses an interesting partial result in this direction which seems more accessible to current methods:

Question (Li Angsheng (see [Sl])). *Are there degrees $\mathbf{a} < \mathbf{b} < \mathbf{c}$ in $\mathbf{E}$ and a formula $\varphi(x)$ in the language of partial ordering without parameters such*

that for all enumerable degrees $\mathbf{x} \in (\mathbf{a}, \mathbf{c})$,

$$(\mathbf{E}, \leq) \models \varphi(\mathbf{x}) \quad \textit{iff} \quad \mathbf{x} = \mathbf{b}?$$

The strongest possible definability result would be the following

Biinterpretability Conjecture (Slaman, Woodin [Sl91]). *There is a map* f *from* $\mathbf{E}$ *into a standard model coded in* $\mathbf{E}$ *such that for all* $\mathbf{a} \in \mathbf{E}$, $\deg W_{f(\mathbf{a})} = \mathbf{a}$.

This conjecture would have implied in particular the *rigidity* of $\mathbf{E}$ (i.e., that the only automorphism of $\mathbf{E}$ is the identity). This consequence, and thus the Biinterpretability Conjecture, was refuted by Cooper as described in the next section.

5. Second-order aspects. The earliest results on automorphisms of $\mathbf{E}$ concerned *automorphism bases* (i.e., sets $\mathbf{S} \subseteq \mathbf{E}$ such that any automorphism which is the identity on $\mathbf{S}$ must be the identity on $\mathbf{E}$). A number of nontrivial automorphism bases were found, mainly in the 1980's. An interesting recent result here is due to Ambos-Spies [Amta] that any nontrivial initial segment of $\mathbf{E}$ forms an automorphism base.

However, the question of whether there are any automorphisms of $\mathbf{E}$ other than the identity remained open until Cooper's recent results [Cota] about the existence of such automorphisms. He also showed that there is an automorphism mapping a low to a nonlow enumerable degree. Thus the low enumerable degrees are not definable in $\mathbf{E}$ from the partial ordering alone, in contrast to all other jump classes as mentioned above.

Once a full proof of Cooper's results is available, a whole number of questions will arise: How many automorphisms are there (e.g., are there continuum many)? Are all automorphisms arithmetical? Is there a finite automorphism base?

6. The Π_2-theory. The main open question about the enumerable Turing degrees at this time, accessible to currently available methods, is in our opinion the decidability of the Π_2-theory of the enumerable degrees. It is not hard to see that the Π_2-theory can be rephrased in purely algebraic terms as follows:

Equivalent formulation of the decidability of the Π_2-theory. *Decide if given any finite partial orders* $P \subseteq Q_0, \ldots, Q_n$ *(for some* $n \geq 0$*), any embedding of* P *into* $\mathbf{E}$ *can be extended to an embedding of* Q_i *into* $\mathbf{E}$ *for some* $i \leq n$. *(Note here that* i *may depend on the embedding of* P *into* $\mathbf{E}$.*)*

A natural subproblem of the above is obtained by setting $i = 0$, i.e., deciding whether any embedding of a finite partial order P into $\mathbf{E}$ can be extended to an embedding of a finite partial order $Q \supset P$ into $\mathbf{E}$. (This is usually called the *extension of embeddings problem.*) A solution to this problem was given by Slaman and Soare, which we state in a modified version due to Lempp and Lerman:

Extension of Embeddings Theorem (Slaman, Soare [SS95], rephrased). *Fix a finite lattice P and a finite upper semilattice Q extending P as an upper semilattice and respecting the lattice structure of P. Then any embedding of P into $\mathbf{E}$ (as a lattice) can be extended to an embedding of Q into $\mathbf{E}$ (as an upper semilattice) iff*

$$\forall a, b \in P \, \forall x \in Q - P(a = \min\{c \in P \mid c > x\} \,\&\, x \not\leq b \rightarrow x \vee b = a \vee b).$$

Lerman calls the above the "Saturation Axiom". It is a generalization of the phenomenon encountered in the superminimal pair mentioned earlier. Lerman [Le96] then suggests the following general approach to deciding the Π_2-theory: Expand the language of partial ordering to include

- the language of bounded upper semilattices (i.e., $\leq$, $\vee$, 0, and 1);
- $(n+2)$-ary meet predicates $M(a, b_0, \ldots, b_n)$ (for all $n \geq 1$): This denotes that all $x \leq b_0, \ldots, b_n$ are also $\leq a$ and takes into account that the meet of two degrees need not always exist;
- saturation predicates generalizing the phenomenon of the Saturation Axiom mentioned above; and
- a unary predicate for the so-called promptly simple degrees. (The class of the promptly simple degrees was shown by Ambos-Spies, Jockusch, Shore, and Soare [AJSS84] to coincide with a number of other interesting classes (such as the degrees which do not form one half of a minimal pair). Prompt simplicity interacts nontrivially with saturation, e.g., Cooper, Slaman, and Yi (unpublished) observed that saturation cannot occur between a promptly simple degree $\mathbf{a}$ and a non-promptly simple degree $\mathbf{b}$, namely, given any such degrees $\mathbf{a}$ and $\mathbf{b}$, there is an enumearable degree $\mathbf{x} < \mathbf{a}$ with $\mathbf{x} \not\leq \mathbf{b}$ and $\mathbf{x} \cup \mathbf{b} < \mathbf{a} \cup \mathbf{b}$.)

Lerman's approach then consists in deciding the Π_1-theory in this expanded language in order to give a decision procedure for the Π_2-theory in the language of partial ordering.

Lerman's approach thus highlights another natural subproblem, the so-called *Lattice Embeddings Problem*, namely, to characterize the finite lattices which are embeddable into $\mathbf{E}$. This is actually a very old open problem going back to the early investigations in the 1960's of the algebraic structure of $\mathbf{E}$ and, in particular, of its finite substructures.

Of course, the minimal pair theorem of Lachlan and Yates mentioned earlier yields an embedding of the diamond lattice into $\mathbf{E}$. This was soon generalized to showing that all finite distributive lattices are embeddable into $\mathbf{E}$ by Lerman (unpublished) and Thomason [Th71]. Lachlan [La72] proved that the two non-distributive five-element lattices M_5 and N_5 are embeddable into $\mathbf{E}$. However, Lachlan and Soare [LS80] exhibited a finite lattice, S_8, which is not embeddable into $\mathbf{E}$. The known embedding and non-embedding techniques up to the late 1980's were then distilled into two conditions (the so-called *Embeddability Condition (EC)* and *Nonembeddability Condition (NEC)*) by Ambos-Spies and Lerman [AL86, AL89].

EC turned out to be a rather unwieldy condition and received little attention. NEC, on the other hand, was a nice, algebraic condition and was conjectured by many to be the correct condition characterizing exactly the non-embeddable finite lattices. In particular, NEC requires the existence of a so-called critical triple in the lattice.

Definition. A triple $\langle a, b, c\rangle$ of elements of a finite lattice L is called a *critical triple* if a, b, and c are pairwise-incomparable, $a \vee b = a \vee c$, and $b \wedge c \leq a$.

Ambos-Spies and Lerman [AL86] observed that in a finite lattice, the absence of critical triples is equivalent to another property which is easier to verify.

Proposition. *A finite lattice L with least element 0 fails to have critical triples if for all $a < d$ in L such that the interval (a, d) is empty, the difference of the intervals $[0, d] - [0, a]$ has a (unique) least element.*

The above proposition allows one to organize the enumeration of elements for an embeddings proof for a lattice without critical triple much more easily, so the absence of critical triples was conjectured by many to ensure the embeddability of a finite lattice. This was recently refuted, however, by Lempp and Lerman [LLta], who exhibited a finite lattice, L_{20}, which is not embeddable into the enumerable degrees but does also not contain a critical triple.

The search for a characterization of the finite lattices embeddable into the enumerable degrees continues and is likely to be hard; it involves analyzing the obstructions to embeddability in a typical pinball machine proof and using them to produce a nonembeddability proof in an effective fashion if possible. Only then, a decision procedure for the Π_2-theory of the enumerable degrees can reasonably be attempted.

REFERENCES

[Amta] K. Ambos-Spies, *Automorphism bases for the recursively enumerable degrees* (to appear).

[AJSS84] K. Ambos-Spies, C. G. Jockusch, Jr., R. A. Shore, and R. I. Soare, *An algebraic decomposition of the recursively enumerable degrees and the coincidence of several degree classes with the promptly simple degrees*, Trans. AMS **281** (1984), 109-128.

[AL86] K. Ambos-Spies and M. Lerman, *Lattice embeddings into the recursively enumerable degrees*, J. Symbolic Logic **51** (1986), 257-272.

[AL89] ———, *Lattice embeddings into the recursively enumerable degrees, II*, J. Symbolic Logic **54** (1989), 735-760.

[AS93] K. Ambos-Spies and R. A. Shore, *Undecidability and 1-types in the recursively enumerable degrees*, Ann. Pure Appl. Logic **63** (1993), 3-37.

[AS89] K. Ambos-Spies and R. I. Soare, *The recursively enumerable degrees have infinitely many one types*, Ann. Pure Appl. Logic **44** (1989), 1-23.

[Bo54-57] W. W. Boone, *Certain simple unsolvable problems of group theory I, II*, Indag. Math. **16** (1954), 231-237, 492-497; *III, IV*, Indag. Math. **17** (1955), 252-256, 571-577; *V, VI*, Indag. Math. **19** (1957), 22-27, 227-232.

[Bo66] ———, *Word problems and recursively enumerable degrees of unsolvability, a sequel on finitely presented groups*, Ann. of Math. (2) **84** (1966), 49-84.

[CD93] P. A. Cholak and R. G. Downey, *Lattice nonembeddings and intervals in the recursively enumerable degrees*, Ann. Pure Appl. Logic **61** (1993), 195-222.

[Co74] S. B. Cooper, *On a theorem of C. E. M. Yates*, handwritten notes.

[Cota] ———, *Beyond Gödel's Theorem – the failure to capture information content*, Logic Colloquium '95 (V. Harnik and J. A. Makowsky European Summer Meeting of the Association for Symbolic Logic, eds.), Springer Lecture Notes in Logic, Springer-Verlag, Heidelberg (to appear).

[Cota] ———, *The Turing degrees are not rigid* (to appear).

[Cl64] C. R. J. Clapham, *Finitely presented groups with word problems of arbitrary degrees of insolubility*, Proc. London Math. Soc. (3) **14** (1964), 633-676.

[ELTT65] Yu. Ershov, L. Lavrov, A. Taimanov, M. Taitslin, *Elementary theories*, Russian Math. Surveys **20** (1965), 35-105.

[ET63] Yu. L. Ershov and M. A. Taitslin, *Undecidability of certain theories*, Alg. i Log. **2** no. 5 (1963), 37-41. (Russian)

[Fr67] A. A. Fridman, *Degrees of unsolvability of the word problem for finitely presented groups*, Nauka, Moscow, 1967.

[Fr57] R.M. Friedberg, *Two recursively enumerable sets of incomparable degrees of unsolvability*, Proc. Natl. Acad. Sci. USA **43** (1957), 236-238.

[HS82] L. Harrington and S. Shelah, *The undecidability of the recursively enumerable degrees*, Bull. AMS (N. S.) **6** no. 1, 79-80.

[Kl36] S. C. Kleene, *General recursive functions of natural numbers*, Math. Ann. **112** (1936), 727-742.

[Kl52] ———, *Introduction to metamathematics*, North-Holland, Amsterdam, 1952.

[La66] A. H. Lachlan, *Lower bounds for pairs of r.e. degrees*, Proc. London Math. Soc. **16** (1966), 537-569.

[La72] ———, *Embedding nondistributive lattices in the recursively enumerable degrees*, Conference in Mathematical Logic, London, 1970 (W. Hodges, ed.), Lecture Notes in Mathematics No. 255, Springer-Verlag, Berlin, 1972, pp. 149-177.

[LS80] A. H. Lachlan and R. I. Soare, *Not every finite lattice is embeddable in the recursively enumerable degrees*, Adv. Math. **37** (1980), 78-82.

[La63] I. Lavrov, *Effective inseparability of the set of identically true formulas and finitely refutable formulas for certain elementary theories*, Alg. i Log. **1** (1963), 5-18. (Russian)

[LLta] S. Lempp and M. Lerman, *The decidability of the existential theory of the poset of recursively enumerable degrees with jump relations*, Adv. Math. (to appear).

[LNSta] S. Lempp, A. Nies, and T. A. Slaman, *The Π_3-theory of the enumerable Turing degrees is undecidable*, Trans. AMS (to appear).

[Le96] M. Lerman, *Embeddings into the recursively enumerable degrees*, Computability, enumerability, unsolvability (S. B. Cooper, T. A. Slaman, and S. S. Wainer, eds.), London Mathematical Society Lecture Notes No. 224, Cambridge University Press, Cambridge, England, 1996, pp. 185-204.

[LSS84] M. Lerman, R. A. Shore, and R. I. Soare, *The elementary theory of the recursively enumerable degrees is not $\aleph_0$-categorical*, Adv. Math. **53** (1984), 301-320.

[Ma70] Yu. V. Matijasevič, *Enumerable sets are diophantine*, Dokl. Akad. Nauk SSSR **191** (1970), 279-282. (Russian)

[Ma93] Yu. V. Matijasevič, *Desyataya problema Gil'berta*, Nauka, Moscow, 1993; Hilbert's Tenth Problem, MIT Press, Cambridge, MA.

[Mu56] A. A. Mučnik, *On the unsolvability of the problem of reducibility in the theory of algorithms*, Dokl. Akad. Nauk SSSR, N.S. **108** (1956), 194-197. (Russian)

[Ni96] A. Nies, *Undecidable fragments of elementary theories*, Alg. Universalis **35** (1996), 8-33.

[NSSta] A. Nies, R. A. Shore, and T. A. Slaman, *Standard models of arithmetic and definability in the enumerable degrees* (to appear).

[No55] P. S. Novikov, *On the algorithmic unsolvability of the word problem in group theory*, Trudy Mat. Inst. Steklov **44** (1955), 1-143. (Russian)

[Od89] P. Odifreddi, *Classical Recursion Theory*, vol. 1, North-Holland, Amsterdam, 1989.

[Odta] ———, *Classical Recursion Theory*, vol. 2 (to appear).

[Po44] E. L. Post, *Recursively enumerable sets of positive integers and their decision problems*, Bull. AMS **50** (1944), 284-316.

[Po48] ———, *Degrees of recursive unsolvability*, Bull. AMS **54** (1948), 641-642.

[Ro67] H. Rogers, Jr., *Theory of recursive functions and effective computability*, McGraw-Hill, New York, 1967.

[Sa63] G. E. Sacks, *Degrees of Unsolvability*, Ann. of Math. Studies No. 55, Princeton University Press, Princeton, N.J., 1963.

[Sa64] ———, *The recursively enumerable degrees are dense*, Annals of Math. (2) **80** (1964), 300-312.

[Sh65] J. R. Shoenfield, *Application of model theory to degrees of unsolvability*, Symposium on the Theory of Models (J. Addison, L. Henkin, A. Tarski, eds.), North-Holland, Amsterdam, 1965, pp. 359-363.

[Sl91] T. A. Slaman, *Degree structures*, Proceedings of the International Congress of Mathematicians, August 21-29, 1990, Kyoto, Japan, vol. 1, Springer-Verlag, Tokyo, 1991, pp. 303-316.

[Sl] ———, *Open Questions in Recursion Theory*, list of open questions on the world-wide web at http://www.nd.edu/~pcholak/computability/computability.html.

[SS95] T. A. Slaman and R. I. Soare, *Algebraic Aspects of the Computably Enumerable Degrees*, Proc. Natl. Acad. Sci. USA **92** (1995), 617-621.

[SWta] T.A. Slaman and H. Woodin, *Definability in degree structures* (to appear).

[So87] R. I. Soare, *Recursively Enumerable Sets and Degrees: The Study of Computable Functions and Computably Generated Sets*, Perspectives in Mathematical Logic, Ω-Series, Springer-Verlag, Berlin, 1987.

[Sota] ———, *Computability and recursion* (to appear).

[Th71] S. K. Thomason, *Sublattices of the recursively enumerable degrees*, Z. math. Logik Grundlag. Math. **17** (1971), 273-280.

[Tr53] B. A. Trakhtenbrot, *Recursive separability*, Dokl. Akad. Nauk SSSR **88** (1953), 953-956.

[Tu36] A. M. Turing, *On computable numbers, with an application to the Entscheidungsproblem*, Proc. London Math. Soc. (2) **42** (1936), 230-265 **43** (1943), 544-546.

[Tu39] ———, *Systems of logic based on ordinals*, Proc. London Math. Soc. (2) **45** (1939), 161-228.

[Ya66] C. E. M. Yates, *A minimal pair of recursively enumerable degrees*, J. Symbolic Logic **31** (1966), 159-168.

THE THEORY OF FINITE MODELS *

LUO Libo

Department of Mathematics
Beijing Normal University, Beijing 100875

(Abstract)

We study the model theory of finite structures. By releasing the requirement of a predefined equal sign and using formulas of infinite lengths we obtain new theorems for preservation of model extensions, submodels, model homomorphisms and inverse homomorphisms. This kind of theorems were discussed in Chang and Keisler's Model Theory systematically for general models but Gurevich obtained some different theorems at this direction for finite models. In our paper the old theorems managed to survive in finite model theory. There are some differences between into homomorphisms and onto homomorphisms in preservation theorems too. We also study reduced models and minimum models. The characterization sentence of a model is given which derives a general result for any theory T to be equivalent to a set of existential - universal sentences. Some results about completeness and model completeness are also given.

SECTION 0. INTRODUCTION

The general model theory is studied for a long time. Chang and Keisler in [1] give a detailed introduction. It seems that the model theory is mainly studying infinite models. For the case of finite model theory some people would think that it was easy to deal with and not as important as the infinite case. The development of computer science gives us a different thought. Database is finite, operating system is finite, even computer itself is considered as a finite machine. The finite model theory becomes more important. Another reason for studying finite model theory separately is that many of the theorems in general models theory cannot be used in the finite cases. For example the Gödel's Incompleteness Theorem, Craig's

*Project supported by the National Natural Science Foundation of China

Interpolation Theorem and Compactness Theorem. Without these theorems the finite model theory looks quite different from the general model theory. The research also finds out that it is not always easy to prove a theorem in the finite model theory in comparing with the same theorem in general model theory. In some cases the proofs are even more difficult. A partial reason for that is the lacking of the theorems mentioned above. Gurevich in [2] and [3] gave a systematic discussion for the finite model theory in connection with the theory of computer science. Then in [4] he investigated the preservation theorems for finite model theory. He gave counter examples in finite model theory to some of the theorems for infinite models and some of the others are still valid in finite model theory. In our paper we prove some different kinds of preservation theorems for finite model theory. In general model theory the equal relation is supposed to be built into the system but we treat all relation equally. Our notations are as follows:

$$\mathfrak{A}, = < A,\ c_1,\ \ldots,\ c_p,\ R_1,\ \ldots,\ R_q >$$

is a model with the universe A, p constants $c_1, \ldots, c_p$ and q relations $R_1, \ldots, R_q$. The universe A is always finite and the language $L = < A, c_1, \ldots, c_p, R_1, \ldots, R_q >$ is also finite. The logical formulas in our paper are usually first order formulas. We also use infinite conjunctions and disjunctions of first order formulas. In the latter case we will use capital Greeks to denote a formula with infinite length. We adopt the idea in [4] to say that two formulas are equivalent if they are equivalent in all finite models. They are logically equivalent if they are equivalent in all finite and infinite models. The notation $\mathfrak{A} \models T$ means that all first order sentences in T are true in $\mathfrak{A}$. T_1 and T_2 are theories. The notation $T_1 \models T_2$ means that for any model $\mathfrak{A}$, $\mathfrak{A} \models T_1$ implies $\mathfrak{A} \models T_2$. T_1 is equivalent to T_2 if and only if $T_1 \models T_2$ and $T_2 \models T_1$. A theory T is consistent if and only if T has a (finite) model.

SECTION 1. ELEMENTARY PROPERTIES

As in the general model theory we can define elementary equivalence between models, Elementary submodels and elementary extensions of models with almost the same definitions. Most of the theorems about these concepts in general model theory are still true in finite model theory. The others are similar. We provide some of them here and skip the proofs.

Theorem 1.1. Two models $\mathfrak{A}$ and $\mathfrak{B}$ are elementary equivalent to each other if and only if the following two conditions are true:

(1) There is a constant keeping function F from A to B such that for any relation R in the language L with arity k and any k-tuple of elements $a_1, \ldots, a_k$ in A,

$\mathfrak{A} \models R[a_1, \ldots, a_k]$ if and only if $\mathfrak{B} \models R[Fa_1, \ldots, Fa_k]$.

(2) There is a constant keeping function G from B to A such that for any relation R in the language L with arity k and any k-tuple of elements $b_1, \ldots, b_k$ in B,

$\mathfrak{B} \models R[b_1, \ldots, b_k]$ if and only if $\mathfrak{A} \models R[Gb_1, \ldots, Gb_k]$.

Proof. (Outline) By the functions F and G we can find

$\mathfrak{A} = \mathfrak{A}_0 \supseteq \mathfrak{A}_1 \supseteq \mathfrak{A}_2 \supseteq \ldots,$

$\mathfrak{B} = \mathfrak{B}_0 \supseteq \mathfrak{B}_1 \supseteq \mathfrak{B}_2 \supseteq \ldots,$

such that $\mathfrak{A}_i \equiv \mathfrak{B}_{i+1}$ and $\mathfrak{B}_i \equiv \mathfrak{A}_{i+1}$. Since they are finite models the two sequences will be the same models in finitely many steps.

Definition 1.1. Model $\mathfrak{A}$ is an elementary submodel of model $\mathfrak{B}$ if $A \subseteq B$ and for any first order formula $\varphi(x_1, \ldots, x_k)$ and for any k-tuple $(z_1, \ldots, z_k)$ from $A \bigcup \{c_1, \ldots, c_p\}$ the sentence $\varphi(z_1, \ldots, z_k\}$ is true in $\mathfrak{A}$ if and only if the same sentence is true and in $\mathfrak{B}$.

Theorem 1.2. Model $\mathfrak{A}$ is an elementary submodel of $\mathfrak{B}$ (or model $\mathfrak{B}$ is an elementary extension of $\mathfrak{A}$) if and only if there is a constant keeping function F from B onto A such that for any relation R in the language L with arity k and any k-tuple of elements $b_1, \ldots, b_k$ in B,

$\mathfrak{B} \models R[b_1, \ldots, b_k]$ if and only if $\mathfrak{A} \models R[Fb_1, \ldots, Fb_k]$.

We also give the definitions of reduced models and minimum models.

Definition 1.2. Two elements a and b are said to satisfy the same relation in a model $\mathfrak{A}$ of language $L = < A, c_1, \ldots, c_p, R_1, \ldots, R_q >$ with respect to elements $a_1, \ldots, a_n$ if for all relation $R(x_1, \ldots, x_k)$ all k-tuples $(z_1, \ldots, z_k)$ from $\{a_1, \ldots, a_n\} \bigcup \{c_1, \ldots, c_p\}$ whenever the relation $R(z_1, \ldots, z_{i-1}, a, z_{i+1}, \ldots, z_k\}$ is true the relation $R(z_1, \ldots, z_{i-1}, b, z_{i+1}, \ldots, z_k\}$ is also true and vise versa.

In a model $\mathfrak{A}$ if two elements a and b satisfy the same relations with respect to all elements of the model $\mathfrak{A}$, then we can reduce one of them to form a new model $\mathfrak{B}$. The new model is an elementary submodel of the original model $\mathfrak{A}$.

Definition 1.3. A model in which any two elements do not satisfy the same relation with respect to all elements of the model is call a reduced model.

Definition 1.4. A model $\mathfrak{A}$ of a theory T is called a minimum model of T if $\mathfrak{A}$ itself is a reduced model and it does not contain any proper submodel of T.

Theorem 1.3. For any model $\mathfrak{A}$ there is a reduced model $\mathfrak{B}$ such that $\mathfrak{B} \prec \mathfrak{A}$. The reduced model $\mathfrak{B}$ is unique up to isomorphism.

Theorem 1.4. A theory T is complete if and only if T has only one reduced model up to isomorphism.

We can define model completeness like in the general model theory, but here the theorem is different.

Theorem 1.5. If a theory T is complete then T is model complete.

Proof. If $\mathfrak{A} \subset \mathfrak{B}$ are T models, then both of them can be reduced to the same reduced model.

Theorem 1.6. For any model $\mathfrak{A}$ of language $L = < A, c_1, \ldots, c_p, R_1, \ldots, R_q >$ there is a characterization formula φ such that for any model $\mathfrak{B}$, $\mathfrak{B} \models \varphi$ if and only if $\mathfrak{B}$ is elementary equivalent to $\mathfrak{A}$. The characterization formula is an existential universal sentence.

Proof. Suppose that model $\mathfrak{A}$ has n elements $\{a_1, \ldots, a_n\}$. We define a conjunction formula

$$\chi_{y_1 = a_{i_1}, \ldots, y_m = a_{i_m}}(y_1, \ldots, y_m)$$

with free variables $(y_1, \ldots, y_m)$ where y_1 represents $a_{i_1}, \ldots, y_m$ represents a_{i_m}. The formula χ is a conjunction of all positive and negative relations $\varepsilon R(z_1, \ldots, z_k)$ where the k-tuple $(z_1, \ldots, z_k)$ runs over all possible combinations (including repetitions) of symbols in

$$Z = \{y_1, \ldots, y_m\} \bigcup \{a_1, \ldots, a_n\} \bigcup \{c_1, \ldots, c_p\}.$$

The symbol ε is chosen according to the truth of the relation $R(z_1, \ldots, z_k)$ in $\mathfrak{A}$. We give the following definition:

$$\chi_{y_1 = a_{i_1}, \ldots, y_n = a_{i_m}}(y_1, \ldots, y_m)$$

$$= \bigwedge_{\substack{R(z_1, \ldots, z_k) \in L \\ \{z_1, \ldots, z_k\} \subset Z}} \Big(\bigwedge_{\mathfrak{A} \models R(z_1, \ldots, z_k)} R(z_1, \ldots, z_k)$$

$$. \wedge . \bigwedge_{\mathfrak{A} \not\models R(z_1, \ldots, z_k)} \neg R(z_1, \ldots, z_k) \Big).$$

Using formula χ we can define the conjunctive formula of the model $\mathfrak{A}$.

$$\chi_{\mathfrak{A}}(x_1, \ldots, x_n) = \chi_{x_1 = a_1, \ldots, x_n = a_n}(x_1, \ldots, x_n)$$

Formula $\chi_{\mathfrak{A}}(x_1, \ldots, x_n)$ is true if and only if elements $x_1, \ldots, x_n$ have the same positive and negative relations as elements $a_1, \ldots, a_n$ in model $\mathfrak{A}$.

Next we define a formula ψ_i for an element y_i to have the same relations as the element a_i with respect to all a's.

$$\psi_i(y_i, x_1, \ldots, x_n) = \chi_{y_i = a_i, x_1 = a_1, \ldots, x_n = a_n}(y_i, x_1, \ldots, x_n).$$

Putting together $\psi_i(y_i, x_1, \ldots, x_n)$, for $i = 1, \ldots, n$ and replacing y for y_i we obtain a disjunction.

$$\bigvee_{i=1}^{n} \psi_i(y_i, x_1, \ldots, x_n).$$

The above formula is true if and only if y satisfies the same relations as one of the elements in $a_1, \ldots, a_n$. with respect to all a's.

If $y_{i_1}, \ldots, y_{i_k}$ have the same relations as elements $a_{i_1}, \ldots, a_{i_k}$ with respect to all elements of $\mathfrak{A}$ we want the list $(y_{i_1}, \ldots, y_{i_k})$ to satisfy the same relations as $(a_{i_1}, \ldots, a_{i_k})$. In the following the arity k is considered as the maximum of the arity of all relations in the language L and the k-tuple $(y_{i_1}, \ldots, y_{i_k})$ runs through all possible strings of length k in $(y_1, \ldots, y_n)$. The formula is as follows:

$$\bigwedge_{\{y_{i_1}, \ldots, y_{i_k}\} \subset \{y_1, \ldots, y_n\}} \left(\bigwedge_{j=1}^{k} \psi_{i_j}(y_{i_j}, x_1, \ldots, x_n) \right.$$

$$\left. \rightarrow \chi_{y_{i_1} = a_{i_1}, \ldots, y_{i_k} = a_{i_k}} (y_{i_1}, \ldots, y_{i_k}) \right).$$

Now we are ready to define our main sentence φ. The meaning of φ is that there exist n elements $x_1, \ldots, x_n$ such that they satisfy the same relations as $a_1, \ldots, a_n$ and for any y, y satisfies the same relations as one of the x's and for any k-tuple $(y_{i_1}, \ldots, y_{i_k})$ chosen from $\{y_1, \ldots, y_n\}$ if each y_{i_j} satisfies the same relations as a_{i_j}, then the k-tuple $(y_{i_1}, \ldots, y_{i_k})$ satisfies the same relations as $(a_{i_1}, \ldots, a_{i_k})$.

$$\varphi = \exists x_1, \ldots, x_n \, (\chi_{\mathfrak{A}}(x_1, \ldots, x_n)$$

$$. \wedge . \forall y (\bigvee_{i=1}^{n} \psi_i(y, x_1, \ldots, x_n))$$

$$. \wedge . \forall y_1, \ldots, y_n \, (\bigwedge_{\{y_{i_1}, \ldots, y_{i_k}\} \subset \{y_1, \ldots, y_n\}} (\bigwedge_{j=1}^{k} \psi_{i_j}(y_{i_j}, x_1, \ldots, x_n)$$

$$\rightarrow \chi_{y_{i_1} = a_{i_1}, \ldots, y_{i_k} = a_{i_k}} (y_{i_1}, \ldots, y_{i_k})))).$$

We will prove that for any model $\mathfrak{B}$, $\mathfrak{B} \models \varphi$ if and only if $\mathfrak{B} \equiv \mathfrak{A}$. Let $\mathfrak{B}$ be a model of φ and $B = \{b_1, \ldots, b_m\}$ be the universe of $\mathfrak{B}$. We can set up a many to many and onto mapping from B to A satisfying

$$\left\{ \begin{array}{ll} F(b_i) = a_j & \textit{if } b_i \textit{ has the same relations as } a_j \\ & \quad \textit{with respect to } a_1, \ldots, a_n. \end{array} \right.$$

For any relation R in the language L with arity k and for any k-tuple $(b_{i_1}, \ldots, b_{i_k})$ in B, $\mathfrak{B} \models R[b_{i_1}, \ldots, b_{i_k}]$ if and only if $\mathfrak{A} \models R[Fb_{i_1}, \ldots, Fb_{i_k}]$. With this property we can prove that the two models $\mathfrak{A}$ and $\mathfrak{B}$ are elementary equivalent to each other.

Observe that there is not a quantifier in formulas ψ and χ. Therefore φ is an existential universal sentence.

The following two theorems are of special interest for finite model theory. We only give a simple proof for Theorem 1.7 and skip the prove of Theorem 1.8.

Theorem 1.7. Any theory T is equivalent to a countable disjunction of first order existential universal sentences.

Proof. For every model $\mathfrak{A}$ with n elements we can write the characterization sentence $\varphi_{\mathfrak{A}}$ as in the proof of Theorem 1.6 such that $\varphi_{\mathfrak{A}}$ is true in a model $\mathfrak{B}$ if and only if $\mathfrak{B}$ is elementary equivalent to $\mathfrak{A}$. Let $\sum$ be the disjunction of $\varphi_{\mathfrak{A}}$ over all models of T. $\sum$ is equivalent to T.

Theorem 1.8. Any theory T is equivalent to a countable set of first order universal existential sentences.

Proof. (It is the dual argument of Theorem 1.7.)

SECTION 2. PRESERVATION THEOREMS

There are a lot of theorems in general model theory in preserving properties among models. Using preservation theorems Chang and Keisler in [1] p.123 give a list of mathematical theories which preserve certain kinds of properties. In the book following Theorems are given:

[MT] Theorem 3.2.2. A theory T is preserved under submodels if and only if T has a set of universal axioms.

[MT] Theorem 3.2.3. A theory T is preserved under union of chains if and only if T has a set of universal existential axioms.

[MT] Theorem 3.2.4. A consistent theory T is preserved under homomorphisms if and only if T has a set of positive axioms.

Can any of the preservation theorems be true in the theory of finite models? The answer was negative for many cases. Gurevich in [4] gave counter examples to obtain the following results for finite models:

(i) There is a first order sentence φ such that φ is preserved by extensions of models but it cannot be equivalent to a first order existential sentence.

(ii) There is a first order sentence φ such that φ is preserved by submodels but it cannot be equivalent to a first order universal sentence.

(ii) There is a first order sentence $\varphi(P)$ such that $\varphi(P)$ is monotone in P but it cannot be equivalent to a first order sentence positive in P.

There are two differences between Gurevich's results and the results in [1]. Gurevich's theorems work for finite models only while the theorems in [1] work for all finite and infinite models. Also Gurevich is discussing the possibility of a first order sentence to be equivalent to another first order sentence while the theorems in [1] are using a set of (possibly) infinitely many sentences to be equivalent to another set of infinitely many sentences. We will show in this section that [MT] Theorem 3.2.2 survives in finite model theory. We also give similar theorems for preservation of homomorphisms of models.

Theorem 2.1. A Theory T is preserved under extensions of models if and only if T is equivalent to a countable disjunction of existential sentences.

Proof. The right to left implication is trivial so we only have to prove the left to right implication. Assume T is preserved under extensions. For every finite model $\mathfrak{A}$ in language $L = < c_1, \ldots, c_p, R_1, \ldots, R_q >$ with n elements we can write an existential sentence φ which describes all relations and negative relations among all elements. Suppose that the n elements are $A = \{a_1, \ldots, a_n\}$. We first define a conjunctive formula

$$\chi_{x_1 = a_1, \ldots, x_n = a_n}(x_1, \ldots, x_n)$$

with free variables $x_1, \ldots, x_n$ where x_1 represents a_1, ..., and x_n represents a_n. The formula χ is a conjunction of all positive and negative atomic formulas for all relations $\varepsilon\ R(z_1, \ldots, z_k)$ where the k-tuple $(z_1, \ldots, z_k)$ runs over all possible combinations (including repetitions) of

$$Z = \{y_1, \ldots, y_m\} \bigcup \{a_1, \ldots, a_n\} \bigcup \{c_1, \ldots, c_p\}.$$

The symbol ε is chosen according to the truth of relation $R(z_1, \ldots, z_k)$ with the corresponding k-tuple in model $\mathfrak{A}$. To specify the details we give the following definition:

$$\chi_{x_1 = a_1, \ldots, x_n = a_n}(x_1, \ldots, x_n) =$$

$$\bigwedge_{\substack{R(z_1, \ldots, z_k) \in L \\ \{z_1, \ldots, z_k\} \subset Z}} \Big(\bigwedge_{\mathfrak{A} \models R(z_1, \ldots, z_k)} R(z_1, \ldots, z_k). \wedge .$$

$$\bigwedge_{\mathfrak{A} \not\models R(z_1, \ldots, z_k)} \neg R(z_1, \ldots, z_k)\Big).$$

The maximum existential conjunctive sentence is defined as follows:

$$\varphi\big(\mathfrak{A}\big) = \exists x_1, \ldots, x_n \quad \chi_{x_1 = a_1, \ldots, x_n = a_n}(x_1, \ldots, x_n).$$

Let $\sum$ be the disjunction of the above sentence $\varphi(\mathfrak{A})$ over all models of the theory T.

$$\sum = \bigvee_{\mathfrak{A} \models T} \varphi(\mathfrak{A}).$$

Every model $\mathfrak{A}$ of the theory T is a model of the formula $\sum$ because $\mathfrak{A}$ satisfies at lease a disjunct of $\sum$. Conversely, every model $\mathfrak{B}$ of the formula $\sum$ satisfies a disjunct of $\sum$ which is written according to a model $\mathfrak{A}$. Model $\mathfrak{B}$ is an extension of the reduced model of $\mathfrak{A}$. $\mathfrak{A}$ is a model of T. Therefore $\mathfrak{B}$ is a model of the theory T.

Theorem 2.2. A Theory T is preserved under submodels if and only if T is equivalent to a set of universal sentences.

Proof. The theory T is considered as a conjunction $\prod$ of (possibly infinitely many) sentences. We consider the theory

$$T \longleftrightarrow \prod = \bigwedge_{1 \leq i \leq \infty} \varphi_i.$$

The negation of it is

$$\sum = \bigvee_{1 \leq i \leq \infty} \neg \varphi_i.$$

It is easy to see that $\sum$ is preserved under extension of models. In the proof of Theorem 2.1 the theory T is very general. We do not require the T be a set of first order sentences or in any other special form. So we can use it for $\sum$. Hence $\sum$ is equivalent to a countable disjunction

$$\sum{}' = \bigvee_{1 \leq i \leq \infty} \psi_i.$$

where each ψ_i is a first order existential sentence. Therefore the original theory T is equivalent to

$$\neg \sum{}' = \bigwedge_{1 \leq i \leq \infty} \neg \psi_i.$$

The left hand side is equivalent to a set of first order universal sentences.

A homomorphism is considered as a mapping h from a model $\mathfrak{A}$ into (including onto) a model $\mathfrak{B}$ such that for any relation R and any elements $a_1, \ldots, a_k \in \mathfrak{A}$ if $R(a_1, \ldots, a_k)$ is true in $\mathfrak{A}$, then $R(ha_1, \ldots, ha_k)$ is true in $\mathfrak{B}$. In [1], p. 126, a result of Lyndon's characterized the theories preserving under onto homomorphisms. We first discuss the preservation property of into homomorphisms in finite models. In our theorem we construct a countable disjunction of first order positive existential sentential sentences to be equivalent to the original theory.

Theorem 2.3. A Theory T is preserved under into homomorphisms if and only if T is equivalent to a countable disjunction of positive existential sentences.

Proof. For every finite model $\mathfrak{A}$ with n elements we can write a first order existential sentence φ which describes all positive relations (but not negative relations) among all elements.

$$\varphi = \exists x_1, \ldots, x_n \bigwedge_{\substack{R(z_1, \ldots, z_k) \in L \\ \{z_1, \ldots, z_k\} \subset Z}} \Big(\bigwedge_{\mathfrak{A} \models R(z_1, \ldots, z_k)} R(z_1, \ldots, z_k)\Big)$$

where x_1 represents $a_1, \ldots, x_n$ represents a_n and

$$Z = \{x_1, \ldots, x_n\} \bigcup \{a_1, \ldots, a_n\} \bigcup \{c_1, \ldots, c_p\}.$$

Let $\sum$ be the disjunction of the above sentences over all models of the theory T. Every model $\mathfrak{B}$ of the theory T is a model of the formula $\sum$ because $\mathfrak{B}$ satisfies at least a disjunct of $\sum$ which is written according to the model $\mathfrak{B}$ itself. Conversely if model $\mathfrak{B}$ satisfies a disjunct of $\sum$ which is written according to the model $\mathfrak{A}$. Model $\mathfrak{A}$ can be reduced to a model $\mathfrak{C}$. $\mathfrak{B}$ also satisfies the positive existential sentence written according to model $\mathfrak{C}$. Hence $\mathfrak{B}$ is a homomorphism image of $\mathfrak{C}$. $\mathfrak{C}$ is a model of T. Therefore $\mathfrak{B}$ is a model of the theory T because T is preserved under into homomorphisms.

Theorem 2.4. A Theory T is preserved under inverse into homomorphisms if and only if T is equivalent to a set of negative universal sentences.

Proof. Similar to the proof of Theorem 2.2.

For the case of onto homomorphisms our result is weaker. We will use countable disjunction of countable conjunctions of first order positive sentences to be equivalent to the original theory. We first give a lemma.

Lemma 2.1. A theory T in the language $L = < A, c_1, \ldots, c_p, R_1, \ldots, R_q >$ is preserved under onto homomorphisms. Let Δ be the set of all positive sentences in L. If $\mathfrak{A}$ is a model of T and every sentence $\delta \in \Delta$ holds in $\mathfrak{A}$ holds in model $\mathfrak{B}$, then $\mathfrak{B}$ is a model of T.

Proof. Suppose that model $\mathfrak{A}$ has n elements $\{a_1, \ldots, a_n\}$. We define a conjunctive formula

$$\chi_{y_1 = a_{i_1}, \ldots, y_n = a_{i_m}} (y_1, \ldots, y_m)$$

with free variables $(y_1, \ldots, y_m)$ where y_1 represents $a_{i_1}, \ldots, y_m$ represents a_{i_m}. The formula χ is a conjunction of all positive relations $R(z_1, \ldots, z_k)$ where k-tuple $(z_1, \ldots, z_k)$ runs over all possible symbols in

$$Z = \{y_1, \ldots, y_m\} \bigcup \{a_1, \ldots, a_n\} \bigcup \{c_1, \ldots, c_p\}.$$

We give the following definition:

$$\chi_{y_1 = a_{i_1}, \ldots, y_n = a_{i_m}} (y_1, \ldots, y_m)$$

$$= \bigwedge_{\substack{R(z_1, \ldots, z_k) \in L \\ \{z_1, \ldots, z_k\} \subset Z}} \Big(\bigwedge_{\mathfrak{A} \models R(z_1, \ldots, z_k)} R(z_1, \ldots, z_k)\Big).$$

Using formula χ we can define the positive conjunctive formula of the model $\mathfrak{A}$.

$$\chi_{\mathfrak{A}} (x_1, \ldots, x_n) = \chi_{x_1 = a_1, \ldots, x_n = a_n} (x_1, \ldots, x_n).$$

Formula $\chi_{\mathfrak{A}}(x_1, \ldots, x_n)$ is true if and only if the n-tuple $(x_1, \ldots, x_n)$ has at least the same positive relations as $(a_1, \ldots, a_n)$ in model $\mathfrak{A}$.

Let $\mathfrak{B} =< B, L >$ where the universe is $B = \{b_1, \ldots, b_m\}$ with $m \le n$. Choose a set of new variables $Z = \{z_1, \ldots, z_l\}$ where $l = m - n$. Let Ω be the set of all functions from Z to the universe of $\mathfrak{A}$.

$$\Omega = \{\omega | \omega \ is \ a \ function \ of \ Z \ to \ A\}.$$

We define a formula ψ_ω with free variables $\{z_1, \ldots, z_l\}$ saying that the l-tuple $(z_1, \ldots, z_l)$ has at least the same positive relations as $(\omega z_1, \ldots, \omega z_l)$ with respect to all elements of $\mathfrak{A}$.

$$\psi_\Omega \ (z_1, \ldots, z_l) \ = \ \chi_{z_1 = \omega z_1, \ldots, z_l = \omega z_l.} \ (x_1, \ldots, x_n)$$

Putting together $\psi_\Omega(z_1, \ldots, z_l)$ for $\omega \in \Omega$ we obtain a formula:

$$\Psi = \bigvee_{\omega \in \Omega} \psi_\Omega(z_1, \ldots, z_l).$$

Formula Ψ is true if and only if $(z_1, \ldots, z_l)$ satisfies at least the same positive relations as one of the l-tuples of elements $(\omega z_1, \ldots, \omega z_l)$ with respect to all elements of $\mathfrak{A}$.

Now we define our main formula φ.

$$\varphi = \exists x_1, \ldots, x_n \chi_{\mathfrak{A}}(x_1, \ldots, x_n). \wedge .\forall y_1, \ldots, y_l \Psi(y_1, \ldots, y_l).$$

φ is a positive sentence which holds in $\mathfrak{A}$. From the condition of the theorem $\mathfrak{B} \models \varphi$. By the above explanations and the structure of sentence φ we know that there are elements $b_1, \ldots, b_n \in \mathfrak{B}$ satisfying at least the same positive relations as $a_1, \ldots, a_n \in \mathfrak{B}$. Now let $d_1, \ldots, d_l \in \mathfrak{B}$ be the elements in $\mathfrak{B}$ different from $b_1, \ldots, b_n$. By the formula Ψ we know that there is a function $\omega \in \Omega$ such that $d_1, \ldots, d_l$ satisfy at least the same positive relations as the l-tuple $(\omega z_1, \ldots, \omega z_l)$ with respect to all a's.

Divide the elements of $\mathfrak{B}$ into subsets $B_1, \ldots, B_n$ in which $b_i \in B_i$ and every element in B_i satisfies at least the same positive relations as a_i with respect to all a's. Construct another model $\mathfrak{D}$ such that for every element $a_i \in \mathfrak{A}$, $\mathfrak{D}$ has a subset D_i with $|B_i|$ different copies of elements having the same (positive and negative) relations as the element $a_i \in \mathfrak{A}$. Although it is possible $b_i = b_j$ for $i \neq j$ but D_i and D_j are disjoint. From Theorem 1.2, $\mathfrak{D}$ is an elementary extension of $\mathfrak{A}$. Hence $\mathfrak{D}$ is a model of T. Define a function F so that it maps all the elements of D_i to element b_i in B. We can check that F is a homomorphism from $\mathfrak{D}$ onto $\mathfrak{B}$. Because the theory T is preserved under onto homomorphisms therefore $\mathfrak{B}$ is a model of T.

Theorem 2.5. A Theory T is preserved under onto homomorphisms if and only if T is equivalent to a countable disjunction of countable conjunction of positive existential sentences.

Proof. For every T model $\mathfrak{A}$ with n elements we can write a countable conjunction of positive sentence

$$\Phi = \bigwedge_{0 \leq l \leq \infty} \varphi_l$$

as in the Proof of Lemma 2.1 such that Φ is true in a model $\mathfrak{B}$ if and only if $\mathfrak{B}$ is an image of an onto homomorphism of an elementary extension of $\mathfrak{A}$. Let $\sum$ be the disjunction of the above formula Φ over all models of T. Every model of T satisfies at least a disjunct of $\sum$ which is written according to itself. Conversely, every model $\mathfrak{B}$ of the sentence $\sum$ satisfies a disjunct which is written according to a model $\mathfrak{A}$. $\mathfrak{B}$ is an image of an onto homomorphism of an elementary extension of $\mathfrak{A}$. Therefore $\mathfrak{B}$ is a model of T.

Theorem 2.6. A Theory T is preserved under inverse onto homomorphisms if and only if T is equivalent to a set of countable disjunction of negative universal sentences.

Proof. (Similar to the proof Theorem 2.2.)

In Section 3 an example is given which shows that a sentence φ is preserved under extensions of models but it can not be equivalent to an existential sentence. Of course from our Theorem 2.1 the sentence φ is equivalent to a countable disjunction of existential sentences.

SECTION 3. EXAMPLES

We give an example to show how we work differently in the finite model theory and the general model theory.

Example 3.1. The language L has two relations $x < y$ and $x \, S \, y$ expressing the relations 'x is smaller than y' and 'x is succeeded by y' respectively and two constants m and M representing the minimum and maximum elements of the linear order.

The symbol $'x \sim y'$ is defined as the simplification of the following:

$x \sim y: \ \forall z((x < z \to y < z). \wedge .(z < x \to z < y))$.

Then we define the following sentences:

(1) $\forall xyz(x < y \wedge y < z :\to x < z)$,

(2) $\forall x \neg(x < x)$,

(3) $\forall xy(x < y \to \neg(y < x))$,

(4) $\forall x(\neg(x < m))$,

(5) $\forall x(\neg(M < x))$,

(6) $\forall xy(x < y \vee y < x \vee x \sim y)$,

(7) $\forall xy(xSy \to x < y)$,

(8) $\forall xyz(xSy \wedge xSz :\to .y \sim z)$,

(9) $\forall xyz(xSy \wedge x < z :\to: y \sim z \vee y < z)$,

(10) $\forall x(x \sim M \vee \exists y(xSy))$.

We need strong axioms here because there is not an equal sign in our language. It is not necessary to discuss the independence of the axioms. The sentences (1) - (6) give a linear order with two end elements. (7) - (10) give a successor relation which is consistent with the linear order. The sentence φ is now given as the conjunction of the above all 10 sentences.

$$\varphi = \bigwedge_{i=1}^{10} (i).$$

The theory T consists of only one sentence $\{\varphi\}$. It is not difficult to see that the theory T is model complete but not complete.

If we change our sentence to the following:

$$\psi = \bigwedge_{i=1}^{9} (i) \rightarrow (10),$$

then the sentence ψ is preserved by extensions of models but it can not be equivalent to a first order existential sentence. This was the original example given by Gurevich in [4].

REFERENCES

[1] Chang, C. C., and Keisler, H. J., Model theory, North-Holland, Amsterdam, 3rd ed., 1989.

[2] Gurevich, Y., Toward logic tailored for computational complexity, Computation and proof theory (Ed. M. Richter et al.), Springer Lecture Notes in Math., v.1104, 1984, pp. 175 - 216.

[3] Gurevich, Y., Logic and the challenge of computer science, Current Trends in Theoretical Computer Science (Ed. E. Börger), Computer Science Press, 1988, pp. 1 - 57.

[4] Gurevich, Y., On finite model theory, Feasible Mathematics, (Ed. S. R. Buss et al.), A mathematical sciences institute workshop, Ithaca, New York, 1989, pp. 211 - 219.

[5] Lyndon, R. C., An interpolation theorem in the predicate calculus, Pacific J. math., v. 9, 1959, 155 - 164.

A Note on Weak Segments of PFA

Tadatoshi MIYAMOTO

Abstract. We consider a family of large cardinals below supercompact and weak segments of PFA. This generalizes Σ_1-reflecting cardinals and BPFA in [Go].

Introduction

In [Go], the Σ_1-reflecting cardinals and the Bounded Proper Forcing Axiom (BPFA) are devised and the equiconsistency between them is established. Since Σ_1-reflecting cardinals are weaker than Mahlo cardinals, the corresponding forcing axiom BFPA sits near the "bottom" of the Proper Forcing Axiom (PFA) whose consistency is gotten by a supercompact cardinal.

In this note, we first define the localized reflecting cardinals which generalize the Σ_1-reflecting cardinals. We then mention a couple of relations between these cardinals and supercompact cardinals and provide a list of equivalent formulations for some cases of these cardinals involving elementary embeddings. It turns out that we can force Laver type diamond sequences to some cases of the localized reflecting cardinals.

We move on to consider the forcing axiom $\Sigma(\lambda)$ for the infinite cardinals λ. The $\Sigma(\lambda)$ are weak segments of the PFA. The BPFA is our $\Sigma(\omega_1)$.

The large cardinals we introduce are sufficient for obtaining levelwise consistency of some of these weak segments of PFA. We actually have an equiconsistency result for $\Sigma(\omega_2)$ along the line of [Go]. We also make use of known results to get very rough estimate on the consistency strength at some of higher levels.

This note is a shortened version of [Mi].

Preliminaries

For a set $P \neq \emptyset$ and a binary relation $\leq$ on P, the pair $(P, \leq)$ is said a *preorder* iff (1) $\forall p \in P\ p \leq p$ (reflexive) ; (2) $\forall p, q, r \in P$ if $q \leq p$ and $r \leq q$, then $r \leq p$ (transitive). For a preorder $(P, \leq)$, $p \in P$, and $A \subseteq P$, A is *predense below* p (in P) iff $\forall x \in P\ x \leq p \exists a \in A$ such that x and a are compatible in P, i.e., $\exists y \in P$ s.t. $y \leq x, a$. If A is predense below every $p \in P$, then A is simply said *predense* (in P). A is a *centered* subset of P iff for any nonempty finite subset S of A, there is $x \in P$ (may or may not $x \in A$) such that for all $y \in S$, $x \leq y$.

For a set x, $|x|$ denotes its cardinality and $\mathrm{TC}(x)$ denotes the transitive closure of x. For an infinite cardinal λ, H_λ denotes the set $\{x : |\mathrm{TC}(x)| < \lambda\}$. We consider $\in$-structures $(H_\lambda, \in)$ and their elementary substructures

$(Y, \in)$ (written as $Y \prec H_\lambda$). A *formula* φ is a formula in the language of set theory. So the only non-logical symbol is the binary relation symbol $\in$. We may assume φ is in H_ω as an object.

An *Easton support* iteration $\langle P_\alpha \mid \alpha \leq \kappa \rangle$ is constructed by taking the direct limit at every strongly inaccessible cardinal $\alpha \leq \kappa$ and inverse limit at any other limit $\alpha \leq \kappa$.

For other notions not found here, see [Ku] and [Je].

§1. Localized Reflecting Cardinals

We consider a hierarchy of large cardinals which generalize the Σ_1-reflecting cardinals in [Go].

1.1 Definition. Let λ be a cardinal and κ be a regular cardinal with $\lambda \geq \kappa \geq \omega_1$. We say κ is H_λ-*reflecting* iff for any $a \in H_\lambda$ and any formula φ, if $H_\chi \models$ "$\varphi(a)$" holds for some cardinal χ, then $\{X \in [H_\lambda]^{<\kappa} \mid (1)\ a \in X \prec H_\lambda$; (2) there is a cardinal $\chi' < \kappa$ s.t. $H_{\chi'} \models$ "$\varphi(\bar{a}^{(X)})$"$\}$ is stationary in $[H_\lambda]^{<\kappa}$, where $\bar{a}^{(X)}$ denotes the image of a under the transitive collapse of X. We say a cardinal κ is a *localized reflecting* cardinal iff κ is H_λ-reflecting for some λ.

We first observe the following relation between this type of large cardinals, Σ_1-reflecting cardinals of [Go], and supercompact cardinals.

1.2 Proposition. *(1) For an uncountable regular cardinal κ, κ is Σ_1-reflecting iff κ is H_κ-reflecting.*

And so if this is the case, then κ is the κ-th strongly inaccessible cardinal.

(2) For cardinals λ_1 and λ_2 with $\lambda_2 \geq \lambda_1 \geq \kappa$, if κ is H_{λ_2}-reflecting, then κ is H_{λ_1}-reflecting.

(3) κ is supercompact iff for all $\lambda \geq \kappa$, κ is H_λ-reflecting.

(4) Let μ be a strongly inaccessible cardinal with $\mu > \lambda \geq \kappa$. If κ is H_λ-reflecting, then $V_\mu \models$ "κ is H_λ-reflecting".

And so,

(5) If λ is a strongly inaccessible cardinal and if κ is H_λ-reflecting with $\lambda > \kappa$, then $V_\lambda \models$ "κ is supercompact".

□

1.3 Note. Therefore, if we are interested in an H_λ-reflecting cardinal κ whose consistency is weaker than supercompact, then we may restrict λ to $\lambda < \kappa^*$, where κ^* denotes the least strongly inaccessible cardinal above κ, if any. In particular, going into V_{κ^*}, if necessary, we may assume that

κ is H_λ-reflecting and κ is the greatest strongly inaccessible cardinal in the universe.

We next give a list of equivalent formulations for some of the localized reflecting cadinals with associated diamond sequences.

1.4 Theorem. *For any cardinal λ and any strongly inaccessible cardinal κ with $\lambda^{<\kappa} = \lambda$, the following are equivalent :*

(1) κ is H_{λ^+}-reflecting with an associated diamond sequence $h : \kappa \longrightarrow H_\kappa$. Namely, we mean that

- *For any $a \in H_{\lambda^+}$ and any formula φ, if $H_\chi \models$ "$\varphi(a)$" holds for some cardinal χ, then $\{X \in [H_{\lambda^+}]^{<\kappa} \mid$ (1) $a \in X \prec H_{\lambda^+}$; (2) $X \cap \kappa = \alpha$ for some $\alpha < \kappa$; (3) there is a cardinal $\chi' < \kappa$ with $H_{\chi'} \models$ "$\varphi\big(h(\alpha)\big)$" ; (4) $h(\alpha) = \bar{a}^{(X)}\}$ is stationary in $[H_{\lambda^+}]^{<\kappa}$.*

(2) There is $g : \kappa \longrightarrow [\kappa]^{<\kappa}$ such that

- *For any $A \subseteq \lambda$ and any formula φ, if $H_\chi \models$ "$\varphi(A)$" holds for some cardinal χ, then $\{X \in [\lambda]^{<\kappa} \mid$ (1) $X \cap \kappa = \alpha$ for some $\alpha < \kappa$; (2) there is a cardinal $\chi' < \kappa$ s.t. $H_{\chi'} \models$ "$\varphi\big(g(\alpha)\big)$" ; (3) $g(\alpha) = \pi_X$"$(A \cap X)\}$ is stationary in $[\lambda]^{<\kappa}$, where π_X is the transitive collapse of X onto its order type.*

(3) There is $f : \kappa \longrightarrow H_\kappa$ such that

For any cardinal ρ, any $a \in H_{\lambda^+}$, and any N s.t.

- *N is a transitive set of size λ and $\kappa, f \in N$.*
- *$\forall x, y \in N$ $\{x, y\} \in N$ and $\forall \alpha < \kappa \forall s \in^\alpha N$ s"$\alpha \in N$.*

there is (j, M) such that

- *M is a transitive set with $j, N, H_\rho \in M$ and ${}^{\leq\rho}M \subseteq M$.*
- *$j : N \longrightarrow M$ is an elementary embedding with crit$(j) = \kappa$ and $\rho < j(\kappa)$.*
- *$j(f)(\kappa) = a$.*

Proof. We just give a proof for (1) implies (3). The rests are left to the readers or can be found in [Mi].

For each $\beta < \kappa$, we define $f(\beta) = x$, if $h(\beta) = (x_1, x_2, x_3, x_4, x)$ for some x_1, x_2, x_3, and x_4. Otherwise, we just set $f(\beta) = \emptyset$.

We show this $f : \kappa \longrightarrow H_\kappa$ works. Notice that $f \in H_{\kappa^+} \subseteq H_{\lambda^+}$. Fix (ρ, a, N) as in the hypothesis of (3). We prove by contradiction. So suppose to the contrary that there were no (j, M) as such. Let us abbreviate this by writting $\neg\exists(j, M)$ $\psi(j, M, \rho, N, \kappa, f, a)$. We would like to derive a contradiction.

To this end, we choose a regular cardinal θ so that for all $\mu < \theta$, $\mu^{\rho\cdot\lambda} < \theta$. Then H_θ satisfies the following :

- There is a cardinal ρ s.t.
  - For all cardinals μ, $\mu^{\rho\cdot\lambda}$ exist.
  - $\kappa \leq |N| = \lambda$, H_{λ^+} exists, and $a \in H_{\lambda^+}$.
  - $\neg\exists(j, M)\ \psi(j, M, \rho, N, \kappa, f, a)$.

  So by (1), we may choose (X, θ', ρ') so that
- $N, \kappa, \lambda, f, a \in X \prec H_{\lambda^+}$, $|X| < \kappa$, and $X \cap \kappa = \bar{\kappa}^{(X)} = \alpha$ for some $\alpha < \kappa$.
- $\rho' < \theta' < \kappa$, ρ' is a cardinal, θ' is a regular cardinal, and for all $\mu < \theta'$, $\mu^{\rho'\cdot\bar{\lambda}^{(X)}} < \theta'$ hold.
- $\alpha \leq |\bar{N}^{(X)}| = \bar{\lambda}^{(X)} < \theta'$ and $\bar{a}^{(X)} \in H_{\bar{\lambda}^{(X)+}}$.
- $H_{\theta'} \models$ "$\neg\exists(j, M)\ \psi(j, M, \rho', \bar{N}^{(X)}, \alpha, \bar{f}^{(X)}, \bar{a}^{(X)})$".
- $h(\alpha) = (\bar{N}^{(X)}, \alpha, \bar{\lambda}^{(X)}, \bar{f}^{(X)}, \bar{a}^{(X)})$.

  And so,
- $f(\alpha) = \bar{a}^{(X)}$.

Claim. *We may construct (Y, M, j) so that*

(1) $(N \cap X) \cup \bar{\lambda}^{(X)} \cup H_{\rho'} \cup H_{\bar{\lambda}^{(X)+}} \cup \bar{N}^{(X)} \cup \{H_{\rho'}, \bar{N}^{(X)}, \rho', f\} \subset Y \prec N$, ${}^{\leq\rho'\cdot\bar{\lambda}^{(X)}}Y \subseteq Y$, *and* $|Y| < \theta'$.

(2) Let $^{-(Y)} : Y \longrightarrow M$ *denote the transitive collapse. Then* ${}^{\leq\rho'\cdot\bar{\lambda}^{(X)}}M \subseteq M$, $\forall x, y \in M\ \{x, y\} \in M$, $\forall\beta \leq \rho'\cdot\bar{\lambda}^{(X)} \forall s \in {}^\beta M\ s``\beta \in M$, *and* $\overline{f(\alpha)}^{(Y)} = \bar{a}^{(X)}$ *holds.*

(3) $j = \{(\bar{x}^{(X)}, \bar{x}^{(Y)}) \mid x \in N \cap X\}$ *is an elementary embedding from* $\bar{N}^{(X)}$ *into M and* $j(\bar{f}^{(X)})(\alpha) = \bar{a}^{(X)} \in M$ *holds.*

(4) $\psi(j, M, \rho', \bar{N}^{(X)}, \alpha, \bar{f}^{(X)}, \bar{a}^{(X)})$.

But since $j, M \in H_{\theta'}$, so $H_{\theta'} \models$ "$\psi(j, M, \rho', \bar{N}^{(X)}, \alpha, \bar{f}^{(X)}, \bar{a}^{(X)})$". This is a contradiction.

Proof of claim. For (1) : We first remark, since for all $x, y \in N$, $\{x, y\} \in N$ and since for all $\beta < \kappa$ and all $s \in {}^\beta N$, $s``\beta \in N$, we have ${}^{<\kappa}N \subseteq N$ and $H_\kappa \subset N$. We also necessarily have that $\rho', \bar{\lambda}^{(X)} < \theta' < \kappa$ and that $H_{\rho'}, H_{\bar{\lambda}^{(X)+}}, \bar{N}^{(X)} \in H_{\theta'} \in H_\kappa \subset N$. So $(N \cap X) \cup \bar{\lambda}^{(X)} \cup H_{\rho'} \cup H_{\bar{\lambda}^{(X)+}} \cup \bar{N}^{(X)} \cup \{H_{\rho'}, \bar{N}^{(X)}, \rho', f\} \subset N$ holds.

But for any $B \subseteq N$ with $|B| < \theta'$, since θ' is a regular cardinal s.t. for all $\mu < \theta'$, $\mu^{\rho' \cdot \bar{\lambda}^{(X)}} < \theta'$, we have Y s.t. $B \subset Y \prec N$, ${}^{\leq \rho' \cdot \bar{\lambda}^{(X)}} Y \subseteq Y$, and $|Y| < \theta'$.

For (2) : It is clear that $^{-(Y)}$ and M are well-defined. Since $\rho' \cdot \bar{\lambda}^{(X)} + 1 \subset Y$ and since ${}^{\leq \rho' \cdot \bar{\lambda}^{(X)}} Y \subseteq Y$, so ${}^{\leq \rho' \cdot \bar{\lambda}^{(X)}} M \subseteq M$ holds. Since $N \models$ "$\forall x, y \exists z\ z = \{x, y\}$" and since $Y \prec N$, so $M \models$ "$\forall x, y \exists z\ z = \{x, y\}$" holds. Since $N \models$ "$\forall \beta < \kappa \forall s \in^{\beta} N \exists z\ z = s``\beta$" and since $\kappa \in N \cap X \subset Y \prec N$, so $M \models$ "$\forall \beta < \bar{\kappa}^{(Y)} \forall s \in^{\beta} M \exists z\ z = s``\beta$". But since $\rho' \cdot \bar{\lambda}^{(X)} + 1 \leq \bar{\kappa}^{(Y)}$, so for any $\beta \leq \rho' \cdot \bar{\lambda}^{(X)}$ and any $s \in^{\beta} M$, since $s \in M$, we have $s``\beta \in M$. Since $f(\alpha) = \bar{a}^{(X)} \in H_{\bar{\lambda}^{(X)+}} \subset Y$, so $\overline{f(\alpha)}^{(Y)} = f(\alpha) = \bar{a}^{(X)}$.

For (3) : Since $N \in X \prec H_{\lambda^+}$, so $N \cap X \prec N$. But $N \cap X \subset Y \prec N$ and so $N \cap X \prec Y$. Also $\bar{N}^{(X)}$ and $N \cap X$ are isomorphic via $^{-(X)} \lceil (N \cap X)$. Therefore j is well-defined and is an elementary embedding from $\bar{N}^{(X)}$ into M. Since $X \models$ "$N \models$ "f is a function with domain κ"", so $\bar{N}^{(X)} \models$ "$\bar{f}^{(X)}$ is a function with domain α" and so $M \models$ "$j(\bar{f}^{(X)})$ is a function with domain $j(\alpha)$". But $\alpha \leq \bar{\lambda}^{(X)} < \bar{\kappa}^{(Y)} = j(\bar{\kappa}^{(X)}) = j(\alpha)$.

Thus $j(\bar{f}^{(X)})(\alpha)$ makes sense and $j(\bar{f}^{(X)})(\alpha) = \bar{f}^{(Y)}(\alpha) = \bar{f}^{(Y)}(\bar{a}^{(Y)}) = \overline{f(\alpha)}^{(Y)} = \bar{a}^{(X)} \in H_{\bar{\lambda}^{(X)+}} \subset M$.

For (4) : It remains to observe that $j \in M$, $\mathrm{crit}(j) = \alpha$, and that $\rho' < j(\alpha)$. Let $e : \bar{\lambda}^{(X)} \longrightarrow N \cap X$ be a bijection. Then for each $i < \bar{\lambda}^{(X)}$, $\overline{e(i)}^{(X)} \in \bar{N}^{(X)} \in M$ and $\overline{e(i)}^{(Y)} \in M$, so $(\overline{e(i)}^{(X)}, \overline{e(i)}^{(Y)}) \in M$. Let $s = \{(i, (\overline{e(i)}^{(X)}, \overline{e(i)}^{(Y)})) \mid i \in \bar{\lambda}^{(X)}\}$. Then $s \in^{\bar{\lambda}^{(X)}} M$ and so we have seen that $j = s``\bar{\lambda}^{(X)} \in M$.

Since $X \cap \kappa = \alpha \leq \bar{\lambda}^{(X)} < \bar{\lambda}^{(X)} + 1 \subset Y$, so for every $\beta < \alpha$, $j(\beta) = j(\bar{\beta}^{(X)}) = \bar{\beta}^{(Y)} = \beta$. Also $j(\alpha) = j(\bar{\kappa}^{(X)}) = \bar{\kappa}^{(Y)} \geq \rho' \cdot \bar{\lambda}^{(X)} + 1 > \rho', \alpha$.

□

□

We may force associated diamond sequences to some of localized reflecting cardinals in the following situation.

1.5 Theorem. *Let κ be an H_{λ^+}-reflecting cardinal for some cardianl λ with $\lambda^{<\kappa} = \lambda$. If there are no strongly inaccessible cardinals above κ, then there is an Easton support iteration $P_\kappa \in H_{\kappa^+}$ such that*

(1) P_κ has the κ-c.c..

(2) In V^{P_κ}, $\lambda^{<\kappa} = \lambda$ remains true and κ is an H_{λ^+}-reflecting cardinal with an associated diamond sequence.

Proof. Let us consider $Q = \{\text{IN, OUT, STOP}, 1_Q\}$ partially ordered by $\leq_Q = \{(x, x) \mid x \in Q\} \cup \{(\text{IN}, 1_Q), (\text{OUT}, 1_Q), (\text{STOP}, 1_Q)\}$. So Q has 4 elements and 1_Q is the greatest element. {IN, OUT, STOP} is a maximal antichain in Q. We may assume that IN, OUT, STOP, 1_Q are absolute objects such as specific integers.

We construct an Easton support iteration $\langle P_\alpha \mid \alpha \leq \kappa \rangle$ with $\langle \dot{Q}_\alpha \mid \alpha < \kappa \rangle$ such that for all $\alpha < \kappa$, $\Vdash_\alpha$"$\dot{Q}_\alpha = Q$". So we just keep forcing with the Q under the Easton support. We may assume that $P_\kappa \in H_{\kappa^+}$. Since κ is H_{λ^+}-reflecting, so κ is at least Mahlo and so P_κ has the κ-c.c.. It is also standard to check that κ remains strongly inaccessible and $\lambda^{<\kappa} = \lambda$ remains true in the generic extensions via P_κ. Now by 1.4 theorem, it suffices to find $g : \kappa \longrightarrow [\kappa]^{<\kappa}$ in V^{P_κ} such that the following hold in V^{P_κ} :

- For any $A \subseteq \lambda$ and any formula φ, if $H_\chi \models$ "$\varphi(A)$" holds for some cardinal χ, then $\{X \in [\lambda]^{<\kappa} \mid (1)\ X \cap \kappa = \alpha$ for some $\alpha < \kappa$; (2) there is a cardinal $\chi' < \kappa$ with $H_{\chi'} \models$ "$\varphi(g(\alpha))$" ; (3) $g(\alpha) = \pi_X$"$(A \cap X)\}$ is stationary in $[\lambda]^{<\kappa}$, where π_X denotes the transitive collapse of X onto its order type.

Let G_κ be an arbitray P_κ-generic filter over V. For each $\alpha < \kappa$, let $G_\alpha = \{p\lceil\alpha \mid p \in G_\kappa\}$. So G_α is a P_α-generic filter over V. We also let $E_\alpha = \{p(\alpha)_{G_\alpha} \mid p \in G_\kappa\}$. Then E_α is a Q-generic filter over $V[G_\alpha]$. Notice Q is so finite that E_α is actually one of either {In, 1_Q}, {OUT, 1_Q}, or {STOP, 1_Q}.

Now we define $\langle g(\alpha) \mid \alpha < \kappa \rangle$ together with $\langle\ l_\alpha \mid \alpha < \kappa \rangle$ as follows :

- l_α is the least $l < \kappa$ s.t. for all $i < l$, STOP$\notin E_{\alpha+i}$ but STOP$\in E_{\alpha+l}$.

Namely, $\alpha + l_\alpha$ is the coordinate $\geq \alpha$ where STOP appears first. It is clear that l_α is well-defined by genericity. Then we define

- $g(\alpha) = \{i < l_\alpha \mid \text{IN} \in E_{\alpha+i}\}$.

So $g(\alpha)$ corresponds to the set of coordinates in the interval $[\alpha, \alpha + l_\alpha)$ where IN gets appear. It can be shown that this $g : \kappa \longrightarrow [\kappa]^{<\kappa}$ works in $V[G_\kappa]$. Details are left to the readers or can be found in [Mi].

□

§2. Weak Segments of the Proper Forcing Axiom

2.1 Definition. For any infinite cardinal λ, let $\Sigma(\lambda)$ denote the following :

For any preorder P with $|P| = \lambda$ and any sequence $\langle B_i \mid i < \omega_1 \rangle$ of subsets of P, if there is a proper preorder Q such that $\Vdash_Q$"there is a filter G in P s.t. for all $i < \omega_1$, $G \cap B_i \neq \emptyset$", then there is a filter G in P s.t. for all $i < \omega_1$, $G \cap B_i \neq \emptyset$.

Notice that in this definition the B_i's are just subsets of P and Q may have any size at all. The following is not hard to observe :

2.2 Proposition. *(1) $\Sigma(\omega)$ is equivalent to the PFA restricted to the Cohen p.o. set $Fn(\omega, 2)$.*

(2) $\Sigma(\omega_1)$ is equivalent to the BPFA of [Go].

(3) Con(ZFC) implies Con($\Sigma(\omega) + \neg\Sigma(\omega_1)$).

(4) For infinite cardinals λ_1 and λ_2 with $\lambda_2 \geq \lambda_1$, $\Sigma(\lambda_2)$ implies $\Sigma(\lambda_1)$.

(5) PFA iff for all infinite cardinals λ, $\Sigma(\lambda)$.

(6) In particular, $\Sigma(\lambda)$ implies the PFA restricted to the proper preorders P with $|P| \leq \lambda$.

□

2.3 Proposition. *For any cardinal $\lambda \geq \omega_1$, the following hold :*

(1) The PFA restricted to the proper preorders P with $|P| \leq \lambda^\omega \cdot \lambda^+$ implies $\neg\Box_\lambda$.

(2) And so, $\Sigma(\lambda^\omega \cdot \lambda^+)$ implies $\neg\Box_\lambda$.

So for example, we have

- $\Sigma(\omega_2) + 2^\omega = \omega_2$ implies $\neg\Box_{\omega_1}$.
- $\Sigma(\omega_3) + 2^\omega = \omega_2$ implies $\neg\Box_{\omega_2}$.
- $\Sigma(\omega_4) + 2^\omega = \omega_2$ implies $\neg\Box_{\omega_3}$.

Proof. By pages 198,199 in [Sc], a p.o. set R concerned is an iteration of :

(1) Forcing a club subset $\dot{C}$ of λ^+ of order type ω_1 by countable conditions.

(2) Specializing a relevant tree on the $\dot{C}$ by finite conditions.

So by considering an appropriate dense subset of R, we may assume $|R| \leq (\lambda^+)^\omega \cdot (\omega_1 \times \omega)^{<\omega}$.

□

2.4 Proposition. *For any cardinal $\lambda \geq \omega_1$, if $\Sigma(\lambda)$ holds, then there is an ω_1-closed and λ^+-dense p.o. set R such that $\Vdash_R$ "$\Sigma(\lambda) + \Box_\lambda + \lambda^\omega \leq \lambda^+$ and so $\neg\Sigma(\lambda^+)$".*

So for example, we have

- Con($\Sigma(\omega_1)$) implies Con($\Sigma(\omega_1) + \neg\Sigma(\omega_2)$).
- Con($\Sigma(\omega_2)$) implies Con($\Sigma(\omega_2) + \neg\Sigma(\omega_3)$).
- Con($\Sigma(\omega_3)$) implies Con($\Sigma(\omega_3) + \neg\Sigma(\omega_4)$).

Proof. Suppose $\Sigma(\lambda)$ holds. We force a $\Box_\lambda$-sequence via an ω_1-closed and λ^+-dense p.o. set ([Bu]). Since $\lambda \geq \omega_1$, it is easy to see $\Sigma(\lambda)$ remains in the generic extensions. By genericity, $[\lambda]^\omega$ can be coded in the $\square_\lambda$-sequence. So we have $\lambda^\omega \leq \lambda^+$. We are done.

□

We give a list of equivalent formulations for the axioms.

2.5 Theorem. *For any cardinal $\lambda \geq \omega_1$, the following are equivalent :*

(1) For any proper preorder P and any sequence $\langle A_i \mid i < \omega_1 \rangle$ of predense subsets of P s.t. for all $i < \omega_1$ $|A_i| \leq \lambda$, there is a centered subset C of P s.t. for all $i < \omega_1$, $C \cap A_i \neq \emptyset$.

(2) For any $a \in H_{\lambda^+}$ and any Σ_0-formula φ, if $\Vdash_P$ "$\exists y \varphi(y, a)$ holds in some transitive $\in$-structure" for some proper preorder P, then $\{N \in [H_{\lambda^+}]^{\omega_1} \mid (1)\ a \in N \prec H_{\lambda^+}$; $(2)\ \exists y \varphi(y, \bar{a}^{(N)})$ holds in some transitive $\in$-structure of size $\omega_1\}$ is stationary in $[H_{\lambda^+}]^{\omega_1}$, where $^{-(N)}$ denotes the transitive collapse of N.

(3) $\Sigma(\lambda)$.

Proof. (1) implies (2) : Fix $a \in H_{\lambda^+}$, a proper preorder P, and a Σ_0-formula φ s.t. $\Vdash_P$"$\exists y \varphi(y, a)$ holds in some transitive $\in$-structure $\dot{M}$". We may assume that $\Vdash_P$"$|\dot{M}| = |\lambda| = \omega_1$" by further forcing with a countably closed forcing. We may also assume that P is a complete Boolean algebra (sans its zero).

The following is Lemma 4.2 and 4.7 in [Go] with minor changes.

Claim 1. *If $\Vdash_P$ "$\dot{E}_1$ is a well-founded binary relation on ω_1", then there are at most ω_1 many predense subsets of size at most ω_1 s.t. if $G \subseteq P$ is a filter which meets all of these predense subsets and if we define $E_1 = \{(\alpha, \beta) \in \omega_1 : \exists p \in G\ p \Vdash_P$ "$\alpha \dot{E}_1 \beta$"$\}$, then*

(1) (ω_1, E_1) is well-founded.

(2) For any formula ψ and any $\vec{\alpha} \in^{<\omega} \omega_1$, we have

- *$(\omega_1, E_1) \models$ "$\psi(\vec{\alpha})$" iff $\exists p \in G\ p \Vdash_P$ "$(\omega_1, \dot{E}_1) \models$ "$\psi(\vec{\alpha})$" ".*

□

We recall that every club $C \subseteq [\lambda]^{\omega_1}$ has a function $h : [\lambda]^{<\omega} \longrightarrow \lambda$ s.t. C is induced by h. Namely, $\{X \in [\lambda]^{\omega_1} \mid (1)\ \omega_1 \subseteq X$; $(2)\ h``[X]^{<\omega} \subseteq X\} \subseteq C$. The following is standard.

Claim 2. *Let $\Vdash_P$ "$\dot{f} : \omega_1 \longrightarrow \lambda$ onto " and let $h : [\lambda]^{<\omega} \longrightarrow \lambda$. Then there are at most ω_1 many predense subsets of P of size at most λ s.t. if G is a filter in P which meets them all and if we define $f = \{(i, \xi) \in \omega_1 \times \lambda : \|\dot{f}(i) = \xi\| \in G\}$, then*

(1) $f : \omega_1 \longrightarrow \lambda$ into.

(2) $\omega_1 \subseteq f``\omega_1$.

(3) $h``[f``\omega_1]^{<\omega} \subseteq f``\omega_1$.

(4) *Furthermore, if* $\Vdash_P$ "$\dot{f}$ *is one-to-one", then so is* f.

□

We make use of the following easy fact on the Mostowski collapses.

Claim 3. *Let E be a binary relation on a set X ($\neq \emptyset$) s.t. (X, E) is transitive collapsed to $(M, \in)$ by π. If $Y \subseteq X$ is an E-initial segment of X (i.e., for any $x \in X$ and any $y \in Y$, if xEy, then $x \in Y$), then $\pi``Y$ is a transitive set and so $(Y, E\lceil Y)$ is uniquely transitive collapsed to $(\pi``Y, \in)$ via $\pi\lceil Y$.*

□

Now we proceed in four steps.

Step 1. Since we want to keep track of a, while taking a copy $(\omega_1, \dot{E}_1)$ of $(\dot{M}, \in)$ in V^P, we first code $(\mathrm{TC}(a) \cup \{a\}, \in)$ in V as follows : We split λ into $A \subset \lambda$ and $\lambda - A$ so that $0 \in A$, $|A| = |\mathrm{TC}(a) \cup \{a\}| \leq \lambda$, and $|\lambda - A| = \lambda$. Take an isomorphic copy (A, E_0) of $(\mathrm{TC}(a) \cup \{a\}, \in)$ via some π_0 with $\pi_0(0) = a$. So for $\xi, \eta \in A$, $\xi E_0 \eta$ iff $\pi_0(\xi) \in \pi_0(\eta)$.

Step 2. In V^P, we may assume $\dot{M} \models$ "$\exists y \varphi(y, a)$" with $|\dot{M} - (\mathrm{TC}(a) \cup \{a\})| = |\lambda| = \omega_1$ by a simple manipulation. So we may take a copy $(\lambda, \dot{E})$ of $(\dot{M}, \in)$ via some $\dot{\pi}$ s.t. $\dot{\pi}\lceil A = \pi_0$. And so $\dot{E}\lceil A = E_0$. Since $\mathrm{TC}(a) \cup \{a\}$ is an $\in$-initial segment of $\dot{M}$, A is an $\dot{E}$-initial segment of λ.

Step 3. In V^P, we now take a copy $(\omega_1, \dot{E}_1)$ of $(\lambda, \dot{E})$ via some $\dot{f} : (\omega_1, \dot{E}_1) \longrightarrow (\lambda, \dot{E})$ with $\dot{f}(0) = 0$. So $(\omega_1, \dot{E}_1)$ is well-founded and $(\omega_1, \dot{E}_1) \models$ " The Axiom of Extensionality and $\exists y \varphi(y, 0)$".

Step 4. Let $C \subseteq [H_{\lambda^+}]^{\omega_1}$ be an arbitrary club. We may assume $C \subseteq \{N \in [H_{\lambda^+}]^{\omega_1} \mid \pi_0 \in N \prec H_{\lambda^+}\}$ by taking the intersection of C and the latter. Notice that for $N \in C$, since $\pi_0 \in N$, $a \in N$. Since $\{N \cap \lambda \in [\lambda]^{\omega_1} \mid N \in C\}$ contains a club in $[\lambda]^{\omega_1}$, we have $h : [\lambda]^{<\omega} \longrightarrow \lambda$ s.t. $\{X \in [\lambda]^{\omega_1} \mid (1)\ \omega_1 \subseteq X\ ;\ (2)\ h``[X]^{<\omega} \subseteq X\} \subseteq \{N \cap \lambda \in [\lambda]^{\omega_1} \mid N \in C\}$.

Since P is a complete Boolean algebra (sans its zero), every centered subset C of P is contained in a filter $G = \{p \in P : \exists c_1, \cdots, c_n \in C\ c_1 \wedge \cdots \wedge c_n \leq p\}$ in P. So by assumption (1) and by applying claims 1 and 2, we have

(a) (ω_1, E_1) is well-founded.

(b) $(\omega_1, E_1) \models$ "The Axiom of Extensionality and $\exists y \varphi(y, 0)$".

(c) $f : \omega_1 \longrightarrow N \cap \lambda$ is one-to-one and onto with $\omega_1 \subseteq N \cap \lambda$ for some $N \in C$.

Now take a copy $(N \cap \lambda, E^*)$ of (ω_1, E_1) via f and transitive collapse $(N \cap \lambda, E^*)$ to $(M, \in)$ via, say, π.

It is easy to check the following :

Claim 4. *(1) $A \cap N$ is an E^*-initial segment of $N \cap \lambda$.*

(2) $E_0 \lceil (A \cap N) = E^* \lceil (A \cap N)$.

□

Since $\pi_0 \in N$, $\pi_0 \lceil (A \cap N)$ is an isomorphism between $(A \cap N, E_0 \lceil (A \cap N))$ and $((\mathrm{TC}(a) \cup \{a\}) \cap N, \in)$. Since $((\mathrm{TC}(a) \cup \{a\}) \cap N, \in)$ is an $\in$-initial segment of $(N, \in)$, ${}^{-(N)} \lceil ((\mathrm{TC}(a) \cup \{a\}) \cap N)$ collapses $((\mathrm{TC}(a) \cup \{a\}) \cap N, \in)$ onto a transitive set. Hence the composition of functions ${}^{-(N)} \circ \pi_0 \lceil (A \cap N)$ is the unique collapse $\pi \lceil (A \cap N)$. In particular, $\pi(0) = \bar{a}^{(N)}$ and so $M \models$ "$\exists y \varphi(y, \bar{a}^{(N)})$" by (b) in step 4. Thus $\{N \in [H_{\lambda^+}]^{\omega_1} \mid (1)\ a \in N \prec H_{\lambda^+}$; (2) $\exists y \varphi(y, \bar{a}^{(N)})$ holds in some transitive $\in$-structure of size $\omega_1\}$ is stationary.

(2) implies (3) : Let P be an arbitrary preorder with $|P| = \lambda$ and $\langle B_i \mid i < \omega_1 \rangle$ be an indexed subsets of P. We may assume $P \in H_{\lambda^+}$. If $\Vdash_Q$"there is a filter G in P s.t. for all $i < \omega_1$, $G \cap B_i \neq \emptyset$" for some proper preorder Q, then by (2) with simple coding, we may fix $N \prec H_{\lambda^+}$ s.t. $|N| = \omega_1$, $\{P, \leq_P, \langle B_i \mid i < \omega_1 \rangle\} \cup \omega_1 \subseteq N$, and fix a filter G in $\bar{P}^{(N)}$ s.t. for each $i < \omega_1$ $G \cap \bar{B}_i^{(N)} \neq \emptyset$. But since $\bar{P}^{(N)}$ and $P \cap N$ are isomorphic, we have a filter G' in $P \cap N$ s.t. for each $i < \omega_1$, $G' \cap (B_i \cap N) \neq \emptyset$. So we are done by considering an upward-closure of G' in P.

(3) implies (1) : Let $j : P \longrightarrow B - \{0_B\}$ be a dense embedding, where B is the completion of P. Let $B_i = j$"A_i for each $i < \omega_1$, then B_i is a predense subset of $B - \{0_B\}$. Let θ be a sufficiently large regular cardinal and let M be an elementary substructure of H_θ s.t. $\{B, \wedge_B\} \cup \bigcup \{B_i \mid i < \omega_1\} \subseteq M$ and $|M| \leq \lambda$. Let $X = (B - \{0_B\}) \cap M$, then $|X| \leq \lambda$, and for each $i < \omega_1$ $B_i \subseteq X$.

In V^P, let $G \subseteq B - \{0_B\}$ be a generic filter and so $G \cap B_i \neq \emptyset$ for all $i < \omega_1$. Let $\bar{G} = G \cap M$, then since $\wedge_B \in M$, $\bar{G}$ is a filter in X s.t. for all $i < \omega_1$, $\bar{G} \cap B_i \neq \emptyset$.

Thus by (3), there is a filter G in X s.t. for each $i < \omega_1$, $G \cap B_i \neq \emptyset$ in V. Fix $\langle c_i \mid i < \omega_1 \rangle$ s.t. for each $i < \omega_1$, $c_i \in A_i$ and $j(c_i) \in G$. Let $C = \{c_i \mid i < \omega_1\}$, then C is a centered subset of P s.t. for each $i < \omega_1$ $C \cap A_i \neq \emptyset$, as j is a dense embedding.

□

§3. The Consistency of $\Sigma(\lambda)$

3.1 Theorem. *Let $\lambda \geq \kappa$ be cardinals such that $\lambda^{<\kappa} = \lambda$ and that κ is H_{λ^+}-reflecting with a Laver type diamond sequence f. Namely, we assume that :*

(1) $f : \kappa \longrightarrow H_\kappa$.

(2) For any $a \in H_{\lambda^+}$, any cardinal ρ, and any N s.t.

- *N is a transitive set of size λ and $f, \kappa \in N$.*
- *$\forall x, y \in N$ $\{x, y\} \in N$ and $\forall \alpha < \kappa \forall s \in^{\alpha} N$ $s``\alpha \in N$.*

there is (j, M) s.t.

- *M is a transitive set with $j, N, H_\rho \in M$ and ${}^{\leq\rho}M \subseteq M$.*
- *$j : N \longrightarrow M$ is an elementary embedding with crit$(j) = \kappa$ and $\rho < j(\kappa)$.*
- *$j(f)(\kappa) = a$.*

Then there is a countable support iteration $\langle P_\alpha \mid \alpha \leq \kappa \rangle$ of proper preordes such that P_κ has the κ-c.c. and $\Vdash_\kappa$ "$\Sigma(\lambda) + 2^\omega = \omega_2 = \kappa$".

For example by 1.3 note and 1.5 theorem, we have

- Con(H_{κ^+}-reflecting κ exists) implies Con($\Sigma(\omega_2) + 2^\omega = \omega_2$).
- Con($H_{\kappa^{++}}$-reflecting κ exists) implies Con($\Sigma(\omega_3) + 2^\omega = \omega_2$).
- Con($H_{(2^\kappa)^+}$-reflecting κ exists) implies Con($\Sigma(2^{\omega_2}) + 2^\omega = \omega_2$).

Proof. The basic idea of the proof can be found on page 102 in [Ma]. We construct a countable support iteration $\langle P_\alpha \mid \alpha \leq \kappa \rangle$ together with $\langle \dot{Q}_\alpha \mid \alpha < \kappa \rangle$ by recursion on α so that

(1) For all $\alpha < \kappa$, $P_\alpha \in H_\kappa$ and $\Vdash_\alpha$"$\dot{Q}_\alpha \in H_\kappa^{V[\dot{G}_\alpha]}$ is a proper preorder".
(2) $P_\kappa \in H_{\kappa^+}$.
(3) P_κ has the κ-c.c..

It suffices to describe what $\dot{Q}_\alpha$ names in V^{P_α}.

Case 1. In V, α is a strongly inaccessible cadinal and

$$f(\alpha) = \big((\lambda_\alpha, \dot{\leq}), \langle \dot{B}_i \mid i < \omega_1 \rangle\big)$$

for some ordinal λ_α, some P_α-name $\dot{\leq}$, and some sequence of P_α-names $\langle \dot{B}_i \mid i < \omega_1 \rangle$.

In V^{P_α}, if $(\lambda_\alpha, \dot{\leq})$ is a preorder, $\langle \dot{B}_i \mid i < \omega_1 \rangle$ is a sequence of subsets of λ_α, and there is a proper preorder $Q \in H_\kappa^{V[\dot{G}_\alpha]}$ with $\Vdash_Q^{V[\dot{G}_\alpha]}$"there is a filter G in $(\lambda_\alpha, \dot{\leq})$ s.t. for all $i < \omega_1$, $G \cap \dot{B}_i \neq \emptyset$", then $\dot{Q}_\alpha$ is one of the Q. Otherwise, set $\dot{Q}_\alpha =^{<\omega_1} \alpha$.

Case 2. Otherwise. Let $\Vdash_\alpha$"$\dot{Q}_\alpha =^{<\omega_1} \alpha$".

The second case is intended to collapse cardinals between ω_1 and κ. This completes the construction of the iteration.

We claim that this iteration works. To see this, let G_κ be an arbitrary P_κ-generic filter over V and suppose in $V[G_\kappa]$ that $(P, \leq_P)$ is a preorder with $P = \lambda$, that $\langle B_i \mid i < \omega_1 \rangle$ is a sequence of subsets of λ, and that there is a proper preorder Q (of any size) with $\Vdash_Q^{V[G_\kappa]}$ "there is a filter G in P s.t. for all $i < \omega_1$, $G \cap B_i \neq \emptyset$". We want to find G as such in $V[G_\kappa]$.

To this end, we fix a P_κ-name $\dot{\leq}_P$ of $\leq_P$ and fix a sequence of P_κ-names $\langle \dot{B}_i \mid i < \omega_1 \rangle$ for $\langle B_i \mid i < \omega_1 \rangle$. Since $P_\kappa \in H_{\kappa^+} \subseteq H_{\lambda^+}$ and $\leq_P, \langle B_i \mid i < \omega_1 \rangle \in (H_{\lambda^+})^{V[G_\kappa]} = H_{\lambda^+}[G_\kappa]$, we may assume that these two objects are in H_{λ^+}.

Claim 1. *We may fix j, N, and M in V so that*

(1) N is a transitive set of size λ, contains the relevant objects f, κ, $\langle P_\alpha \mid \alpha \leq \kappa \rangle$, $\langle \dot{Q}_\alpha \mid \alpha < \kappa \rangle$, $\dot{\leq}_P$, and $\langle \dot{B}_i \mid i < \omega_1 \rangle$, and satisfies as many Axioms of ZFC as we want.

(2) M is a transitive set, $j : N \longrightarrow M$ is an elementary embedding with $crit(j) = \kappa$, M can be as fat as we want, and $j(\kappa)$ is made as large as we want.

(3) $j(f)(\kappa) = \big((\lambda, \dot{\leq}_P), \langle \dot{B}_i \mid i < \omega_1 \rangle\big)$.

Proof. Let ρ be any strong limit cardinal such that H_ρ reflects enough Axioms of ZFC in V. Now since $\lambda^{<\kappa} = \lambda$, we may consider an elementary substructure X of H_ρ with ${}^{<\kappa}X \subseteq X$ and $|X| = \lambda$. We may assume that X contains the relevant objects. We are done by taking the transitive collapse N of X and by applying the hypothesis on κ and f. We regard the fatness of M as $j, N, H_\rho \in M$ and ${}^{\leq\rho}M \subseteq M$. Also we take the largeness of $j(\kappa)$ as $\rho < j(\kappa)$.

□

Claim 2. *We force over N and M via P_κ. Then*

(1) In both $N[G_\kappa]$ and $M[G_\kappa]$, $(\lambda, \leq_P)$ is a preorder and $\langle B_i \mid i < \omega_1 \rangle$ is a sequence of subsets of λ.

(2) In $M[G_\kappa]$, Q is a member of $H_{j(\kappa)}^{M[G_\kappa]}$ and is a proper preorder with $\Vdash_Q^{M[G_\kappa]}$ "there is a filter G in P which hits the B_i".

Proof. We consider (2). Since we assume $H_\rho \in M$ for some cardinal ρ which is sufficiently large, we may have $Q \in H_\rho^{V[G_\kappa]} = H_\rho[G_\kappa] = H_\rho^{M[G_\kappa]}$. In particular, we may assume that the power set of Q in $V[G_\kappa]$ and $M[G_\kappa]$ are the same, that $[2^{|Q|}]^\omega$ is absolute between $V[G_\kappa]$ and $M[G_\kappa]$, and that $V[G_\kappa]$ and $M[G_\kappa]$ share the exactly same stationary subsets of $[2^{|Q|}]^\omega$. Since the preservation of the stationary subsets of $[2^{|Q|}]^\omega$ is enough to imply Q is proper (see pages 76,77, and 83 in [Sh]), we may conclude (2).

□

Now we define $\langle P_\alpha^M \mid \alpha \leq j(\kappa) \rangle = j(\langle P_\alpha \mid \alpha \leq \kappa \rangle)$ and define $\langle \dot{Q}_\alpha^M \mid \alpha < j(\kappa) \rangle = j(\langle \dot{Q}_\alpha \mid \alpha < \kappa \rangle)$.

Claim 3. *We may assume that*

(1) *In M, $\langle P^M_\alpha \mid \alpha \le j(\kappa)\rangle$ and $\langle \dot{Q}^M_\alpha \mid \alpha < j(\kappa)\rangle$ are defined using $j(f)$ in the corresponding manner as $\langle P_\alpha \mid \alpha \le \kappa\rangle$ and $\langle \dot{Q}_\alpha \mid \alpha < \kappa\rangle$ are defined using f in V (and N).*

(2) *For all $\alpha < \kappa$, $P_\alpha = P^M_\alpha$ and $P_\kappa = P^M_\kappa$ hold.*

Proof. For (1) : Since N is gotten by collapsing some elementary substructure of H_ρ which reflects V well, $\langle P_\alpha \mid \alpha \le \kappa\rangle$ and $\langle \dot{Q}_\alpha \mid \alpha < \kappa\rangle$ are defined using f in N in the same manner as in V. So by the elementarity of j, we have (1).

For (2) : Since $P_\alpha \in H_\kappa$ and $j\lceil H_\kappa = \mathrm{id}$, so $P_\alpha = P^M_\alpha$ for all $\alpha < \kappa$. Since we may assume κ is a strongly inaccessible cardinal in M, so P^M_κ is the direct limit of $\langle P_\alpha \mid \alpha < \kappa\rangle$. But since M is fat, we may assume $P^M_\kappa = P_\kappa$.

□

Claim 4. *We next force over $V[G_\kappa]$ via $P^M_{\kappa j(\kappa)}$. So let $G^M_{\kappa j(\kappa)}$ be a $P^M_{\kappa j(\kappa)}$-generic filter over $V[G_\kappa]$ and denote the $P^M_{j(\kappa)}$-generic filter $G_\kappa * G^M_{\kappa j(\kappa)}$ over M by $G^M_{j(\kappa)}$. Then we have the following in $V[G_\kappa][G^M_{\kappa j(\kappa)}]$:*

(1) *$j : N \longrightarrow M$ is extended to an elementary embedding $j_1 : N[G_\kappa] \longrightarrow M[G^M_{j(\kappa)}]$ such that $j_1(G_\kappa) = G^M_{j(\kappa)}$.*

(2) *In $M[G^M_{j(\kappa)}]$, there is a directed subset F of $j_1(P)$ s.t. for all $i < \omega_1$, $F \cap j_1(B_i) \ne \emptyset$.*

And so by the elementarity of j_1, we are done.

Proof. For (1) : For any $p \in P_\kappa$, since the support of p is bounded below κ, we have $j``G_\kappa \subset G^M_{j(\kappa)}$. Therefore it is standard that $j_1 = \{(\tau_{G_\kappa}, j(\tau)_{G^M_{j(\kappa)}}) \mid \tau \in N^{P_\kappa}\}$ works.

For (2) : We have arranged so that in M, κ is a strongly inaccessible cardinal, that $j(f)(\kappa) = ((\lambda, \dot{\le}_P), \langle \dot{B}_i \mid i < \omega_1\rangle)$, that in $M[G_\kappa]$, $(\lambda, \le_P)$ is a preorder, that $\langle B_i \mid i < \omega_1\rangle$ is a sequence of subsets of λ, and that $Q \in H^{M[G_\kappa]}_{j(\kappa)}$ is a proper preorder with $\Vdash^{M[G_\kappa]}_Q$ "there is a filter in P which hits the B_i". Therefore $(\dot{Q}^M_\kappa)_{G_\kappa}$ is one of the Q by definition. So in $M[G^M_{j(\kappa)}\lceil(\kappa+1)]$, there is a filter G in P s.t. G hits the B_i.

Let $F = j``G$. Since $j \in M$, we have $F \in M[G^M_{j(\kappa)}\lceil(\kappa+1)] \subset M[G^M_{j(\kappa)}]$ Now it is easy to check that this F works.

□

§4. The Strength of $\Sigma(\lambda)$

We give some account concerning the consistency strength of $\Sigma(\lambda)$.

4.1 Theorem([Go]). *The following are equiconsistent :*

(1) There is an H_κ-reflecting cardinal κ.

(2) $\Sigma(\omega_1)$ holds.

Proof. A Σ_1-reflecting cardinal κ in [Go] is called H_κ-reflecting in our hierarchy of localized reflecting cardinals. BPFA in [Go] is equivalent to $\Sigma(\omega_1)$ as we have seen. So this theorem is a restatement of a theorem in [Go].

□

4.2 Theorem. *The followng are equiconsistent :*

(1) There is an H_{κ^+}-reflecting cardinal κ.

(2) $\Sigma(\omega_2)$ holds.

The proof is given below.

4.3 Proposition. *(1) $\Sigma(\omega_3)+2^\omega = \omega_2$ implies $\neg\Box_{\omega_2} + \neg\exists\omega_2$-Aronszajn tree.*

(2) $\neg\Box_{\omega_2} + \neg\exists\omega_2$-Aronszajn tree implies $0^\#$.

Proof. For (1) : Since $\Sigma(\omega_3^\omega)$ implies $\neg\Box_{\omega_2}$ (see pages 198, 199 in [Sc]) and since $\Sigma(\omega_2)$ implies $\neg\exists\omega_2$-Aronszajn tree (see page 110 in [Ma]), we are done.

For (2) : Let $\kappa = \omega_2$. Since $\Box_\kappa$ holds in L, we have $(\kappa^+)^L < \omega_3$. The rest is the same as page 216 in [Ab].

□

4.4 Note. We may consider a stronger version $\Sigma(\lambda$, semiproper) of $\Sigma(\lambda)$ by expanding the class of preorders from proper to semiproper (more precisely, $\{\omega_1\}$-semiproper) preorders. By paying attention to the sizes of relevant p.o. sets, we may note from pages 57–60 in [Be] :

(1) $\Sigma(2^{\omega_2}$, semiproper) implies $2^\omega = \omega_2$.

In particular, $\Sigma(2^{\omega_2}$, semiproper) iff $\Sigma(2^{2^{\omega_1}}$, semiproper). Actually, on page 63 in [Go], it is claimed that $\Sigma(\omega_1$, semiproper) implies $2^\omega = \omega_2$.

(2) $\Sigma(2^{2^{\omega_1}}$, semiproper) implies NS_{ω_1} is ω_2-saturated. And so there are inner models of set theory with many measurable cardinals (see page 430 in [Je]).

(3) $\Sigma(2^\lambda$, semiproper) implies $\lambda^{\omega_1} = \lambda^\omega = \lambda$ for regular cardinals $\lambda \geq \omega_2$. ([Be] and [Fo])

We now give a proof to 4.2 theorem which is a reworking of [Go] adjustments made to fit into our context.

Proof of 4.2 theorem. We need to show (2) implies (1). Let us put $\kappa = \omega_2$. We first mention that the following weak form of $\Sigma(\kappa)$ is sufficient to conclude that κ is an $L_{(\kappa^+)^L}$-reflecting cardinal in L.

- For any $A_0, \cdots, A_n \subseteq \kappa$, any $a \in H_\omega$, and any Σ_0-formula ψ, if $\Vdash_P$ "$\exists y\psi(y, A_0, \cdots, A_n, a)$ holds in some transitive $\in$-structure" for some finite iteration of c.c.c. and ω_1-closed preorders P, then

$$\{\alpha < \kappa \mid \exists y\psi(y, A_0 \cap \alpha, \cdots, A_n \cap \alpha, a)$$

holds in some transitive $\in$-structure of size $\omega_1\}$ is stationary in κ.

If $0^\#$ exists, then every Silver indiscernible μ is an $L_{(\mu^+)^L}$-reflecting cardinal in L. In particular, the uncountable cardinal κ is $L_{(\kappa^+)^L}$-reflecting in L. So we may assume $0^\#$ does not exist. In V, let $\theta = 2^{\omega_1}$. By applying the covering lemma, we have $A \subseteq \theta$ such that for any $X \in [\mathrm{ON}]^{\leq\omega_1}$, $X \in L[A]$. In particular, we have

- For any $X \in [\mathrm{ON}]^{\omega_0}$, $X \in L[A]$ and so $\omega_1 = \omega_1^{L[A]}$.

  To show κ is $L_{(\kappa^+)^L}$-reflecting in L, it suffices to show the following :

- For any subset $X \in L$ of κ and any formula φ, if $L_\chi \models$ "$\varphi(X)$" holds for some cardinal χ in V with $\theta < \chi$, then $\{\alpha < \kappa \mid$ there is a cardinal $\chi' < \kappa$ in L with $L_{\chi'} \models$ "$\varphi(X \cap \alpha)$"$\}$ is stationary in V (and so in L).

So fix $X \subseteq \kappa$ with $X \in L$, a formula φ, and a cardinal χ in V with $\theta < \chi$ such that $L_\chi \models$ "$\varphi(X)$". We find a (meta) Σ_1-formula $\Phi(\omega_1, \kappa, X, \varphi)$ and a finite iteration P of c.c.c. and ω_1-closed preorders with $\Vdash_P$ "$\Phi(\omega_1, \kappa, X, \varphi)$". Then by the weak $\Sigma(\kappa)$ we mentioned, we have $S = \{\alpha < \kappa \mid \Phi(\omega_1, \alpha, X \cap \alpha, \varphi)$ in some transitive $\in$-structure of size $\omega_1\}$ is stationary in V. We choose Φ in such a way that for any $\alpha \in S$, there is a cardinal $\chi' < \kappa$ in L with $L_{\chi'} \models$ "$\varphi(X \cap \alpha)$". This will complete the proof.

Step 1. We begin by describing what P is and what objects we pay attention to in V^P.

⟪ **Defining** P ⟫

We first force with ${}^{<\omega_1}\theta$ over V to make $|\theta| = \omega_1$. Then since χ is a cardinal in V with $\theta^\omega = \theta < \chi$, χ remains a cardinal in $V^{{}^{<\omega_1}\theta}$. In $V^{{}^{<\omega_1}\theta}$, since $|\theta| = \omega_1$, we may fix $B \subseteq \omega_1$ which codes both the collapse of θ and $A \subseteq \theta$. So we have $A \in L[B] \models$ "$|\theta| = \omega_1 \wedge \chi$ remains a cardinal". Since ${}^{<\omega_1}\theta$ is ω_1-closed, for any $i < \omega_1$, $B \cap i \in V$. So by the assumption on A, we have for any $i < \omega_1$, $B \cap i \in L[A]$. Notice that $\omega_1^{L[B]} = \omega_1$.

We next force with ${}^{<\omega_1}\chi$ over $V^{<\omega_1\theta}$ to make $|\chi| = \omega_1$. In $V^{<\omega_1\theta * <\omega_1\chi}$, since $|\chi| = \omega_1$, we may fix $C \subseteq \omega_1$ which codes both the collapse of χ and $B \subseteq \omega_1$. So we have $B \in L[C] \models$ "$|\chi| = \omega_1$". Notice that we also have $\omega_1^{L[C]} = \omega_1$.

Now going down to $L[B]$, we set T in $L[B]$ as follows : $t \in T$ iff

- $t = (\langle \theta_i^t \mid i \leq \beta^t \rangle, \langle f_{ij}^t \mid i < j \leq \beta^t \rangle)$.
- $\beta^t < \omega_1$.
- For all $i \leq \beta^t$, $\theta_i^t < \chi$.
- For all i, j with $i < j \leq \beta^t$, $f_{ij}^t : \theta_i^t \longrightarrow \theta_j^t$ s.t. for any $\eta, \xi < \theta_i^t$, if $\eta < \xi$, then $f_{ij}^t(\eta) < f_{ij}^t(\xi)$ (i.e., f_{ij}^t is an order-preserving function).
- For all i, j, k with $i < j < k \leq \beta^t$, $f_{jk}^t \circ f_{ij}^t = f_{ik}^t$ (i.e., commutative).

We then consider the following binary relation $\leq_T$ on T : For $t, s \in T$, $t \leq_T s$ iff

- $\beta^t \leq \beta^s$.
- For all $i \leq \beta^t$, $\theta_i^t = \theta_i^s$.
- For all i, j with $i < j \leq \beta^t$, $f_{ij}^t = f_{ij}^s$.

Then $(T, \leq_T)$ is a tree of height ω_1 such that each node is contained in a cofinal branch through T and that there are at least ω_1-many cofinal branches through T.

Notice this remains in the bigger universe $V^{<\omega_1\theta * <\omega_1\chi}$. So by § 3 in [Go], we have a finite iteration of c.c.c. and ω_1-closed preorders Q in $V^{<\omega_1\theta * <\omega_1\chi}$ which forces the "sealing devices" $\langle B_\alpha \mid \alpha < \omega_1 \rangle$ and $g : T \longrightarrow \omega$ to this tree T. Namely,

- The B_α are pairwise disjoint end-segments of cofinal branches through T with $T = \bigcup \{B_\alpha \mid \alpha < \omega_1\}$.
- For any $s, t \in T$ with $s \leq_T t$, if $g(s) = g(t)$, then there is $\alpha < \omega_1$ with $\{s, t\} \subset B_\alpha$.

Back in V, we denote the whole iteration ${}^{<\omega_1}\theta * {}^{<\omega_1}\chi * Q$ by P. This completes the construction of P.

⟪ **Objects and Properties in V^P** ⟫

Now we choose a sufficiently large regular cardinal $\delta > \chi$ in V^P so that

- $T \in H_\delta^{L[B]}$ and for any $\beta < \chi$, $([\beta]^\omega)^{L[B]} \in H_\delta^{L[B]}$.
- $\langle B_\alpha \mid \alpha < \omega_1 \rangle, g \in H_\delta^{V^P}$.

Then since $A \subseteq \theta < \delta$ and since $B, C \subseteq \omega_1 < \delta$, via condensation arguments (see page 105 in [De]), we have

- $A \in L_\delta[A] = H_\delta^{L[A]}$.
- $B, T \in L_\delta[B] = H_\delta^{L[B]}$.
- For all $i < \omega_1$, $B \cap i \in L_\delta[B \cap i] = H_\delta^{L[B\cap i]}$.
- $C \in L_\delta[C] = H_\delta^{L[C]}$.

  Since $H_\delta^{L[B\cap i]} \subseteq H_\delta^{L[A]} \subset H_\delta^{L[B]} \subset H_\delta^{L[C]} \subset H_\delta^{V^P}$, so

- For all $i < \omega_1$, $L_\delta[B \cap i] \subseteq L_\delta[A] \subset L_\delta[B] \subset L_\delta[C] \subset H_\delta^{V^P}$.
- For all $i < \omega_1$, $B \cap i \in L_\delta[A]$, $A \in L_\delta[B]$, and $B \in L_\delta[C]$.

Notice that for every $i < \omega_1$, $L_\delta[B \cap i] = (L[B \cap i])^{L_\delta[A]}$, $L_\delta[A] = (L[A])^{L_\delta[B]}$, $L_\delta[B] = (L[B])^{L_\delta[C]}$, and $L_\delta[C] = (L[C])^{H_\delta^{V^P}}$.

Let $M = H_\delta^{V^P}$ and we list the following rather ad hoc properties in V^P :

(1) M is a transitive set and $\omega_1, \kappa, \theta, \chi, A, B, C, X \in M$.

(2) $X \subseteq \kappa$, $X \in L_\chi$, and $L_\chi \models$ "$\varphi(X)$".

(3) $A \subseteq \theta$, $B \subseteq \omega_1$, $C \subseteq \omega_1$, $\omega_1 < \kappa \leq \theta < \chi < \delta = M \cap \mathrm{ON}$, and δ is a limit ordinal.

Once again,

(4) $A \in L_\delta[A]$, $B \in L_\delta[B]$, $C \in L_\delta[C]$, $B \cap i \in L_\delta[B\cap i] \subseteq L_\delta[A] \subset L_\delta[B] \subset L_\delta[C] \subset M$ $(i < \omega_1)$.

Since $\omega_1 = \omega_1^{L[A]} = \omega_1^{V^P}$, so $\omega_1 = \omega_1^{L[B]} = \omega_1^{L[C]}$. Since $\omega_1^{L[A]} = \omega_1^{L_\delta[A]}$, so we have

(5) $\omega_1^{L_\delta[A]} = \omega_1^{L_\delta[B]} = \omega_1^{L_\delta[C]} = \omega_1^M = \omega_1$.

Since $L[B] \models$ "$|\theta| = \omega_1$ and χ remains a cardinal", $L[C] \models$ "$|\chi| = \omega_1$", $\theta < \chi < \delta$, $L_\delta[B] = H_\delta^{L[B]}$, and $L_\delta[C] = H_\delta^{L[C]}$, we have

(6) $L_\delta[B] \models$ "$|\theta| = \omega_1$ and χ is a cardinal" and $L_\delta[C] \models$ "$|\chi| = \omega_1$".

Since $B \subseteq \omega_1$, we have $L[B] \models$ " for any cardinal μ with $\mathrm{cf}(\mu) = \omega$, $|[\mu]^\omega| = \mu^+$" via condensation. But since δ has chosen so that $L[B] \models$ " for every cardinal $\mu < \chi$, $[\mu]^\omega \in H_\delta$" and that $L_\delta[B] = H_\delta^{L[B]}$, so we have

(7) $L_\delta[B] \models$ "$\forall \mu < \chi$ if μ is a cardinal with $\mathrm{cf}(\mu) = \omega$, then $|[\mu]^\omega| = \mu^+$".

By the assumption on A and the fact that ${}^{<\omega_1}\theta * {}^{<\omega_1}\chi$ is ω_1-closed, for any limit ordinal $\beta < \chi$, as $|\chi| = \omega_1$ in $L[C] \subset V^{{}^{<\omega_1}\theta * {}^{<\omega_1}\chi}$, $\mathrm{cf}^{L[C]}(\beta) = \omega$ or ω_1, so either there is $h \in L[A]$ s.t. $h : \omega \longrightarrow \beta$ increasing and cofinal or $h \in L[C]$ s.t. $h : \omega_1 \longrightarrow \beta$ increasing, cofinal, and for all $i < \omega_1$, $h \restriction i \in L[A]$. Therefore, we certainly have

(8) For any limit ordinal $\beta < \chi$, either (a) there is $h \in L_\delta[A]$ s.t. $h : \omega \longrightarrow \beta$ is increasing and cofinal, or (b) there is $h \in L_\delta[C]$ s.t. $h : \omega_1 \longrightarrow \beta$ is increasing, cofinal, and for all $i < \omega_1$, $h\lceil i \in L_\delta[A]$.

Since $L_\delta[B] = H_\delta^{L[B]}$ and $L_\delta[C] = H_\delta^{L[C]}$, and δ is regular in both $L[B]$ and $L[C]$, so we have

(9) $L_\delta[C] \models$ "$\forall\gamma \in$ON $\forall y \subset$ON $\exists z\ z = L_\gamma[y]$".

(10) $L_\delta[B] \models$ "$\forall\bar{\mu} \in$ON $\forall\bar{\beta} \in$ON $\forall(\langle\theta_\xi \mid \xi \leq \bar{\mu}\rangle, \langle f_{\xi\eta} \mid \xi < \eta \leq \bar{\mu}\rangle)\ \forall h : \bar{\beta}+1 \longrightarrow \bar{\mu}+1$ increasing $\exists y\ y = (\langle\theta_{h(i)} \mid i \leq \bar{\beta}\rangle, \langle f_{h(i)h(j)} \mid i < j \leq \bar{\beta}\rangle)$".

(11) $L_\delta[C] \models$ "$\forall x\ |x| < \omega$ or $|x| = \omega$ or $\omega_1 \leq |x|$".

Since $|\chi| = \omega_1$ in $L[C]$, we have $|L_\chi| = \omega_1$ in $L_\delta[C] = H_\delta^{L[C]}$. So we have

(12) $L_\delta[C] \models$ "$\exists\langle Y_i, \bar{Y}_i, \pi_i \mid i < \omega_1\rangle$ s.t. $\langle Y_i \mid i < \omega_1\rangle$ is a sequence of continuously increasing countable subsets of L_χ with $\bigcup\{Y_i \mid i < \omega_1\} = L_\chi$ and that for all $i < \omega_1$, $\pi_i : Y_i \longrightarrow \bar{Y}_i$ is the transitive collapse and $\forall z \exists y\ y = \{x \in Y_i \mid \pi_i(x) \in z\}$".

Since $L[A]$, V, and $V^{{}^{<\omega_1}\theta * {}^{<\omega_1}\chi}$ calculate the same $[\mathrm{ON}]^\omega$, so do $L[B]$ and $L[C]$. Since we have chosen δ so that for any $\beta < \chi$, $L[B] \models$ "$[\beta]^\omega \in H_\delta$" and so $L[C] \models$ "$[\beta]^\omega \in H_\delta$", $([\beta]^\omega)^{L[B]} = ([\beta]^\omega)^{H_\delta^{L[B]}}$, and $([\beta]^\omega)^{L[C]} = ([\beta]^\omega)^{H_\delta^{L[C]}}$. So we have

(13) $\forall\beta < \chi\ ([\beta]^\omega)^{L_\delta[B]} = ([\beta]^\omega)^{L_\delta[C]}$.

And finally, since $\langle B_\alpha \mid \alpha < \omega_1\rangle$, $g \in M = H_\delta^{V^P}$, we have

(14) There is T in $L_\delta[B]$ and $\langle B_\alpha \mid \alpha < \omega_1\rangle$, g, and $\langle((\langle\theta_i^\alpha \mid i < \omega_1\rangle, \langle f_{ij}^\alpha \mid i < j < \omega_1\rangle) \mid \alpha < \omega_1\rangle$ in M such that

- $T = \{(\langle\theta_i \mid i \leq \beta\rangle, \langle f_{ij} \mid i < j \leq \beta\rangle) \in L_\delta[B] \mid$ (1) $\beta < \omega_1$; (2) $\forall i \leq \beta$, $\theta_i < \chi$; (3) $\forall i, j$ with $i < j \leq \beta$, $f_{ij} : \theta_i \longrightarrow \theta_j$ is an order-preserving map ; (4) commutative $\}$.
- $\forall\alpha < \omega_1 \forall\beta < \omega_1\ (\langle\theta_i^\alpha \mid i \leq \beta\rangle, \langle f_{ij}^\alpha \mid i < j \leq \beta\rangle) \in T$.
- For all $\alpha < \omega_1$, there is $\bar{\beta} < \omega_1$ s.t. $B_\alpha = \{(\langle\theta_i^\alpha \mid i \leq \beta\rangle, \langle f_{ij}^\alpha \mid i < j \leq \beta\rangle) \mid \bar{\beta} \leq \beta < \omega_1\}$ (i.e., B_α is an end-segment of the cofinal branch $\{(\langle\theta_i^\alpha \mid i \leq \beta\rangle, \langle f_{ij}^\alpha \mid i < j \leq \beta\rangle) \mid \beta < \omega_1\}$ through T).
- For all $\alpha, \alpha' < \omega_1$, if $\alpha \neq \alpha'$, then $B_\alpha \cap B_{\alpha'} = \emptyset$ (i.e., disjoint).
- $T = \bigcup\{B_\alpha \mid \alpha < \omega_1\}$.
- $g : T \longrightarrow \omega$ and for any $t, s \in T$ with $t < s$ in T, if $g(t) = g(s)$, then there is $\alpha < \omega_1$ with $t, s \in B_\alpha$.

Since every cofinal branch through T in V^P has the direct limit which is a well-order,

- $\forall\alpha<\omega_1\exists\gamma<\delta$ s.t. γ is the direct limit of $(\langle\theta_i^\alpha\mid i<\omega_1\rangle,\langle f_{ij}^\alpha\mid i<j<\omega_1\rangle)$.

Step 2. Let $\Phi(\omega_1,\kappa,X,\varphi)$ be a Σ_1-formula which is equivalent to (in, say, ZFC) the following formula ; $\exists M,\delta,A,B,C,\theta,\chi$ s.t. (1) through (14) hold. So $\Vdash_P$"$\Phi(\omega_1,\kappa,X,\varphi)$" holds. By our assumption $\Sigma(\kappa)$, we have $S=\{\alpha<\kappa\mid\Phi(\omega_1,\alpha,X\cap\alpha,\varphi)$ holds in some transitive $\in$-structure of size $\omega_1\}$ is stationary in κ.

We list explicitely what $\Phi(\omega_1,\alpha,X\cap\alpha,\varphi)$ says :

For each $\alpha\in S$, we have M', δ', A', B', C', θ', $\chi'\in H_\kappa$ such that

(1)′ M' is a transitive set, $\omega_1,\alpha,\theta',\chi',A',B',C',X\cap\alpha\in M'$, and so $|M'|=\omega_1$.

(2)′ $X\cap\alpha\subseteq\alpha$, $X\cap\alpha\in L_{\chi'}$, $L_{\chi'}\models$ "$\varphi(X\cap\alpha)$".

(3)′ $A'\subseteq\theta'$, $B'\subseteq\omega_1$, $C'\subseteq\omega_1$, $\omega_1<\alpha\leq\theta'<\chi'<\delta'=M'\cap\mathrm{ON}$, $\delta'<\kappa$ is a limit ordinal.

(4)′ $A'\in L_{\delta'}[A']$, $B'\in L_{\delta'}[B']$, $C'\in L_{\delta'}[C']$, and for all $i<\omega_1$, $B'\cap i\in L_{\delta'}[B'\cap i]\subseteq L_{\delta'}[A']\subset L_{\delta'}[B']\subset L_{\delta'}[C']\subset M'$.

(5)′ $\omega_1^{L_{\delta'}[A']}=\omega_1^{L_{\delta'}[B']}=\omega_1^{L_{\delta'}[C']}=\omega_1^{M'}=\omega_1$.

(6)′ $L_{\delta'}[B']\models$ "$|\theta'|=\omega_1$ and χ' is a cardinal" and $L_{\delta'}[C']\models$ "$|\chi'|=\omega_1$".

(7)′ $L_{\delta'}[B']\models$ " For any $\mu<\chi'$, if μ is a cardinal with $\mathrm{cf}(\mu)=\omega$, then $|[\mu]^\omega|=\mu^+$".

(8)′ For any limit ordinal $\beta<\chi'$, either (a) there is $h\in L_{\delta'}[A']$ s.t. $h:\omega\longrightarrow\beta$ is increasing and cofinal, or (b) there is $h\in L_{\delta'}[C']$ s.t. $h:\omega_1\longrightarrow\beta$ is increasing, cofinal, and for all $i<\omega_1$, $h\lceil i\in L_{\delta'}[A']$.

(9)′ $L_{\delta'}[C']\models$ "$\forall\gamma\in\mathrm{ON}\exists y\subset\mathrm{ON}\ \exists z\ z=L_\gamma[y]$".

(10)′ $L_{\delta'}[B']\models$ "$\forall\bar\mu\in\mathrm{ON}\ \forall\bar\beta\in\mathrm{ON}\ \forall(\langle\theta_\xi\mid\xi\leq\bar\mu\rangle,\langle f_{\xi\eta}\mid\xi<\eta\leq\bar\mu\rangle)$ $\forall h:\bar\beta+1\longrightarrow\bar\mu+1$ increasing $\exists y\ y=(\langle\theta_{h(i)}\mid i\leq\bar\beta\rangle,\langle f_{h(i)h(j)}\mid i<j\leq\bar\beta\rangle)$".

(11)′ $L_{\delta'}[C']\models$ "$\forall x\ |x|<\omega$ or $|x|=\omega$ or $\omega_1\leq|x|$".

(12)′ $L_{\delta'}[C']\models$ "$\exists\langle Y_i,\bar Y_i,\pi_i\mid i<\omega_1\rangle$ s.t. $\langle Y_i\mid i<\omega_1\rangle$ is a sequence of continuously increasing countable subsets of $L_{\chi'}$ with $\bigcup\{Y_i\mid i<\omega_1\}=L_{\chi'}$ and that for all $i<\omega_1$, $\pi_i:Y_i\longrightarrow\bar Y_i$ is the transitive collapse and $\forall z\exists y\ y=\{x\in Y_i\mid\pi_i(x)\in z\}$".

(13)′ $\forall\beta<\chi'\ ([\beta]^\omega)^{L_{\delta'}[B']}=([\beta]^\omega)^{L_{\delta'}[C']}$.

(14)′ There are $T'\in L_{\delta'}[B']$, $\langle B'_\eta\mid\eta<\omega_1\rangle\in M'$, $\langle((\langle\theta_i'^\eta\mid i<\omega_1\rangle,\langle f_{ij}'^\eta\mid i<j<\omega_1\rangle)\mid\eta<\omega_1\rangle\in M'$, $g'\in M'$ such that

- $T'=\{(\langle\theta_i\mid i\leq\beta\rangle,\langle f_{ij}\mid i<j\leq\beta\rangle)\in L_{\delta'}[B']\mid$ (1) $\beta<\omega_1$; (2) $\forall i\leq\beta$, $\theta_i<\chi'$; (3) $\forall i,j$ with $i<j\leq\beta$, $f_{ij}:\theta_i\longrightarrow\theta_j$ is an order-preserving map ; (4) commutative $\}$.

- $\forall\eta<\omega_1\forall\beta<\omega_1$ $(\langle\theta_i'^\eta \mid i\leq\beta\rangle,\langle f_{ij}'^\eta \mid i<j\leq\beta\rangle)\in T'$.
- $\forall\eta<\omega_1\exists\bar\beta<\omega_1$ s.t. $B'_\eta=\{(\langle\theta_i'^\eta \mid i\leq\beta\rangle,\langle f_{ij}'^\eta \mid i<j\leq\beta\rangle)\mid\bar\beta\leq\beta<\omega_1\}$.
- For any $\eta,\eta'<\omega_1$, if $\eta\neq\eta'$, then $B'_\eta\cap B'_{\eta'}=\emptyset$.
- $T'=\bigcup\{B'_\eta\mid\eta<\omega_1\}$.
- $g':T'\longrightarrow\omega$ and for any $t,s\in T'$ with $t<s$ in T', if $g'(t)=g'(s)$, then there is $\eta<\omega_1$ with $t,s\in B'_\eta$.
- $\forall\eta<\omega_1\exists\gamma<\delta'$ s.t. γ is the direct limit of $(\langle\theta_i'^\eta \mid i<\omega_1\rangle,\langle f_{ij}'^\eta \mid i<j<\omega_1\rangle)$.

Step 3. It suffices to show, by $(2)'$, that χ' is a cardinal in $L[B']$. We establish this by comparing $L_{\delta'}[B']$ and $L[B']$ as follows :

Lemma 1. *(1) For any $\beta<\chi'$, $(cf\beta=\omega)^{L[B']}$ iff $(cf\beta=\omega)^{L_{\delta'}[B']}$ iff $cf^V\beta=\omega$.*

(2) For any $\beta<\chi'$, $([\beta]^\omega)^{L[B']}=([\beta]^\omega)^{L_{\delta'}[B']}$.

(3) For any $\mu<\chi'$, if $cf^{L_{\delta'}[B']}(\mu)=\omega$ and μ is a cardinal in $L[B']$, then $(|[\mu]^\omega|=\mu^+)^{L[B']}$ and $(|[\mu]^\omega|=\mu^+)^{L_{\delta'}[B']}$.

And so,

(4) For any $\mu<\chi'$, if $cf^V\mu=\omega$ and μ is a cardinal in $L[B']$, then $(\mu^+)^{L[B']}=(\mu^+)^{L_{\delta'}[B']}<\delta'$.

Proof. For (1) : This follows from $(4)'$ and $(8)'$.

For (2) : It suffices to show $([\beta]^\omega)^{L[B']}\subseteq([\beta]^\omega)^{L_{\delta'}[\beta']}$.

Fix $y\in([\beta]^\omega)^{L[B']}$ arbitrarily. Since $\beta<\chi'<\delta'<\kappa$, $B'\subseteq\omega_1$, and $L_{\chi'}\in L_\kappa[B']=H_\kappa^{L[B']}$, there is $\gamma<\kappa$ s.t. $y\in L_\gamma[B']$ and $L_{\chi'}\subset L_\gamma[B']$. So in V, we may fix a continuously increasing sequence $\langle X_i\mid i<\omega_1\rangle$ of countable elementary substructures of $L_\gamma[B']$ s.t. $y\cup\{y\}\subset X_0$.

Since $L_{\chi'}\subset L_\gamma[B']$, so $\langle X_i\cap L_{\chi'}\mid i<\omega_1\rangle$ continously increases to $L_{\chi'}$. So by $(12)'$, we may find $i<\omega_1$ s.t. $X_i\cap L_{\chi'}=Y_i$ and $X_i\cap\omega_1=i$. And so $y\subset Y_i$. Let $\pi:X_i\longrightarrow L_{\hat\gamma}[\hat B]$ be the transitive collapse. Then since $L_{\chi'}$ is transitive, $\pi\lceil Y_i=\pi_i$. By $(4)'$, $\hat B=\pi``(X_i\cap B')=\pi``(B'\cap i)=B'\cap i\in L_{\delta'}[B']$. Since $|X_i|=\omega$, so $\hat\gamma<\omega_1<\delta'$. By $(4)'$ again, $L_{\delta'}[B']\subset L_{\delta'}[C']$ and so $\hat B\in L_{\delta'}[C']$. So by $(9)'$, $L_{\hat\gamma}[\hat B]\in L_{\delta'}[C']$ and so $\pi(y)\in L_{\delta'}[C']$. By $(12)'$, $y=\{\xi\in Y_i\mid\pi_i(\xi)\in\pi(y)\}\in L_{\delta'}[C']$. Since $|y|=\omega$ in $L[B']$, so in V and so $L_{\delta'}[C']\models$ "$|y|=\omega$" by $(11)'$. But by $(13)'$, we conclude $y\in([\beta]^\omega)^{L_{\delta'}[B']}$.

For (3) : Suppose $\mu<\chi'$, $\mathrm{cf}^{L_{\delta'}[B']}(\mu)=\omega$ and μ is a cardinal in $L[B']$. By $(7)'$, we have $(|[\mu]^\omega|=\mu^+)^{L_{\delta'}[B']}$. Since $\mathrm{cf}^{L[B']}(\mu)=\omega$, $L[B']\models$ "$|[\mu]^\omega|\geq\mu^+$". Since $B'\subseteq\omega_1=\omega_1^{L[B']}$ by $(5)'$, we have $L[B']\models$ "$|[\mu]^\omega|=\mu^+$" by a condensation argument.

For (4) : By (1), $\mathrm{cf}^{L_{\delta'}[B']}(\mu) = \omega$. So by (3), $(|[\mu]^\omega| = \mu^+)^{L[B']}$ and $(|[\mu]^\omega| = \mu^+)^{L_{\delta'}[B']}$. But by (2), $([\mu]^\omega)^{L_{\delta'}[B']} = ([\mu]^\omega)^{L[B']}$. Thus in $L[B']$, we have $\mu^+ = |[\mu]^\omega| = |(\mu^+)^{L_{\delta'}[B']}|$ and so $(\mu^+)^{L[B']} \leq (\mu^+)^{L_{\delta'}[B']}$. But $(\mu^+)^{L[B']} \geq (\mu^+)^{L_{\delta'}[B']}$ holds trivially.

□

Lemma 2. *(1) For any cardinal $\mu < \chi'$ in $L[B']$, we have $H_\mu^{L[B']} \subset L_{\delta'}[B']$.*

(2) For any $\beta < \chi'$, $cf^V \beta = \omega_1$ iff $\exists h \in L_{\delta'}[C']$ s.t. $h : \omega_1 \longrightarrow \beta$ is increasing, cofinal, and for each $i < \omega_1$, $h \lceil i \in L_{\delta'}[B']$.

(3) For any cofinal branch B through T', there is $\eta < \omega_1$ and $\gamma < \delta'$ s.t. $B'_\eta \subseteq B$ and γ is the direct limit of B.

(4) For any cardinal $\mu < \chi'$ in $L[B']$ with $cf^V \mu = \omega_1$ and any γ with $\mu \leq \gamma < (\mu^+)^{L[B']}$, there is a cofinal branch B through T' s.t. γ is the direct limit of B, so $\gamma < \delta'$. And so $(\mu^+)^{L[B']} \leq \delta'$.

Proof. For (1) : Fix a cardinal $\mu < \chi'$ in $L[B']$. If $\mu = \omega$, then $H_\mu^{L[B']} = L_\omega[B'] \in L_{\delta'}[B']$. So we may assume $\omega_1^{L[B']} \leq \mu$. But since $\omega_1 = \omega_1^{L[B']}$ and $B' \subseteq \omega_1$, we may consider in two cases.

Case 1. $\mu = \omega_1$: For any $x \in H_\mu^{L[B']}$, since $B' \subseteq \omega_1$, we have $i, j < \omega_1$ s.t. $x \in L_j[B' \cap i]$ via a condensation argument. And so $x \in L_{\omega_1}[B' \cap i] \subset L_{\delta'}[B']$ by $(3)'$ and $(4)'$.

Case 2. $\omega_1 < \mu$: For any $x \in H_\mu^{L[B']}$, since μ is a cardinal in $L[B']$, we may find $j < \mu$ s.t. $x \in L_j[B']$ via a condensation argument in $L[B']$. Since $\mu < \chi' < \delta'$, we have $x \in L_\mu[B'] \subset L_{\chi'}[B'] \subset L_{\delta'}[B']$. Thus in this case, we actually have $H_\mu^{L[B']} = L_\mu[B']$.

For (2) : This is immediate from $(4)'$ and $(8)'$.

For (3) : Suppose B is a cofinal branch through T'. By $(14)'$, we have $g' : T' \longrightarrow \omega$. Since $|B| = \omega_1$, there are $n < \omega$ and $B' \in [B]^{\omega_1}$ s.t. for all $t \in B'$ $g'(t) = n$. Therefore for all $t, s \in B'$ with $t < s$ in T', there is $\eta < \omega_1$ s.t. $t, s \in B'_\eta$. But since $\langle B'_\eta \mid \eta < \omega_1 \rangle$ is disjoint, there is $\eta < \omega_1$ s.t. $B' \subset B'_\eta$. Since B is a cofinal branch and B'_η is an end-segment of a cofinal branch, $B'_\eta \subseteq B$ holds. Now by $(14)'$ again, we may get $\gamma < \delta'$ such that γ is the direct limit of the cofinal branch B.

For (4) : Fix a cardinal $\mu < \chi'$ in $L[B']$ with $\mathrm{cf}^V \mu = \omega_1$ and fix γ with $\mu \leq \gamma < (\mu^+)^{L[B']}$. Since $|\gamma| = \mu$ in $L[B']$, we have a one-to-one and onto map $f \in L[B']$ s.t. $f : \mu \longrightarrow \gamma$. For each $\xi < \mu$, let θ_ξ be the order type of $f``\xi$ and let $f_\xi : \theta_\xi \longrightarrow f``\xi$ be the isomorphism. For each $\xi < \eta < \mu$, let $f_{\xi\eta} = f_\eta^{-1} \circ f_\xi$. Then $\theta_\xi < \mu$ $(< \chi')$, $f_{\xi\eta} : \theta_\xi \longrightarrow \theta_\eta$ is an order-preserving map, and $\langle f_{\xi\eta} \mid \xi < \eta < \mu \rangle$ are commutative. And γ is the direct limit of this system. By (2), there is $h \in L_{\delta'}[C']$ s.t. $h : \omega_1 \longrightarrow \mu$ is increasing, cofinal and for each $i < \omega_1$, $h\lceil i \in L_{\delta'}[B']$. For each $i < j < \omega_1$, define $\bar{\theta}_i = \theta_{h(i)}$ and $\bar{f}_{ij} = f_{h(i)h(j)}$. We claim that for each $\beta < \omega_1$, $\langle \bar{\theta}_i \mid i \leq \beta \rangle, \langle \bar{f}_{ij} \mid i < j \leq \beta \rangle \in L_{\delta'}[B']$ and so $B = \{(\langle \bar{\theta}_i \mid i \leq \beta \rangle, \langle \bar{f}_{ij} \mid i < j \leq \beta \rangle) \mid \beta < \omega_1\}$ is a cofinal branch through T'. Since h is cofinal in μ, γ is the direct limit of B. And so by (3), $\gamma < \delta'$ holds. Now to see $\langle \bar{\theta}_i \mid i \leq \beta \rangle, \langle \bar{f}_{ij} \mid i < j \leq \beta \rangle \in L_{\delta'}[B']$ for each $\beta < \omega_1$, let $\bar{\mu} = h(\beta) < \mu < \chi' < \delta'$. Since $\langle \theta_\xi \mid \xi \leq \bar{\mu} \rangle, \langle f_{\xi\eta} \mid \xi < \eta \leq \bar{\mu} \rangle \in H_\mu^{L[B']} \subset L_{\delta'}[B']$ by (1) and $h\lceil \beta + 1 : \beta + 1 \longrightarrow \bar{\mu} + 1$ is in $L_{\delta'}[B']$, so by (10)', we have $(\langle \bar{\theta}_i \mid i \leq \beta \rangle, \langle \bar{f}_{ij} \mid i < j \leq \beta \rangle) = (\langle \theta_{h(i)} \mid i \leq \beta \rangle, \langle f_{h(i)h(j)} \mid i < j \leq \beta \rangle) \in L_{\delta'}[B']$.

□

Lemma 3. *(1)* $L_{\delta'}[B'] \models$ *"χ' is a cardinal",*

(2) For any cardinal $\mu < \chi'$ in $L[B']$, $(\mu^+)^{L[B']} \leq \delta'$ holds.

And so,

(3) For any cardinal $\mu < \chi'$ in $L[B']$, $(\mu^+)^{L[B']} \leq \chi'$ holds. So χ' is a cardinal in $L[B']$ and so in L.

Proof. For (1) : This is immediate from (6)'.

For (2) : Fix a cardinal $\mu < \chi'$ in $L[B']$. Since $\mu < \chi' < \delta' < \kappa = \omega_2^V$, we have either $\mathrm{cf}^V \mu = \omega$ or $\mathrm{cf}^V \mu = \omega_1$. But by lemma 1 and 2, we have $(\mu^+)^{L[B']} \leq \delta'$.

For (3) : If $\mu = \omega$, then $(\mu^+)^{L[B']} = \omega_1 < \chi'$. So assume $\omega_1^{L[B']} = \omega_1 \leq \mu$. For any γ with $\mu < \gamma < (\mu^+)^{L[B']}$, there is a one-to-one and onto map $f : \mu \longrightarrow \gamma$ with $f \in L[B']$. But by a condensation argument in $L[B']$, since $B' \subseteq \omega_1$, we may find $j < (\mu^+)^{L[B']}$ such that $f \in L_j[B']$. So $f \in L_{\delta'}[B']$ by (2) and so γ is not a cardinal in $L_{\delta'}[B']$. Thus by (1), $(\mu^+)^{L[B']} \leq \chi'$ must hold.

□

□

References

[Ab] U. Abraham, Aronszajn trees on $\aleph_2$ and $\aleph_3$, *Annals of Pure and Applied Logic*, vol. 24 (1983), pp. 213-230.

[Be] M. Bekkali, *Topics in Set Theory*, Lecture Notes in Mathematics, Vol. 1476, Springer-Verlag, 1991.

[Bu] D. Burke, Generic Embeddings and the failure of Box, *Proceedings of the American Mathematical Society*, vol. 123 (1995), pp. 2867-2871.

[De] K. Devlin, *Constructibility*, Perspectives in Mathematical Logic, Springer-Verlag, 1984.

[Fo] M. Foreman, M. Magidor, and S. Shelah, Martin's Maximum, saturated ideals, and non-regular ultrafilters. Part I, *Annals of Mathematics*, vol. 127 (1988), pp. 1-47.

[Go] M. Goldstern and S. Shelah, The Bounded Proper Forcing Axiom, *Journal of Symbolic Logic*, vol. 60 (1995), pp. 58-73.

[Je] T. Jech, *Set Theory*, New York, Academic Press, 1978.

[Ku] K. Kunen, *Set Theory*, An Introduction to Independence Proofs, Studies in Logic and the Foundations of Mathematics, North-Holland, 1980.

[Ma] A. Mathias, *Surveys in Set Theory*, London Mathematical Society Lecture Note Series 87, Cambridge University Press, 1983.

[Mi] T. Miyamoto, Localized Reflecting Cardinals and Weak Segments of PFA, a typed note, 1996.

[Sc] A. Schimmerling, Combinatorial Principles in the Core Model for One Woodin Cardinal, *Annals of Pure and Applied Logic*, vol. 74 (1995), pp. 153-201.

[Sh] S. Shelah, *Proper Forcing*, Lecture Notes in Mathematics, vol. 940, Springer-Verlag, 1982.

Department of Mathematics
Nanzan University
18, Yamazato-cho, Showa-ku
Nagoya, 466, Japan

e-mail : miyamoto@iq.nanzan-u.ac.jp

ON STRUCTURAL INFERENCE RULES FOR GENTZEN-STYLE NATURAL DEDUCTION, PART I

Koji Nakatogawa
Takeshi Ueno
Department of Philosophy, Department of Mathematics,
Hokkaido University, Sapporo, Japan
e-mail: koji@synthe.hokudai.ac.jp
m074310@math.hokudai.ac.jp

Abstract

For a system of Gentzen-style natural deduction, we present characterizations of two structural rules, namely, weakening and contraction, in terms of the numbers of assumptions discharged at an application of introduction rule for logical implication. It is shown that the number is zero for the weakening rule, while the number is reduced to exactly two for the contraction rule.

1 Introduction

The purpose of the present paper is to characterize two structural inference rules, namely, weakening (Theorem 2) and contraction (Theorem 3), for a system D (defined below in section 3) of Gentzen-style natural deduction. Our characterizations of these two rules are given in terms of the number of assumptions discharged at an application of the introduction rule for implication. For the weakening rule, the number is zero, while it is reduced to exactly two for the contraction rule.

We start with introducing a propositional language equipped with the logical connectives, *implication* $\supset$, *conjunction* $\wedge$, *disjunction* $\vee$, and *negation* $\neg$. Then, the system D is defined as a system without weakening, contraction, and exchange rules. (Oide[9] introduced sequent calculus systems where the number of formulas allowed to occur on each sides of a sequent is restricted to zero or one. This system provided us of motivations and stimulations leading to the definition of D.) In this base system D, the elimination rule for implication is the ordinary modus ponens. For the introduction rule of implication , it is customary not to put any restriction on the number of discharged assumptions. In our D, however, the number of assumptions discharged at each application of the introduction rule $\supset I^1$ for implication is restricted to exactly one.

It will turn out that if an introduction rule $\supset I^0$ for implication, where the number of discharged assumptions is restricted to zero, is added to D as a new rule, it works as the weakening rule. (See *Theorem* 2 below) If the introduction rules $\supset I^n(n \geq 2)$, where the number of discharged assumptions is any positive integer strictly greater that or equal to two, are added to D, they work as the contraction rule.(See *Theorem* 3 below) For the contraction rule, the number of discharged assumptions can be reduced to two (*Lemma* 4).

Inference rules for conjunction, disjunction, and negation are also formulated by using the standard symbols $\wedge$, $\vee$, $\neg$ for those logical connectives . However, due to the restrictions on introduction rules, the inference rules for these logical connectives have to be formulated in such a way that the "contexts" in a proof can be taken into consideration. Thus, the substitutions of $\supset$ with $\multimap$, $\wedge$ with multiplicative conjunction $\otimes$, $\vee$ with multiplicative disjunction $\wp$, and $\neg$ with $(\ )^{\perp}$ in D readily yield a system for a certain weak substructural logic which may provide a basis for non-standard logics such as linear logics. Those who are interested in the linear logic will find in, e.g., Troelstra [4] extensive studies on systems of natural deduction for linear logic which are modifications of the Gentzen-style natural deduction. Those who are interested in substructural logics may find Ono [11] helpful in this regard.

Our primary concern in this paper is on neither substractural logics nor linear logics. It is mostly concerned with seeing what sort of forms the structural rules such as weakening and contraction take in a natural deduction system which follows closely as possible the original formulation of Gentzen [1]. The scope of the present paper is restricted to propositional connectives.

2 Language L

The propositional connectives of our basic language L are the implication symbol $\supset$, the negation symbol $\neg$, the conjunction symbol $\wedge$, and the disjunction symbol $\vee$. We also use the constant symbols 1 (one) to denote the unit element for the conjunction, $\perp$ to denote the unit element for the disjunction. The formulas of L are constructed by combining the constant symbols 1 , $\perp$ and the propositional constant symbols, $A, B, C, \cdots$

together with the connectives $\supset$, $\neg$, $\wedge$, $\vee$, ((left parenthesis),) (right parenthesis). Some lower case Greek alphabets ϕ, ψ , θ, $\cdots$ denote a formula. We use capital letters such as Γ, Δ, Ψ, Φ, $\cdots$ to represent a multi-sets of formulas. In what follows we usually omit "multi-" when it is clear from the context.

The sequent of L have the following form

$$\Gamma \to \Delta \,.$$

The left hand side of a sequent may be empty. The right hand side of a sequent is either empty or consists of a single formula. To specify the element of Γ and Δ we write

$$\gamma_0, \cdots, \gamma_{m-1} \to \delta.$$

When both sides of a sequent are empty, we write

$$\to$$

and call it the empty sequent.

3 Natural Deduction System D

We now introduce a formal system D of natural deductions which is obtained by restricting and modifying Gentzen's natural deduction system NK. Our system D has the following inference rules. The non-negative integer n used as the superscript of I, E indicates that the number of assumptions cancelled at the application of an inference rule is restricted to exactly n.

The introduction of $\supset$ with the number of cancelled assumptions being restricted to exactly one

$$\frac{\begin{matrix}[A]\\ \vdots \\ B\end{matrix}}{A \supset B} \supset I^1$$

The elimination of $\supset$

$$\frac{A \quad A \supset B}{B} \supset E$$

The introduction of $\neg$ with the number of cancelled assumptions being restricted to one

$$\dfrac{\begin{array}{c}[A]\\ \vdots\\ \bot\end{array}}{\neg A}\neg I^1$$

The elimination of $\neg$

$$\dfrac{A \quad \neg A}{\bot}\neg E$$

The introduction of $\wedge$

$$\dfrac{A \quad B}{A \wedge B}\wedge I$$

The elimination of $\wedge$ with the number of the cancelled assumptions being restricted to one for A, and one for B respectively

$$\dfrac{A \wedge B \qquad \begin{array}{c}[A] \quad [B]\\ \vdots\\ C\end{array}}{C}\wedge E^{1,1}$$

The 1 on the left side of the superscript of E indicates that the A used as an assumption is cancelled just once at the application of $\wedge E^{1,1}$. The 1 on the right side of the superscript of E indicates the similar restriction for the assumption B.

The first introduction of $\vee$ with the number of cancelled assumptions being restricted to one

$$\dfrac{\begin{array}{c}[\neg A]\\ \vdots\\ B\end{array}}{A \vee B}\vee I_1^1$$

The second introduction of $\vee$ with the the number of cancelled assumptions being restricted to one

$$\dfrac{\begin{array}{c}[\neg B]\\ \vdots\\ A\end{array}}{A \vee B}\vee I_2^1$$

The above two rules are modifications of the introduction rules of the multiplicative disjunction to suite to the present circumstances . Instead of trying to give the meanings to these rules within an already available framework, their meanings are to be sought in the actual use of these rules themselves.

The first elimination of $\vee$

$$\frac{\neg A \quad A \vee B}{B} \vee E_1$$

The second elimination of $\vee$

$$\frac{\neg B \quad A \vee B}{A} \vee E_2$$

The above two rules are adopted from the elimination rules of the multiplicative disjunction to suite to the restrictions under which we have been working. They may be regarded as expressing a sense of logical incompatibility.

The introduction of $\supset$ for the propositional constant 1 where the number of cancelled assumptions is restricted to zero

$$\frac{A}{1 \supset A} \supset I^0(1)$$

The superscript 0 of I indicates that the number of cancelled assumptions at the application of $\supset I^0$ is restricted to zero.

The symbol 1, regarded as a formula, is the axiom of D.

There are other inference rules not belonging to D.

The introduction of $\supset$ with the number of cancelled assumptions being restricted to zero

$$\frac{B}{A \supset B} \supset I^0$$

The superscript 0 of I indicates that the number of cancelled assumptions at the application of $\supset I^0$ is restricted to zero.

The introduction of $\supset$ with the number of cancelled assumptions being restricted to $n (n \geq 2)$

$$\dfrac{\begin{array}{c}\overbrace{[A]\cdots[A]}^{n\text{ times}}\\ \vdots \\ B\end{array}}{A \supset B} \supset I^n$$

The superscript n of I indicates that the number of cancelled assumptions at the application of $\supset I^n$ is restricted to two or more than two.

4 Sequent Calculus S

Next, we introduce the sequent calculus S as follows. We say that $(\Gamma_1 \rightarrow X_1, \Gamma_2 \rightarrow X_2, \cdots, \Gamma_n \rightarrow X_n \;/\; \Gamma \rightarrow X)$ is an instance of a certain inference rule if it has the form indicated by the corresponding figure. If $(\Gamma_1 \rightarrow X_1, \Gamma_2 \rightarrow X_2, \cdots, \Gamma_n \rightarrow X_n \;/\; \Gamma \rightarrow X)$ is an instance of an inference rule α , we call $\Gamma_i \rightarrow X_i$ the i-th upper sequent of α , and $\Gamma \rightarrow X$ the lower sequent of α.

The sequent $A \rightarrow A$ is called an axiom.

The sequent $\rightarrow 1$ is called the unit axiom of $\wedge$.

The sequent $\mathsf{I} \rightarrow$ is called the unit axiom of $\vee$.

The left rule of $\supset$

$$\frac{\Gamma \rightarrow A \quad B, \Delta \rightarrow C}{\Gamma, A \supset B, \Delta, \rightarrow C} \supset left$$

The right rule of $\supset$

$$\frac{A, \Gamma \rightarrow B}{\Gamma \rightarrow A \supset B} \supset right$$

The left rule of $\neg$

$$\frac{\Gamma \rightarrow A}{\neg A, \Gamma \rightarrow} \neg left$$

The right rule of $\neg$

$$\frac{A, \Gamma \rightarrow}{\Gamma \rightarrow \neg A} \neg right$$

The right hand side upper sequent of $\neg right$ have to be empty.

The left rule of $\wedge$

$$\frac{A, B, \Gamma \rightarrow C}{A \wedge B, \Gamma \rightarrow C} \wedge left$$

The right rule of $\wedge$

$$\frac{\Gamma \rightarrow A \quad \Delta \rightarrow B}{\Gamma, \Delta \rightarrow A \wedge B} \wedge right$$

The first left rule of $\vee$

$$\frac{A, \Gamma \rightarrow \quad B, \Delta \rightarrow C}{A \vee B, \Gamma, \Delta \rightarrow C} \vee left_1$$

The right hand side of first upper sequent of $\vee left_1$ has to be empty.

The second left rule of $\vee$

$$\frac{A, \Gamma \rightarrow C \quad B, \Delta \rightarrow}{A \vee B, \Gamma, \Delta \rightarrow C} \vee left_2$$

The right hand side of second upper sequent of $\vee left_2$ has to be empty.

The first right rule of $\vee$

This and the next rules are adapted from the introduction rule of multiplicative disjunction. Our main concern here is neither the subformula property nor the nomarization of proof procedures. We regard these rules as expressing a certain meaning of logical compatibility in its own forms.

$$\frac{\neg A, \Gamma \rightarrow B}{\Gamma \rightarrow A \vee B} \vee right_1$$

The second right rule of $\vee$

$$\frac{\neg B, \Gamma \rightarrow A}{\Gamma \rightarrow A \vee B} \vee right_2$$

The left rule of 1

$$\frac{\Gamma \to A}{1, \Gamma \to A} 1left$$

The right rule of $\bot$

$$\frac{\Gamma \to}{\Gamma \to \bot} \bot right$$

The right hand side of upper sequent of $\bot right$ have to be empty.

The left exchange rule

$$\frac{\Gamma, A, B, \Delta \to C}{\Gamma, B, A, \Delta \to C} \text{ex.}left$$

The cut rule

$$\frac{\Gamma \to A \quad A, \Delta \to C}{\Gamma, \Delta \to C} \text{cut}$$

There are other structural inference rules not belong to S.

The left contraction rule

$$\frac{A, A, \Gamma \to C}{A, \Gamma \to C} \text{c. } left$$

The left weakening rule

$$\frac{\Gamma \to C}{A, \Gamma \to C} \text{w. } left$$

5 Provability

Our notations about provability in D and S (and in systems resulting from them by certain enlargements) are fairly standard. Thorough definitions will be found in Gentzen [1], Prawitz [2], and Troelstra [4]. A deduction Σ in D is either a single formula or a ("tree-like") figure constructed by applying inference rules of D finitely many times. ("in D" is often deleted if it is clear from the context.) When ϕ is the formula located at the bottom ("root") of a deduction Σ , we say that Σ is a deduction of ϕ in D.

We distinguish proofs from deductions. A proof of ϕ in *D* is a deduction where all the assumptions of Σ are cancelled. ϕ is provable in *D* if we can construct a proof of ϕ in *D*. When Σ is a deduction of ϕ in *D*, and Γ is the set of uncancelled assumptions in Σ , we say Σ is a deduction of ϕ from Γ in *D*. In this case, following the common practice, we often say that Σ is a "proof" of ϕ from Γ in *D*, even if some assumptions Γ of ϕ are remaining uncancelled in the deduction Σ . If the context clearly tells which Σ is meant, we use the notation, $\Gamma \vdash \phi$ in *D*. Since the formula 1 is an axiom of *D*, the formula 1 appearing in Σ is regarded as being cancelled. Thus, one might feel inclined to think that 1 is not listed in the assumptions Γ of Σ. But we explicitly display 1 among the assumptions Γ whenever 1 appears in Σ.

A formula ϕ is provable from Γ in *D* if we can write a deduction Σ of ϕ from Γ in *D*.

Provability for a sequent is also defined in a usual way.

A proof Π in *S* is a list of sequents defined inductively as follows.

1. A sequent axioms $A \to A$, $\to 1$, $\bot \to$ are a proof in *S*.
2. If Π is a proof in *S*, then the figure obtained from Π by applying an inference rule of *S* to the last sequent of Π is also a proof in *S*.
3. The proofs in *S* are only those obtained by applying of (1) and (2) finitely many times.

Let α be an inference rule or axiom. Let X be a formal system. $X \cup \{\alpha\}$ means a new enlarged system resulting from the addition of an inference rule or axiom to a system X . For example, $D \cup \{\supset I^0\}$ is a system obtained from D by adding a new inference rule $\supset I^0$. The notions and phrases about provability carry over to those enlarged systems.

6 Characterization of Structural Inference Rules

In Gentzen [1], structural inference rules are formulated in classical and intuitionistic sequent calculuses called *LK* and *LJ*, respectively . In the corresponding Gentzen-style systems of natural deductions, namely *NK* and *NJ* in Gentzen [1], there is no structural inference rules. The powers of the structural inference rules in sequent calculuses are distributed unevenly over the logical inference rules of natural deduction systems, and can not easily be detected. Our strategy here is then, first, to introduce a system *S* of sequent calculus which is small enough not to include any structural inference rules and any effect resulting from their presence, but, large enough to capture basic characteristics of logical connectives . (The idea of removing and restricting structural inference rules in sequent calculuses is found in Oide [9].) Second, we carefully construct a natural deduction system, here called D, in such a way that the provability in D is the same as S, so that D is free of the effect of structural inference rules. Third, the effect of structural rules such as weakening and contraction will be captured in the corresponding axioms, and then the equivalence of the systems enlarged by these axioms are proved. In this way, the effects stemming from structural rules in sequent calculuses are isolated and added to natural deduction systems in such a way that these two systems posses the equal proving capabilities. Equivalence proof of S and D is obtained by adapting to our situation, the standard equivalence proof of LK and NK , which is carried out in detail in Gentzen [1]. Since the lack of structural inference rules require some care, we present the inductive proof of the equivalence of S and D.

Lemma 1
Let ϕ be a formula of the language L. Let Γ be a set of formulas of L. Then, ϕ is provable from Γ in D if and only if the sequent $\Gamma \to \phi$ is provable in S.

Proof First, we assume that ϕ is provable from Γ in D. Then, we can construct a deduction of ϕ from Γ in D

Let $A, Band\ C$ be formulas of L.

Let Σ be a deduction of ϕ from Γ in D.

We use induction on the number $\sharp\Sigma$ of the application of inference rules in Σ .

If the number of the application of inference rules in Σ is zero, then Σ consisted of only one formula ϕ. So we make a proof in S as follows .

$$\phi \to \phi$$

We assume the result for the case where $\sharp\Sigma$ is less than n, and prove it for n.

Case 1 The bottom inference rule in Σ is $\supset I^1$ and the form of ϕ is $A \supset B$.

By induction hypothesis, the sequent $A, \Gamma \to B$ can be proved in S. So we deduce the sequent $\Gamma \to A \supset B$ by $\supset$ *right* rule.

Case 2 The bottom inference rule in Σ is $\supset E$ and ϕ is B.

By induction hypothesis, the sequent $\Gamma_1 \to A$ and $\Gamma_2 \to A \supset B$ can be proved in S $(\Gamma_1 \cup \Gamma_2 = \Gamma)$.

We deduce the sequent $A \supset B, A \to B$ from $A \to A$ and $B \to B$ by $\supset$ *left* rule. Then we deduce $\Gamma_2, A \to B$ from

$A \supset B, A \to B$ and $\Gamma_2 \to A \supset B$ by *cut* rule. So we deduce $\Gamma_1, \Gamma_2 \to B$ from $\Gamma_1 \to A$ and $\Gamma_2, A \to B$ by *cut* rule.

Case 3 The cases where the bottom inference rules in Σ are $\neg I^1$, $\neg E$, $\wedge I$ are omitted.

Case 4 The bottom inference rule in Σ is $\wedge E^{11}$ and ϕ is C.

By induction hypothesis, the sequent $\Gamma_1 \to A \wedge B$ and $A, B, \Gamma_2 \to C$ can be proved in S $(\Gamma_1 \cup \Gamma_2 = \Gamma)$.

We deduce the sequent $A \wedge B, \Gamma_2 \to C$ from $A, B, \Gamma_2 \to C$ by $\wedge left$ rule. Then we deduce $\Gamma_1, \Gamma_2 \to C$ from $\Gamma_1 \to A \wedge B$ and $A \wedge B, \Gamma_2 \to C$ by *cut* rule.

Case 5 The bottom inference rule in Σ is $\vee I_1^1$ and ϕ is C.

By induction hypothesis, the sequent

$\neg A, \Gamma \to B$ can be proved in S .

Then we deduce the sequent $\Gamma \to A \vee B$ from $\neg A, \Gamma \to B$ by $\vee right_1$ rule.

Case 6 The cases where the bottom inference rules in Σ are $\vee I_2^1$, $\vee E_1$, $\vee E_2$ are omitted.

Case 7 The bottom inference rule in Σ is $\supset I^0(1)$ and ϕ is $1 \supset A$.

By induction hypothesis, the sequent $\Gamma \to A$ can be proved in S .

We deduce the sequent $1, \Gamma \to A$ by $1left$ rule.

Then we deduce the sequent $\Gamma \to 1 \supset A$ from $1, \Gamma \to A$ by $\supset right$ rule.

Next we assume that we can prove a sequent $\Gamma \to \phi$ in S.
$\Gamma \to$ in S is defined to correspond to $\Gamma \vdash \bot$ in D.
Let Π be proof of the sequent $\Gamma \to \phi$ in S.
We use induction on the number of the applications of inference rules in S. This number is denoted by the notation $\sharp\Pi$.
If $\sharp\Pi$ is zero, then Π consists of one of the axioms, $A \to A$, $\to 1$, $\bot \to$. These proofs in S correspond to deductions or proof; A, 1, $\bot$ in D respectively.
We assume the result for the cases where $\sharp\Pi$ is less than n , and prove it for n.

Case 1 the bottom inference rule in Π is $\supset right$ rule and the form of $\Gamma \to \phi$ is $\Gamma \to A \supset B$.

By induction hypothesis, the formula B is provable from $\Gamma \cup \{A\}$ in D. So we deduce the formula $A \supset B$ from B by applying the inference rule $\supset I^1$. Then $A \supset B$ is provable from Γ in D.

Case 2 The bottom inference rule in Π is $\supset left$ rule and the form of $\Gamma \to \phi$ is $\Gamma_1, A \supset B, \Gamma_2 \to C$.

By induction hypothesis, the formula A is provable from Γ_1 in D and the formula C is provable from $\Gamma_2 \cup \{B\}$ in D. So we deduce B from A and $A \supset B$ by $\supset E$ rule. Then C is provable from $\Gamma_1 \cup \Gamma_2 \cup \{A \supset B\}$ in D.

Case 3 The cases where the bottom inference rules in Π are $\neg right$, $\neg left$, $\wedge right$, $\wedge left$, $\vee right_1$, $\vee right_2$ are omitted.

Case 4 The bottom inference rule in Π is $\vee left_1$ rule and the form of $\Gamma \to \phi$ is $A \vee B, \Gamma_1, \Gamma_2 \to C$.

By induction hypothesis, the formula $\perp$ is provable from $\Gamma_1 \cup \{A\}$ in D and the formula C is provable from $\Gamma_2 \cup \{B\}$ in D. We deduce $\neg A$ from $\perp$ by applying $\neg I^1$ rule. So we deduce B from $\neg A$ and $A \vee B$ by applying $\vee E_1$ rule. Then C is provable from $\Gamma_1 \cup \Gamma_2 \cup \{A \vee B\}$ in D.

Case 5 The bottom inference rule in Π is $1left$ rule and the form of $\Gamma \to \phi$ is $1, \Gamma_1 \to A$.

By induction hypothesis, the formula A is provable from Γ_1 in D. We deduce $1 \supset A$ from A by applying $\supset I^0(1)$ rule. So we deduce A from 1 and $1 \supset A$ by applying $\supset E$ rule. Then A is provable from $\Gamma_1 \cup \{1\}$ in D.

Case 6 The cases where the bottom inference rules in Π are $\vee left_2$, $\perp right$, ex.$left$, *cut* are omitted. This completes the proof of *Lemma* 1.

To prove the following two theorems, we do not have to introduce notions such as labeling for the formulas appearing in a proof and the anti-lexicographic ordering on the labels . However, for an inductive argument to provide a concrete guide in re-writing a given proof to obtain another proof possessing an intended properties, we need labels and anti-lexicographic ordering. (Our definitions of these notions are not given for the general cases, but, limited to the cases relevant to our proofs.)

Let R be the set of all finite sequences consisting of zero's and one's. This R is a countably infinite set and is called the set of labels. If two sequences s and t in R are distinct form each other, there must be the first place, namely the i-th terms s_i and t_i, where the two sequences take different values for the first time, so that $s_i \neq t_i$. If $s_i < t_i$ (in the order of the usual magnitude of natural numbers), we put $s >_R t$. If $s_i > t_i$, we set $s <_R t$. Thus, we put the following definition of the anti-lexicographic ordering $>_R$ on R. For any two mutually distinct labels s and t, $s >_R t$

if and only if there is some i in $\mathbf{N}$ (the set of natural numbers including zero) such that $s_i < t_i$ and that $s_j = t_j$ for any $j < i$, where $<$ is the usual order of magnitude among the natural numbers. It is not difficult to see that this $<_R$ is a well ordering on the countable set R. We call $<_R$ the anti-lexicographic ordering on R.

Now, let Σ be any deduction of ϕ from Γ in D, and let $F(\Sigma)$ be the set of all formulas in Σ. We construct a labeling function $lb()$ from $F(\Sigma)$ to R as follows. First the conclusions ϕ (the "root") of the deduction Σ is assigned to $< 0 >$, the sequence of length one, consisting of just one zero. Next, suppose ϕ is the conclusion of (the application of) a certain inference rule, say IR, which has one assumption ψ. Then, ψ is assigned to $< 0, 0 >$. In case IR has two assumptions ϕ_1 on the left and ϕ_2 on the right, then ψ_1 is assigned to $< 0, 0 >$, while ψ_2 is assigned to $< 0, 1 >$. In general, if a formula θ in Σ is assigned to s, i.e., $lb(\theta) = s$, then the assumptions χ of the inference rule IR having θ as its conclusion is assigned to $s\hat{}< 0 >$, provide that IR has one assumption. In case IR has two assumptions χ_1 on the left and χ_2 on the right, then $lb(\chi_1) = s\hat{}< 0 >$, and $lb(\chi_2) = s\hat{}< 1 >$. The symbol $\hat{}$ indicates the concatenation of two sequences. The set of all labels assigned to the formulas in Σ is ordered by $<_R$ as a subset of R. In a similar way, we can define labeling function from the sequences in a given proof of S to$R$.

Theorem 2 (Characterization of weakening)
Let ϕ, X and Y be formulas of L.
Let Γ be a set of formulas of L.
Then, the following four statements are logically equivalent.

(1) We can prove ϕ from Γ in $D \cup \{\supset I^0\}$.

(2) We can prove ϕ from Γ in $D \cup \{$axiom $X \supset (Y \supset X)\}$.

(3) We can prove a sequent $\Gamma \to \phi$ in $S \cup \{$sequent axiom $\to X \supset (Y \supset X)\}$.

(4) We can prove a sequent $\Gamma \to \phi$ in $S \cup \{$the left weakening rule$\}$.

Proof

(1) $\Rightarrow$ (2)

Let Σ be a deduction of ϕ in $D \cup \{\supset I^0\}$.
We use induction on the number of the applications of $\supset I^0$ in Σ.
If Σ contains no application of $\supset I^0$, then we are done.
We assume the result for the case where the number of the applications of $\supset I^0$ in Σ is less than n, and we prove it for n.

The set P of all formulas appearing in Σ are ordered by the anti-lexicographic ordering $<_R$, which we defined earlier.

Let Q be the set of formulas which is the consequence of application of $\supset I^0$ in Σ. Then, Q , as a subset of P, is ordered by $<_R$.

We denote the elements of Q by,

$$q_1 <_R q_2 <_R \cdots <_R q_n.$$

Now, we consider a deduction Ξ of q_1 inside Σ , which has q_1 as its last formula. By the leastness of q_1 , Ξ contains no application of $\supset I^0$ except at the very bottom. Then Ξ can be re-written to Ξ' in the following way, so that Ξ' does not contain any application of $\supset I^0$ at all.

$$\cfrac{\begin{matrix}\Xi\\ B\end{matrix}}{A \supset B}\supset I^0 \qquad q_1 = A \supset B$$

The above deduction Ξ can be re-written as follows.

$$\cfrac{\begin{matrix}\Xi\\ B\end{matrix} \qquad B \supset (A \supset B)}{A \supset B}\supset E$$

Let Σ' the deduction resulting from Σ by replacing Ξ with Ξ'.

We then notice that the number of applications of $\supset I^0$ in Σ' is strictly less than n (in fact $n-1$). By our induction hypothesis, we obtain the proof Σ_0 of ϕ having no application of $\supset I^0$.

(2) $\Rightarrow$ (1)

Let Σ be a deduction of ϕ in $D \cup \{axiom\ X \supset (Y \supset X)\}$.

We use induction on the number of the axioms of $X \supset (Y \supset X)$ in Σ.

If Σ contain no axiom of $X \supset (Y \supset X)$ in Σ , then we are done.

We assume the result for the case where the number of the axioms of $X \supset (Y \supset X)$ in Σ is less than n, and we prove it for n.

The set P of all formulas appearing in Σ are ordered by the anti-lexicographic ordering $<_R$, which we defined earlier.

Let Q be the set of axioms $X \supset (Y \supset X)$ in Σ. Q , as a subset of P, is ordered by $<_R$.

We denote the elements of Q by

$$q_1 <_R q_2 <_R \cdots <_R q_n.$$

Now, the axiom q_1 in Σ can be proved as follows:

$$\cfrac{\cfrac{A}{B \supset A}\supset I^0}{A \supset (B \supset A)}\supset I^1$$

where q_1 is $A \supset (B \supset A)$.

Let Σ' be the deduction resulting from Σ by replacing the axiom q_1 by the above proof.

We then notice that the number of the axioms $X \supset (Y \supset X)$ in Σ' is strictly less than n (in fact $n-1$). By our induction hypothesis, we obtain the deduction Σ_0 of ϕ having no axioms of $X \supset (Y \supset X)$.

(2) $\Leftrightarrow$ (3)

This is an immediate consequence of *Lemma* 1, together with the fact that we can prove the axiom of $X \supset (Y \supset X)$ if and only if we can prove the sequent axiom of $\rightarrow X \supset (Y \supset X)$.

(3) $\Rightarrow$ (4)

Let Π be a proof of $\Gamma \rightarrow \phi$ in $S \cup \{\text{sequent axiom } \rightarrow X \supset (Y \supset X)\}$.

We use induction on the number of the sequent axioms of $\rightarrow X \supset (Y \supset X)$.

If Π contain no sequent axiom of $\rightarrow X \supset (Y \supset X)$, then we are done.

We assume the result for the case where the number of the sequent axioms of $\rightarrow X \supset (Y \supset X)$ in Π is less than n , and we prove it for n.

The set T of all sequents in Π are ordered by the anti-lexicographic ordering $<_R$, which we define earlier.

Let U be the set of sequent axioms of $\rightarrow X \supset (Y \supset X)$ in Π. Then, U , as a subset of T, is ordered by $<_R$.

We denote the elements of U by,

$$u_1 <_R u_2 <_R \cdots <_R u_n.$$

The axiom u_1 can be proved in $S \cup$ { the left weakening rule } as follows.

$$\cfrac{\cfrac{\cfrac{A \rightarrow A}{B, A \rightarrow A}\ \text{w.}left}{A \rightarrow B \supset A}\ \supset right}{\rightarrow A \supset (B \supset A)}\ \supset right$$

Let Π' be the proof resulting from Π by replacing u_1 by its above proof .

We then notice that the number of the sequent axioms $\rightarrow X \supset (Y \supset X)$ in Π is strictly less than n (in fact $n-1$). By our induction hypothesis, we obtain the proof Π_0 of $\Gamma \rightarrow \phi$ having no sequent axiom $\rightarrow X \supset (Y \supset X)$.

(4) $\Rightarrow$ (3)

Let Π be a proof of $\Gamma \to \phi$ in $S \cup \{\text{the left weakening rule}\}$.

We use induction on the number of the applications of w.$left$ in Π.

If Π contain no application of w.$left$, then we are done.

We assume the result for the case where the number of the applications of w.$left$ rule in Π is less than n , and we prove it for n.

The set T of all sequents in Π are ordered by the anti-lexicographic ordering $<_R$, which we define earlier.

Let U be the set of sequents which is the lower sequent of w.$left$ rule in Π. U , as a subset of T, is ordered by $<_R$.

We denote the elements of U by,

$$u_1 <_R u_2 <_R \cdots <_R u_n.$$

Now, we consider a proof Θ of u_1 inside Π , which has u_1 as its last sequent. By the leastness of u_1 , Θ contains no application of w.$left$ rule except at the very bottom. Then Θ can be re-written to Θ' in the following way so that Θ' does not contain any application of w.$left$ rule at all.

$$\dfrac{\begin{matrix}\Theta\\ A,\Gamma_1, \to C\end{matrix}}{B,A,\Gamma_1 \to C}\,\text{w.}left \qquad u_1 = B,A,\Gamma_1 \to A$$

The above proof Θ can be re-written as follows.

$$\dfrac{\dfrac{\to \chi \qquad \dfrac{A \to A \qquad \dfrac{B \to B \quad A \to A}{B \supset A, B \to A}\supset left}{A \supset (B \supset A), A, B \to A}\supset left}{A, B \to A}\,cut \qquad \dfrac{\Theta}{A,\Gamma_1 \to C}}{\dfrac{A,B,\Gamma_1 \to C}{B,A,\Gamma_1 \to C}\,\text{ex.}left}\,\text{cut}$$

where χ is a formula $A \supset (B \supset A)$.

When the set $\{A\} \cup \Gamma_1$ of formulas is empty, we use 1 as the cut formula instead. This proof is as follows.

$$\dfrac{\dfrac{\to 1 \supset (B \supset 1) \qquad \dfrac{\to 1 \qquad \dfrac{B \to B \quad 1 \to 1}{B \supset 1, B \to 1}\supset left}{1 \supset (B \supset 1), B \to 1}\supset left}{B \to 1}\,cut \qquad \dfrac{\dfrac{\Theta}{\to C}}{1 \to C}}{B \to C}\,\text{cut}$$

We then notice that the number of applications of w.$left$ in Π' is strictly less than n (in fact $n-1$). By our induction hypothesis, we obtain the proof Π_0 of $\Gamma \to \phi$ having no application of w.$left$. This completes the proof of Theorem 2.

Theorem 3 (Characterization of contraction)
Let ϕ, X *and* Y *be formulas of* L.
Let Γ *be a set of formula of* L.
Then the following four statements are logically equivalent.

(1) *We can prove* ϕ *from* Γ *in* $D \cup \{\supset I^n (n \geq 2)\}$.

(2) *We can prove* ϕ *from* Γ *in* $D \cup \{\text{axiom } (X \supset (X \supset Y)) \supset (X \supset Y)\}$.

(3) *We can prove a sequent* $\Gamma \rightarrow \phi$ *in* $S \cup \{\text{sequent axiom } \rightarrow (X \supset (X \supset Y)) \supset (X \supset Y)\}$.

(4) *We can prove a sequent* $\Gamma \rightarrow \phi$ *in* $S \cup \{$ the left contraction rule $\}$.

Lemma 4
Let ϕ, X *and* Y *be formulas of* L.
Let Γ *be a set of formula of* L.
We can prove ϕ *from* Γ *in* $D \cup \{\supset I^2\}$ *if and only if we can prove* ϕ *from* Γ *in* $D \cup \{\supset I^n (n \geq 2)\}$.

Proof of Theorem **3**

(1) $\Rightarrow$ (2)

We assume *Lemma* 4.

By *Lemma* 4, we can prove ϕ from Γ in $D \cup \{\supset I^2\}$ if and only if we can prove ϕ from Γ in $D \cup \{\supset I^n (n \geq 2)\}$. So, it is sufficient to show that ϕ from Γ in $D \cup \{\supset I^2\}$ can be proved if and only if ϕ can be proved from Γ in $D \cup \{\text{axiom}(A \supset (A \supset B)) \supset (A \supset B)\}$. Let Σ be a deduction of ϕ in $D \cup \{\supset I^2\}$.

We use induction on the number of the applications of $\supset I^2$ in Σ.

If Σ contain no application of $\supset I^2$, then we are done.

We assume the result for the case where the number of the applications of $\supset I^2$ in Σ is less than n, and we prove it for n.

The set P of all formulas appearing in Σ are ordered by the anti-lexicographic ordering $<_R$, which we defined earlier.

Let Q be the set of formulas which is the consequence of application of $\supset I^2$ in Σ. Q, as a subset of P, is ordered by $<_R$.

We denote the elements of Q by,

$$q_1 <_R q_2 <_R \cdots <_R q_n.$$

Now, we consider a deduction Ξ of q_1 inside Σ , which has q_1 as its last formula. By the leastness of q_1 , Ξ contains no application of $\supset I^2$ except at the very bottom. Then Ξ can be re-written to Ξ' in the following way, so that Ξ' does not contain any application of $\supset I^2$ at all.

$$\dfrac{\begin{matrix}[A]\ [A] \\ \Xi \\ B\end{matrix}}{A \supset B} \supset I^2 \qquad q_1 = A \supset B$$

The above proof Ξ can be re-written as follows.

$$\dfrac{\dfrac{\dfrac{\begin{matrix}[A]\ [A] \\ \Xi \\ B\end{matrix}}{A \supset B} \supset I^1}{A \supset (A \supset B)} \supset I^1 \qquad (A \supset (A \supset B)) \supset (A \supset B)}{A \supset B} \supset E$$

Let Σ' be the poof resulting from Σ by replacing Ξ with Ξ'.

We then notice that the number of applications of $\supset I^2$ in Σ' is strictly less than n (in fact $n-1$). By our induction hypothesis, we obtain the deduction Σ_0 of ϕ having no application of $\supset I^2$.

(2) $\Rightarrow$ (1)

Let A and B be formulas of L.

Let Σ be a deduction of ϕ in $D \cup \{\text{axiom } (X \supset (X \supset Y)) \supset (X \supset Y)\}$.

We use introduction on the number of the axioms of $(X \supset (X \supset Y)) \supset (X \supset Y)$ in Σ.

If Σ contain no axiom of $(X \supset (X \supset Y)) \supset (X \supset Y)$ in Σ, then we are done.

We assume the result for the case where the number of the axioms of $(X \supset (X \supset Y)) \supset (X \supset Y)$ in Σ is less than n, and we prove it for n.

The set P of all formulas appearing in Σ are ordered by the anti-lexicographic ordering $<_R$, which we defined earlier.

Let Q be the set of axiom formulas of $(X \supset (X \supset Y)) \supset (X \supset Y)$ in Σ. Q , as a subset of P, is ordered by $<_R$.

We denote the elements of Q by,

$$q_1 <_R q_2 <_R \cdots <_R q_n.$$

The axiom of q_1 can be re-written as follows

$$\dfrac{\dfrac{\dfrac{A \qquad \dfrac{A \quad A \supset (A \supset B)}{A \supset B} \supset E}{B} \supset E}{A \supset B} \supset I^2}{(A \supset (A \supset B)) \supset (A \supset B)} \supset I^1$$

Let Σ' be the deduction resulting from Σ by re-writing the axiom q_1.

We then notice that the number of the axioms of $(X \supset (X \supset Y)) \supset (X \supset Y)$ in Σ' is strictly less than n (in fact $n-1$). By our induction hypothesis, we obtain the deduction Σ_0 of ϕ having no axioms of $(X \supset (X \supset Y)) \supset (X \supset Y)$.

(2) $\Leftrightarrow$ (3)

This is an immediate consequence of *Lemma* 1, together with the fact that we can prove the axiom of $(X \supset (X \supset Y)) \supset (X \supset Y)$ in $D \cup \{\text{axiom } (X \supset (X \supset Y)) \supset (X \supset Y)\}$ if and only if we can prove the sequent axiom of $\to (X \supset (X \supset Y)) \supset (X \supset Y)$ in $S \cup \{\text{sequent axiom } \to (X \supset (X \supset Y)) \supset (X \supset Y)\}$.

(3) $\Rightarrow$ (4)

Let Π be a proof of $\Gamma \to \phi$ in $S \cup \{\text{sequent axiom } \to (X \supset (X \supset Y)) \supset (X \supset Y)\}$.

We use induction on the number of the sequent axioms of $\to (X \supset (X \supset Y)) \supset (X \supset Y)$.

If Π contain no sequent axiom of $\to (X \supset (X \supset Y)) \supset (X \supset Y)$, then we are done.

We assume the result for the case where the number of the sequent axioms of $\to (X \supset (X \supset Y)) \supset (X \supset Y)$ in Π is less than n , and we prove it for n.

The set T of all sequents in Π are ordered by the anti-lexicographic ordering $<_R$, which we define earlier.

Let U be the set of sequent axioms of $\to (X \supset (X \supset Y)) \supset (X \supset Y)$ in Π. U , as a subset of T, is ordered by $<_R$.

We denote the elements of U by,

$$u_1 <_R u_2 <_R \cdots <_R u_n.$$

The axiom of u_1 can be re-written as follows

$$\dfrac{A \to A \qquad \dfrac{A \to A \quad B \to B}{A \supset B, A \to B}\ \supset left}{A \supset (A \supset B), A, A \to B}\ \supset left$$

$$\dfrac{A \supset (A \supset B), A, A \to B}{A, A, A \supset (A \supset B) \to B}\ \text{ex.}left\ \ n \text{ times}$$

$$\dfrac{A, A, A \supset (A \supset B) \to B}{A, A \supset (A \supset B) \to B}\ \text{c.}left$$

$$\dfrac{\dfrac{A, A \supset (A \supset B) \to B}{A \supset (A \supset B) \to A \supset B}\ \supset right}{\to A \supset (A \supset B) \supset (A \supset B)}\ \supset right$$

Let Π' be the proof resulting from Π by re-writing the sequent axiom u_1.

We then notice that the number of the sequent axioms of $\to (X \supset (X \supset Y)) \supset (X \supset Y)$ in Π' is strictly less than n (in fact $n-1$). By our induction hypothesis, we obtain the proof Π_0 of $\Gamma \to \phi$ having no sequent axiom of $\to (X \supset (X \supset Y)) \supset (X \supset Y)$.

(4) $\Rightarrow$ (3)

Let Π be a proof of $\Gamma \to \phi$ in $S \cup$ { the left contraction rule }.

We use induction on the number of the applications of c.$left$ in Π.

If Π contain no application of c.$left$, then we are done.

We assume the result for the case where the number of the applications of c.$left$ rule in Π is less than n , and we prove it for n.

The set T of all sequents in Π are ordered by the anti-lexicographic ordering $<_R$, which we define earlier.

Let U be the set of sequents which is the lower sequent of c.$left$ rule in Π. U , as a subset of T, is ordered by $<_R$.

We denote the elements of U by,

$$u_1 <_R u_2 <_R \cdots <_R u_n.$$

Now, we consider a proof Θ of u_1 inside Π , which has u_1 as its last sequent. By the leastness of u_1 , Θ contains no application of c.$left$ rule except at the very bottom. Then Θ can be re-written to Θ' in the following way so that Θ' does not contain any application of c.$left$ rule at all.

$$\dfrac{\begin{array}{c}\Theta\\ A, A, \Gamma \to C\end{array}}{A, \Gamma_1 \to C}\,\text{c.}left \qquad (u_1 = A, \Gamma_1 \to C)$$

The above proof Θ can be re-written as follows.

$$\dfrac{\dfrac{\to \chi \qquad \dfrac{\dfrac{\dfrac{\dfrac{\begin{array}{c}\Theta\\ A, A, \Gamma_1 \to C\end{array}}{A, \Gamma_1 \to A \supset C}\supset right}{\Gamma_1 \to A \supset (A \supset C)}\supset right \qquad \dfrac{A \to A \quad C \to C}{A \supset C, A \to C}\supset left}{\Gamma_1, (A \supset (A \supset C)) \supset (A \supset C), A \to C}\supset left}{(A \supset (A \supset C)) \supset (A \supset C), \Gamma_1, A \to C}\,\text{ex.}left\ \text{n}\ times}{\Gamma_1, A \to C}\,\text{cut}}{A, \Gamma_1 \to C}\,\text{ex.}left\ \text{n}\ times$$

where χ is a formula $(A \supset (A \supset C)) \supset (A \supset C)$.

When the formula of C is not present, we use $\perp$ as the cut formula instead. The proof in this case is as follows.

$$
\dfrac{
\rightarrow \chi
\qquad
\dfrac{
\dfrac{
\dfrac{
\dfrac{
\dfrac{
\dfrac{
\begin{matrix}\Theta \\ A, A, \Gamma_1 \rightarrow\end{matrix}
}{A, A, \Gamma_1 \rightarrow \bot}\ \bot right
}{A, \Gamma_1 \rightarrow A \supset \bot}\ \supset right
}{\Gamma_1 \rightarrow A \supset (A \supset \bot)}\ \supset right
\qquad
\dfrac{A \rightarrow A \quad \bot \rightarrow}{A \supset \bot, A \rightarrow}\ \supset left
}{\Gamma_1, (A \supset (A \supset \bot)) \supset (A \supset \bot), A \rightarrow}\ \supset left
}{(A \supset (A \supset \bot)) \supset (A \supset \bot), \Gamma_1, A \rightarrow}\ \text{ex.}left \ \ \text{n} \ \ times
}
}{
\dfrac{\Gamma_1, A \rightarrow}{A, \Gamma_1 \rightarrow}\ \text{ex.}left \ \ \text{n} \ \ times
}\ \text{cut}
$$

where χ is a formula $(A \supset (A \supset \bot)) \supset (A \supset \bot)$.

We then notice that the number of applications of $c.left$ in Π' is strictly less than n (in fact $n-1$). By our induction hypothesis, we obtain the proof Π_0 of $\Gamma \rightarrow \phi$ having no application of $c.left$. This completes the proof of Theorem 3.

Proof of Lemma **4** First, we assume that ϕ is provable from Γ in $D \cup \{\supset I^2\}$. Then it is obvious that we can prove ϕ from Γ in $D \cup \{\supset I^2(n \geq 2\}$.

Next, we shall assume that we prove ϕ from Γ in $D \cup \{\supset I^2(n \geq 2)\}$.

Let Σ a deduction of ϕ in $D \cup \{\supset I^2(n \geq 2)\}$.

We use induction on the number of the applications of $\cup\{\supset I^n(n \geq 2)$ in Σ .

If Σ contain no application of $\supset I^n(n \geq 3)$, then we are done.

We assume the result for the case where the number of the applications of $\supset I^n(n \geq 3)$ in Σ is less than n, and we prove it for n.

The set P of all formulas appearing in Σ are ordered by the anti-lexicographic ordering $<_R$, which we defined earlier.

Let Q be the set of formulas which is the consequence of application of $\supset I^n(n \geq 3)$ in Σ. Q, as a subset of P, is ordered by $<_R$.

We denote the elements of Q by,

$$q_1 <_R q_2 <_R \cdots <_R q_n.$$

Now, we consider a deduction Ξ of q_1 inside Σ , which has q_1 as its last formula. By the leastness of q_1 , Ξ contains no application of $\supset I^n(n \geq 3)$ except at the very bottom. Then Ξ can be re-written to Ξ' in the following way, so that Ξ' does not contain any application of $\supset I^n(n \geq 3)$ at all.

$$
\dfrac{
\begin{matrix}
\overbrace{[A] \cdots [A]}^{\text{n}} \\
\Xi \\
B
\end{matrix}
}{A \supset B}\ \supset I^n(n \geq 3)
$$

The above deduction of Ξ can be re-written as follows.

$$
\begin{array}{ccl}
 & \overbrace{[A]\cdots[A]}^{\text{n}} & \\
 & \Xi & \\
 & B & \\
\cline{2-2}
A & A \supset B & \supset I^2 \\
\hline
 & B & \supset E \\
\cline{2-2}
 & A \supset B & \supset I^2 \\
 & & \supset E \\
 & \vdots & \vdots \\
A & A \supset B & \\
\hline
 & B & \supset E \\
\cline{2-2}
 & A \supset B & \supset I^2
\end{array}
$$

After these re-writing, the number of the application of $\supset I^2$ becomes $n-1$, and the number of the application of $\supset E$ becomes $n-2$.

Let Σ' be the deduction resulting from Σ by replacing Ξ with Ξ'. We then notice that the number of applications of $\supset I^n (n \geq 3)$ in Σ' is strictly less than n (in fact $n-1$). By our induction hypothesis, we obtain the deduction Σ_0 of ϕ having no application of $\supset I^n (n \geq 3)$. So we can prove ϕ from Γ in $D \cup \{\supset I^2\}$.

7 Summary

In the present paper, we characterized two structural inference rules, weakening and contraction, in Gentzen-style natural deduction. We plan to investigate the exchange rule by using the anti-lexicographic ordering on the set of labels for a proof. We also would like to investigate the natural deduction systems which include additive conjunction and disjunction in a sequel to the present paper.

References

[1] Gerhard Gentzen, *The collected papers of Gerhard Gentzen*, ed. by M. E. Szabo, Noth-Holland, 1969.

[2] D. Prawitz, *Natural Deduction, A Proof-Theoretical Study*, Almqvist & Wiksell, 1965.

[3] Gunnar Stålmarck, Normalization theorems for full First-order Classical natural Deduction, *The Journal of Symbolic Logic*, Volume 56, Number 1, March 1991.

[4] A. S. Troelstra, *Lectures on Linear Logic*, CSLI Lecture Note Number 29, Center for the Study of Language and Information, Stanford University, 1992.

[5] A. S. Troelstra, Natural deduction for intuitionistic linear logic, *Annals of Pure and Applied Logic*, Volume 73, 1995.

[6] S. Maehara, *Suri Ronri Gaku Jyosetu* (Introduction to Mathematical Logic), Kyoritu Syuppan, 1966.

[7] S. Maehara, *Suri Ronri Gaku* (Mathematical Logic), Baihu-kan, 1974.

[8] G. Takeuti, *Senkei Ronri Nyumon* (Linear Logic), Nihon Hyoron-sya, 1995.

[9] A. Oide, Quantum Logic and Related Systems, *Annals of Japan for Philosophy of Science*, Volume 6, 1983.

[10] H. Ono, Semantics for Substructural Logics in: *Substructural logics*, K.Dosen and P. Schroeder-Heister eds., Oxford University Press, 1993.

[11] H. Ono, *Joho Kagaku ni okeru Ronri* (Logic in Information Science), Nihon Hyoron-sya, 1994.

Linear set theory with strict comprehension

Masaru Shirahata*

Abstract

In this paper, we study the extensionality axiom in the set theory with the unrestricted comprehension based on linear logic. We first review Grishin's result which shows the imcompatibility of the extensionality axiom and the unrestricted comprehension in linear set theory. As one way to remedy this situation, we introduce the notion of "strict comprehension" and formulate a system of linear set theory which contains the extensionality and the strict comprehension. The consistency of such a system is then proved by a cut-elimination argument. Finally, we sketch the development of arithmetic in our system.

1 Introduction

The set theory based on contraction-free logics [2, 8] has been studied for the reason that the use of unrestricted comprehension does not necessarily yield the contradiction in such logics [3, 4, 7, 9, 10, 13]. However, it has been also noticed that the standard extensionality axiom cannot be added to such a set theory without causing inconsistency [5, 14]. In this paper, we show one way to remedy this situation, using the notion of "strict comprehension."

The basic idea is as follows. Contraction of a formula in a sequent calculus corresponds to *copying* resources, or proofs of that formula (terms of that type). In linear and affine logics [2], the contraction rule is not available and, as a result, we can discharge (abstract) at most one occurrence of the assumption A (variable $x : A$) in forming the implication $A \multimap B$ ($(\lambda x : A)\ M$). We apply the similar restriction to the formation of set terms $\{x : A\}$ so that we have the logical equivalence between the formulas $A[s/x]$ and $s \in \{x : A\}$ only when the variable x has at most one occurrence in the formula A. Such a comprehension rule will be called *strict (or linear) comprehension.* A similar idea to this was already found in Fitch's work on combinatory logic [1].

*This work was partially supported by JSPS research fellowship for young scientists.

We first review Grishin's result which shows the imcompatibility of the extensionality axiom and the unrestricted comprehension in linear set theory. We then introduce the notion of strict comprehension and formulate a system of linear set theory which contains the extensionality and the strict comprehension. The consistency of such a system is then proved by a cut-elimination argument. Finally, we sketch the development of arithmetic in our system.

2 The extensionality and linear set theory

The extensionality principle says that two sets are identical if they have the same members. This is a very natural principle, given our intuitive notion of sets. However, it turned out that the set theory with the unrestricted comprehension based on affine (BCK) logic becomes inconsistent with the extensionality added.

The proof is due to Grishin [5]. We assume the standard axioms for the equality:

$$\vdash s = s \qquad\qquad s = t, A \vdash A[t/s]$$

where the formula $A[t/s]$ is obtained from the formula A by replacing the term t for some of the occurrences of the term s. We work in the set theory with the unrestricted comprehension

$$\frac{\Gamma, A[s/x] \vdash \Delta}{\Gamma, s \in \{x : A\} \vdash \Delta} \qquad \frac{\Gamma \vdash A[s/x], \Delta}{\Gamma \vdash s \in \{x : A\}, \Delta}$$

which is based on affine logic, *i.e.* the logic without contraction but with weakening.

First, let's recall that the left contraction rule for the formula A is equivalent to the axiom $A \vdash A \otimes A$, which may be called *duplication* or *copying*. Then, note that we can easily recover the contraction rule for the equations $s = t$ as follows:

$$\frac{\dfrac{\vdash s = s \quad \vdash s = s}{\vdash s = s \otimes s = s} \qquad s = t, s = s \otimes s = s \vdash s = t \otimes s = t}{s = t \vdash s = t \otimes s = t}$$

Secondly, we define the terms $s \equiv \{x : x = q\}$ and $t \equiv \{x : x = q \otimes A\}$ for the formula A and show that A is logically equivalent to the equation $s = t$. For this, we use the extensionality written as the rule

$$\frac{\Gamma, x \in s \vdash x \in t, \Delta \quad \Gamma, x \in t \vdash x \in s, \Delta}{\Gamma \vdash s = t, \Delta}$$

where the variable x has no free occurrence in the multi-sets Γ and Δ. We state the proof of the logical equivalence in the informal fashion, from which one can easily recover the derivation in the sequent calculus.

Let's assume $x \in t$. We unwind the term and obtain $x = q \otimes A$. Discarding (left weakening) A, we have $x = q$ and hence, $x \in s$. This amounts to the derivation of the sequent $x \in t \vdash x \in s$.

On the other hand, let's assume A and $x \in s$. We can immediately obtain $x = q \otimes A$. Winding this up, we have $x \in t$. Hence, we have the derivation of $A, x \in s \vdash x \in t$. Then, combining the two directions by the weakening and extensionality, we have $A \vdash s = t$.

For the other direction, assume $s = t$. Since $q = q$ always holds, we have $q \in s$. Hence, $q \in t$. Unwinding this, we obtain $q = q \otimes A$. Discarding $q = q$, we have A. Hence, $s = t \vdash A$ is derivable.

Since the contraction is admissible for $s = t$, it is admissible for the formula A as well. Therefore, we can recover the contraction for any formula. This implies that one can again derive the set-theoretical paradoxes.

Note that there are two critical points in the above proof:

- One of the standard axioms for the equality allows the substitution of the multiple occurrences of one term by another term, and makes it possible to recover the contraction for equations. This seems to be against the spirit of linear logic, which is supposed to be resource conscious.

- The weakening is used in establishing the logical equivalence between $s = t$ and A. It is not clear if one can do the same in the logic without weakening, *i.e.* in linear logic.

In this paper, we pursue the way to modify the system with respect to the first point, which leads to the notion of "strict comprehension." The second point should also be investigated, but we postpone it for another occasion.

3 The system STCOM of linear set theory

Our guiding idea is to restrict substitution $A[t/s]$ under the equation $s = t$ so that at most one occurrence of s is replaced by t. To achieve this effect, however, we need to modify the comprehension principle as well. To see the point, let's consider the formula $A \equiv q \in s \otimes s \in p$. Under the assumption $s = t$, we can perform the substitution to obtain $q \in t \otimes s \in p$, but we should not be able to obtain $q \in t \otimes t \in p$. However, A is logically equivalent to $s \in \{x : q \in x \otimes x \in p\}$ by the comprehension, and one can use

$$s = t, s \in \{x : q \in x \otimes x \in p\} \vdash t \in \{x : q \in x \otimes x \in p\}$$

to derive

$$s = t, q \in s \otimes s \in p \vdash q \in t \otimes t \in p.$$

We therefore restrict the comprehension rule so that we can form the formula $s \in \{x : A\}$ from the formula $A[s/x]$ if and only if the variable x occurs at most once in A. This type of comprehension will be called *strict comprehension.*

This condtion can be relaxed when additives are involved. The formulas $A\&B$ and $A \oplus B$ in linear logic represent "choices" between A and B in the computational interpretation. Therefore, even when a term occurs in both A and B, only one of the occurrence of the term is used in computation (*i.e.* cut-elimination). Hence, two occurrences of the variable x in $A(x)$ and $B(x)$ can be identified in $A(x)\&B(x)$ and $A(x) \oplus B(x)$, and we allow the strict comprehension to create the term $\{x : A(x)\&B(x)\}$ and $\{x : A(x) \oplus B(x)\}$.

Note that the strict comprehension does not prevent the set-theoretical paradoxes in classical logic, since one can easily write a term such as $\{x : \exists y(x = y \otimes y \notin y\}$ which has the same effect as the standard Russell set in the presence of contraction. In general, the strict comprehension and ordinary comprehension do not make any difference if contraction is allowed for equations, since the term $\{x : A\}$ with the multiple occurrences of x in A can be written as $\{x : \exists y(x = y \otimes A')\}$ where A' is the formula obtained from A by replacing all the occurrences of x by y.

Definition 1 *The terms and formulas are defined by simultanous induction with the notion of free and strict variables:*

1. *The variables, denoted x, y, z ..., are terms. They are free and strict in themselves.*

2. *If s and t are terms, then $s \in t$ and $s = t$ are (atomic) formulas. They inherit the free variables from the terms. The free variables x are strict in them when x is strict in s or t, but not in both.*

3. *If A is a formula with a strict variable x or without any occurrence of x, then $\{x : A\}$ is a term. This term inherits the free and strict variables of A except for x.*

4. *If A and B are formulas, then $A \otimes B$ and $A \parr B$ are formulas. They inherit the free variables from A and B. The variables x are strict in them when x is strict in A or B, but not in both.*

5. *If A and B are formulas, then $A\&B$ and $A \oplus B$ are formulas. They inherit the free variables from A and B. The variable x is strict in them when x is strict in A or B, and possibly in both.*

6. *If A is a formula and x is a variable, then $A^\perp$, $\forall xA$ and $\exists xA$ are formulas. They inhrerit the free and strict variables from A except for x.*

We write $s \notin t$ and $s \neq t$ for $(s \in t)^\perp$ and $(s = t)^\perp$, respectively.

In the formulation of the deductive system, we use the Gentzen style two-sided sequent calculus. We write $e[t/x]$ for the expression obtained from the expression e by replacing the term t for a strict variable x in e. If e does not contain x at all, $e[t/x]$ is e itself.

Axioms:

$$s \in t \vdash s \in t$$

$$s = t, (p \in q)[s/x] \vdash (p \in q)[t/x] \qquad s = t, (p \in q)[t/x] \vdash (p \in q)[s/x]$$

Rules of inference:

$$\frac{\Gamma, A \vdash \Delta}{\Gamma \vdash A^{\perp}, \Delta} \qquad \frac{\Gamma \vdash A, \Delta}{\Gamma, A^{\perp} \vdash \Delta}$$

$$\frac{\Gamma, A, B \vdash \Delta}{\Gamma, A \otimes B \vdash \Delta} \qquad \frac{\Gamma \vdash A, \Delta \quad \Theta \vdash B, \Sigma}{\Gamma, \Theta \vdash A \otimes B, \Delta, \Sigma}$$

$$\frac{\Gamma, A \vdash \Delta \quad \Theta, B \vdash \Sigma}{\Gamma, \Theta, A \parr B \vdash \Delta, \Sigma} \qquad \frac{\Gamma \vdash A, B, \Delta}{\Gamma \vdash A \parr B, \Delta}$$

$$\frac{\Gamma, A \vdash \Delta}{\Gamma, A \& B \vdash \Delta} \qquad \frac{\Gamma, B \vdash \Delta}{\Gamma, A \& B \vdash \Delta} \qquad \frac{\Gamma \vdash A, \Delta \quad \Gamma \vdash B, \Delta}{\Gamma \vdash A \& B, \Delta}$$

$$\frac{\Gamma, A \vdash \Delta \quad \Gamma, B \vdash \Delta}{\Gamma A \oplus B \vdash \Delta} \qquad \frac{\Gamma \vdash A \Delta}{\Gamma \vdash A \oplus B, \Delta} \qquad \frac{\Gamma \vdash B \Delta}{\Gamma \vdash A \oplus B, \Delta}$$

$$\frac{\Gamma, A[s/x] \vdash \Delta}{\Gamma, \forall x A \vdash \Delta} \qquad \frac{\Gamma \vdash A, \Delta}{\Gamma \vdash \forall x A, \Delta}$$

where the variable x has no free occurrence in Γ or Δ in the righthand rule.

$$\frac{\Gamma, A \vdash \Delta}{\Gamma, \exists x A \vdash \Delta} \qquad \frac{\Gamma \vdash A[s/x], \Delta}{\Gamma \vdash \exists x A, \Delta}$$

where the variable x has no free occurrence in Γ or Δ in the lefthand rule.

$$\frac{\Gamma, A[s/x] \vdash \Delta}{\Gamma, s \in \{x : A\} \vdash \Delta} \qquad \frac{\Gamma \vdash A[s/x], \Delta}{\Gamma \vdash s \in \{x : A\}, \Delta}$$

where A is a formula with the strict variable x or without any occurrence of x.

$$\frac{\Gamma, x \in s \vdash x \in t, \Delta \quad \Gamma, x \in t \vdash x \in s, \Delta}{\Gamma \vdash s = t, \Delta}$$

where the variable x has no free occurrence in Γ or Δ.

$$\frac{\Gamma \vdash A, \Delta \quad \Theta, A \vdash \Sigma}{\Gamma, \Theta \vdash \Delta, \Sigma} \qquad \frac{\Gamma \vdash \Delta}{\Gamma, A \vdash \Delta} \qquad \frac{\Gamma \vdash \Delta}{\Gamma \vdash A, \Delta}$$

Proposition 2 *The sequents*

1. $\vdash s = s$

2. $s = t \vdash t = s$

3. $s = t, (p = q)[s/x] \vdash (p = q)[t/x]$

4. $s = t, (p = q)[t/x] \vdash (p = q)[s/x]$

are derivable.

Proof
We have the derivations:

$$\frac{x \in s \vdash x \in s \quad x \in s \vdash x \in s}{\vdash s = s} \qquad \frac{s = t, x \in t \vdash x \in s \quad s = t, x \in s \vdash x \in t}{s = t \vdash t = s}$$

For 3 and 4, suppose x is strict in p but not in q. We have the derivations:

$$\frac{s = t, y \in p[t/x] \vdash y \in p[s/x] \quad p[s/x] = q, y \in p[s/x] \vdash y \in q}{s = t, p[s/x] = q, y \in p[t/x] \vdash y \in q}$$

and

$$\frac{p[s/x] = q, y \in q \vdash y \in p[s/x] \quad s = t, y \in p[s/x] \vdash y \in p[t/x]}{s = t, p[s/x] = q, y \in q \vdash y \in p[t/x]}$$

¿From them, the sequent $s = t, p[s/x] = q \vdash p[t/x] = q$ is obtained by the extensionality. Other cases are similarly proved. ∎

In order to prove the properties of the system, we need to introduce the *size* of proofs. ;

Definition 3 *The size $\sigma(\pi)$ of a proof π is defined inductively:*

1. *If π is an axiom, then its size is* 1;

2. *If π is obtained from τ by one of the one-premise rules, then $\sigma(\pi)$ is $\sigma(\tau) + 1$;*

3. *If π is obtained from π_1 and π_2 by cut, the right rule for $\otimes$ or the left rule for $\invamp$, then $\sigma(\pi) = \sigma(\pi_1) + \sigma(\pi_2)$;*

4. *If π is obtained from π_1 and π_2 by the extensionality, the right rule for & or the left rule for $\oplus$, then $\sigma(\pi) = max(\sigma(\pi_1), \sigma(\pi_2)) + 1$.*

Proposition 4 *For any proof π of the sequent $\Gamma, A[s/x] \vdash \Delta$ or $\Gamma \vdash A[s/x], \Delta$, there exists a proof $\hat{\pi}$ of the sequent $\Gamma, s = t, A[t/x] \vdash \Delta$ or $\Gamma, s = t \vdash A[t/x], \Delta$, respectively, such that the size of $\hat{\pi}$ is at most $\sigma(\pi) + 1$. The same statement holds with $t = s$ in place of $s = t$.*

Proof

We only consider the case when the equation is $s = t$, since the other case is entirely similar. The proof is by induction on the size of π. When π is obtained from π_1 (and π_2) by an application of one of the inference rules, we call π_1 (and π_2) the immediatel subproof(s) of π. If A is not the main formula of the last inference of π, we apply the inductve hypothesis to the immediate subproof in which A occurs. For the right $\forall$ and left $\exists$ rules, we assume the appropriate renaming of the eigenvariable. If $A[s/x]$ is obtained by one of the weakenings, so is $A[t/x]$. Otherwise, we have five cases.

Case 1 $A[s/x]$ has the form $(B \otimes C)[s/x]$ or $(B \parr C)[s/x]$. Then, it suffices to perform the substitution in exactly one of the formulas $B[s/x]$ and $C[s/x]$ since the strict variable x appears in at most one of them. Hence, we only need to apply the inductive hypothesis to the appropriate immediate subproof.

Case 2 $A[s/x]$ has the form $(B \& C)[s/x]$ or $(B \oplus C)[s/x]$. For the left $\&$ rule and the right $\oplus$ rule, it suffices to perform the substitution in one of the formulas $B[s/x]$ or $C[s/x]$. Hence, we use the inductive hypothesis to the immediate subproof. For the right $\&$ rule and the left $\oplus$ rule, the strict variable x may appear in both B and C. However, we can use the inductive hypothesis to both of the immediate subproofs as well without creating additional occurrence of the formula $s = t$ in the end sequent. By the definition of size, this increases the size of a proof at most by one.

Case 3 $A[s/x]$ has the form $B[s/x]^{\perp}$, $\forall y B[s/x]$ or $\exists y B[s/x]$. We then apply the inductive hypothesis to the immediate subproof.

Case 4 $A[s/x]$ has the form $(p \in q)[s/x]$. Then, we will use the axiom $s = t, (p \in q)[t/x] \vdash (p \in q)[s/x]$ or $s = t, (p \in q)[s/x] \vdash (p \in q)[t/x]$ to perform a cut. This increases the size of a proof by one.

Case 5 $A[s/x]$ has the form $(p = q)[s/x]$. If $s = t$ is obtained as the main formula of an appplication of extensionality, then apply the inductive hypothesis to both of the immediate subproofs of π. This increases the size of the proof at most by one. If the formula $(p = q)[s/x]$ is on the lefthand side, then it is from one of the axioms:

$$(p = q)[s/x], (a \in b)[p[s/x]/y] \vdash (a \in b)[q[s/x]/y]$$

or

$$(p = q)[s/x], (a \in b)[q[s/x]/y] \vdash (a \in b)[p[s/x]/y]$$

where x in fact occurs in at most one of p or q. Without the loss of generality, we can concentrate on the first case with x occurs in p but not in q. We then have the following.

$$\frac{s = t, (a \in b)[p[s/x]/y] \vdash (a \in b)[p[t/x]/y] \quad p[t/x] = q, (a \in b)[p[t/x]/y] \vdash (a \in b)[q/y]}{s = t, p[t/x] = q, (a \in b)[p[s/x]/y] \vdash (a \in b)[q/y]}$$

Other cases are entirely similar. In any on those cases, we only add one application of cut with an axiom. Hence, the size of a proof increases by one. ∎

4 The cut-elimination theorem

We use three types of reduction rules, *i.e.* 1) permutative reductions, 2) symmetric reductions and 3) axiom reductions. For the full exposition of those rules which do not involve the equality, we refer the reader to [9, 10].

Permutative reductions This type of reductions are applied when one of the cut formulas is not the main formula of the immediate subproof. They are mostly standard and we only give the rules for the extensionality:

$$\frac{\Gamma \vdash A, \Delta \quad \dfrac{A, \Theta, x \in s \vdash x \in t, \Sigma \quad A, \Theta, x \in t \vdash x \in s, \Sigma}{A, \Theta \vdash s = t, \Sigma}}{\Gamma, \Theta \vdash s = t, \Delta, \Sigma}$$

is tranformed to ($\leadsto$)

$$\frac{\dfrac{\Gamma \vdash A, \Delta \quad A, \Theta, y \in s \vdash y \in t, \Sigma}{\Gamma, \Theta, y \in s \vdash y \in t, \Sigma} \quad \dfrac{\Gamma \vdash A, \Delta \quad A, \Theta, y \in t \vdash y \in s, \Sigma}{\Gamma, \Theta, y \in t \vdash y \in s, \Sigma}}{\Gamma, \Theta \vdash s = t, \Delta, \Sigma}$$

where y is a fresh new variable. The case when the cut-formula is on the righthand side in the extensionality is similarly handled. Since $(max(m,n)+1)+l = max(m+l, n+l)+1$, this transformation does not change the size of a proof. In general, one can establish by observation that the permutative reductions preserve the size of a proof. In particular, we can freely permute two consecutive applications of cut without changing the size.

Symmetric reductions This type of reductions are applied when both of the cut-formulas are the main formulas of the respective immediate subproofs, except for the weakening reduction which is applied when one of the cut-formulas is the main formula. They are all standard. Hence, we only state two exemplary cases.

$$\frac{\dfrac{\Gamma_1 \vdash A, \Delta_1 \quad \Gamma_2 \vdash B, \Delta_2}{\Gamma_1, \Gamma_2 \vdash A \otimes B, \Delta_1, \Delta_2} \quad \dfrac{\Theta, A, B \vdash \Sigma}{\Theta, A \otimes B \vdash \Sigma}}{\Gamma_1, \Gamma_2, \Theta \vdash \Delta_1, \Delta_2, \Sigma} \leadsto \frac{\Gamma_2 \vdash B, \Delta_2 \quad \dfrac{\Gamma_1 \vdash A, \Delta_1 \quad \Theta, A, B \vdash \Sigma}{\Gamma_1, \Theta, B \vdash \Delta_1, \Sigma}}{\Gamma_1, \Gamma_2, \Theta \vdash \Delta_1, \Delta_2, \Sigma}$$

$$\dfrac{\dfrac{\Gamma \vdash A[s/x], \Delta}{\Gamma \vdash s \in \{x : A\}, \Delta} \quad \dfrac{\Theta, A[s/x] \vdash \Sigma}{\Theta, s \in \{x : A\} \vdash \Sigma}}{\Gamma, \Theta \vdash \Delta, \Sigma} \quad \leadsto \quad \dfrac{\Gamma \vdash A[s/x], \Delta \quad \Theta, A[s/x] \vdash \Sigma}{\Gamma, \Theta \vdash \Delta, \Sigma}$$

We note that the size of a proof is strictly decreased by an application of any of the symmetric reductions.

Axiom reductions This type of reductions are used where one of the cut-formulas is in an axiom. Similar reductions were defined for the normalization of a natural deduction system of set theory [6]. For this type of cut, there are cases where no reduction is defined. When we substitute the term t for the term s in $p \in s$ to obtain $p \in t$, we call such a substitution *critical.* Otherwise, the substitution is *non-crtical.* When there are two reduction rules which only differ in the position (*i.e.* the lefthand side or righthand side of the symbol $\vdash$) of cut formulas or the order of terms in equation (*i.e.* $s = t$ and $t = s$), we only state one of them explicitly.

$$\dfrac{s \in t \vdash s \in t \quad s \in t, \Gamma \vdash \Delta}{s \in t, \Gamma \vdash \Delta} \quad \leadsto \quad s \in t, \Gamma \vdash \Delta$$

The size of a proof is strictly decreased by this reduction. For non-critical cuts, we have the reduction:

$$\dfrac{s = t, (p \in \{x : A\})[s/y] \vdash (p \in \{x : A\})[t/y] \quad \dfrac{A[p/x][t/y], \Gamma \vdash \Delta}{(p \in \{x : A\})[t/y], \Gamma \vdash \Delta}}{s = t, (p \in \{x : A\})[s/y], \Gamma \vdash \Delta}$$

is tranformed to ($\leadsto$)

$$\dfrac{s = t, A[p/x][s/y], \Gamma \vdash \Delta}{s = t, (p \in \{x : A\})[s/y], \Gamma \vdash \Delta}$$

where the new proof is obtained using $\hat{\pi}$ for the subproof π of the sequent $A[p/x][t/y], \Gamma \vdash \Delta$. Since $\sigma(\hat{\pi}) \leq \sigma(\pi) + 1$, this transformation preserves the size of a proof. For critical cuts, we have the following:

$$\dfrac{\dfrac{\Gamma, x \in s \vdash x \in t, \Delta \quad \Gamma, x \in t \vdash x \in s, \Delta}{\Gamma \vdash s = t, \Delta} \quad s = t, p \in s \vdash p \in t}{\Gamma, p \in s \vdash p \in t, \Delta} \quad \leadsto \quad \Gamma, p \in s \vdash p \in t, \Delta$$

where the new proof is obtained by replacing the term p for the variable x in the subproof of the sequent $\Gamma, x \in s \vdash x \in t, \Delta$. By this transformation, the size of a proof is strictly decreased.

Each reduction rule gives a one-step reduction. We say that a proof π *reduces to* a proof τ if there is a finite sequence of one-step reductions which transforms π into τ.

In order to prove the (partial) cut-elimination theorem, we need to define what the results of reductions are. For this, we first describe a certain configuration of cuts in a given proof π. The subproof τ of π will be called a *chain* if τ is a consecutive application of cuts with the axioms for substitution, *i.e.* it has the form

$$\dfrac{s_0 = t_0, p_0 \in q_0 \vdash p_1 \in q_1 \qquad \dfrac{\dfrac{s_n = t_n, p_n \in q_n \vdash p_{n+1} \in q_{n+1}}{\vdots}}{s_1 = t_1, \ldots, s_n = t_n, p_1 \in q_1 \vdash p_{n+1} \in q_{n+1}}}{s_0 = t_0, \ldots, s_n = t_n, p_0 \in q_0 \vdash p_{n+1} \in q_{n+1}}$$

where $p_{i+1} \in q_{i+1}$ is obtained from $p_i \in q_i$ by the substitution of s_i (or t_i) for t_i (s_i).

The equations involved in the chain are called the *gates*. They are either *critical* or *non-critical* according to whether the substitutions with them are critical or not. We call a chain *non-critical* if it consisits entirely of non-critical substitutions. The formulas $p_o \in q_o$ and $p_{n+1} \in q_{n+1}$ are the *end-formulas* of the chain τ. They are called *critical* as well if the substitutions for them are critical.

Now suppose that the chain in a proof π is followed by a consecutive application of cuts such that

1. the cut formulas for each cut are
   (a) one of the non-critical gates and the main formula of a subproof ended with the extensionality, or
   (b) one of the end-formulas, which is critical, and the main formula of a subproof ended with one of the comprehension rules,
2. none of the critical gates and non-critical end-formulas are used as cut-formulas in π.

Such a configuration of cuts will be called a *cluster* in the proof π. To be precise, a configuration consists of a chain and the cut-formulas involved in the cuts following the chain. A cluster α is *closed* if one of the end-formulas of the chain is used as a cut-formula in α. If both of the end-formulas are used as cut-formulas in α, it is *closed at both ends*. If a cluster is not closed, then it is *open*.

Lemma 5 *Suppose that the proof π is obtained from π_1 and π_2 by a cut, where the last inference rule of π_1 is the right (left) comprehension to create* *the cut formula $p \in q$ ($p' \in q'$) and π_2 is a non-critical chain with the end-formulas $p \in q$ and $p' \in q'$. Then, π reduces to a proof τ such that the last inference rule of τ is the right (left) comprehension to create $p' \in q'$ ($p \in q$) and $\sigma(\tau) \leq \sigma(\pi)$.*

Proof
We only consider the case where the cut-formula is created by the right comprehension since the other case is entirely similar. The proof is by induction on the size of the non-ctitical chain. Note that the cut-formula $p \in q$ originated from the substitution axiom $s = t, p \in q \vdash p'' \in q''$. We first use the permutation to create the cut with π_1 and the substitution axiom, followed by the cut with the rest of the chain. We then apply the non-critical axiom reduction to the upper cut. The result τ'again has the configuration described in the condition of the lemma, with $p'' \in q''$ replaced for $p \in q$ and the size of the chain decreased, so that we can apply the inductive hypothesis. Furthermore, the size of the entire proof does not change by the non-critical axiom reduction so that $\sigma(\tau) \leq \sigma(\tau') \leq \sigma(\pi)$. ∎

Definition 6 *The proof π is called normal if the cuts in π appears only as part of clusters in π.*

Theorem 7 *For any proof π, there exists a normal proof τ such that π reduces to τ and $\sigma(\tau) \leq \sigma(\pi)$.*

Proof
The proof is by induction on the size of a proof. If π does not end with a cut, we apply the inductive hypothesis to the immediate subproof(s) of π. If π ends with a cut, we can use the inductive hypothesis to assume that each of the immediate subproof is already normal. Hence, we only consider the proofs π such that π is obtained from the normal subproofs π_1 and π_2 by a cut with the cut-formula C.

We divide the form of each subproof into five types: 1) the identity axiom, 2) the substitution axiom, 3) ended with a rule of inference other than cut and C is its main formula, 4) ended with a rule of inference other than cut but C is not its main formula, and 5) ended with a cut. Accordingly, we need to consider 25 possibilities in total. We will cut down the number of those possibilities, step by step.

Step 1 If one of the subproof is the identity axiom (type 1), we can use the axiom reduction to decrease the size of a proof and apply the inductive hypothesis. This cuts down the number to 16. If one of the subproof ends with a rule of inference but C is not its main formula (type 4), we can use the permutative reduction and apply the inductive hypothesis. Now, we have only 9 possibilities.

Step 2 If both of the subproofs are substitution axioms (type 2/type2), then π itself is already a chain. If both of the subproofs end with rules of inference with C as their main formulas (type 3/type 3), then we can use the symmetric reduction to decrease the size and apply the inductive hypothesis. The number is now 7.

Step 3 Let's consider the cases type 2/type 3 and type 3/type 2. For the former, the last inference rule of π_2 must be the left comprehension. If π_1 is critical, then π already forms a cluster. If π_1 is not critical, then we can use the non-critical axiom reduction and use the inductive hypothesis to the immediate subproof of the result. For the latter, the last inference rule of π_1 must be the right comprehension or the extensionality. If it is the comprehension, then we can use the same argument as we have just seen for type 2/type 3. Hence suppose that the rule is the extensionality. If π_2 is critical, then we can use the critical axiom reduction to decrease the size and apply the inductive hypothesis. If π_2 is non-critical, then π already forms a cluster. The number is then 5.

For the rest of 5 possibilities, the subproof π_1 or π_2 ended with a cut (type 5) is involved. We only argue for the case when π_1 is of type 5 since the other case is entirely similar. Since π_1 is normal, the cut must be part of a cluster. Then, the cut-formula C of the cut with π_1 and π_2 comes from the chain for the cluster or from one of the subproofs τ of π_1 which is used in one of the consecutive applications of cut with the chain. Suppose that C comes from τ. Since τ is part of the cluster, τ is obtained from its immediate subproof τ' by the comprehension or the extensionality. We then use the permutative reductions for cut to move the cut with π_2 right after τ', followed by the comprehension or the extensionalty. Since $\sigma(\tau') + \sigma(\pi_2) < \sigma(\pi_1) + \sigma(\pi_2) = \sigma(\pi)$, we can apply the inductive hypothesis to the proof obtained from τ' and π_2 by a cut. Note that it is still followed by the comprehension or the extensionality after the reduction. Hence, the result forms a cluster. Therefore, we can assume that C comes from the chain.

Step 4 We now consider the cases (type 5/type 5), (type 2/type 5) and (type 5/type 2). For the first case, we can assume that C comes from the chains of π_1 and π_2. Then, we combine the two chains by the permutative reductions to create a new chain. For the second and the third cases, we similarly use the permutative reductions to combine the substitution axiom and the chain in the other subproof to create a new chain.

Step 5 The last two cases are (type 3/type 5) and (type 5/type 3). For the former, the last inference rule of π_1 is either the comprehension or the extensionality. Suppose that it is the extensionality. Then C is one of the gates in the chain of π_2. If the gate is critical, then we can use the critical reduction to decrease the size and apply the inductive hypothesis. If the gate is non-critical, π already forms a cluster. Hence suppose that the last inference rule is the comprehension. Then C is one of the end-formulas of the chain of π_2. If the end-formula is critical, then π already forms a cluster. If the end-formula is non-critical, it is part of a non-critical segment of the chain. We take the maximal such segment in size. We then use the permutative reductions and Lemma 5 to move the non-critical segment to the inside of π_1. The last inference rule of this new subproof, call it τ_1, is the comprehension according to Lemma 5. Note that there may be subproofs of π_2 which are used in cuts with the gates of the non-critical chain. By the permuative reductions for cut, we move such subproofs above the last rule of inference of τ_1, resulting in a subproof τ_2. Since all those transformations do not increase the size, we have $\sigma(\tau_2) \leq \sigma(\pi)$. Hence, we can apply the inductive hypothesis to the immediate subproof of τ_2. The result is a subproof τ_3 which consists of a normal proof followed by the comprehension. The subproof τ_3 is therefore normal. If the non-critical segment is the entire chain, τ_3 is the entire proof. Otherwise, the chain can be divided into the non-critical segment and the segment whose lefthand end-formula is critical. The entire proof then consists of τ_3 followed by a cut with the cluster for the latter segment, whose cut-formula comes from the critical end-formula. Hence, the entire proof forms a cluster as well. This transfomation process can be illustrated by the following diagram:

$$\dfrac{\begin{array}{c}\vdots\, \pi_1 \\ \Gamma \vdash p_0 \in q_0, \Delta\end{array} \qquad \dfrac{\Theta \vdash s = t, \Sigma \qquad \dfrac{\dfrac{\Psi, p \in q \vdash p_{n+1} \in q_{n+1}}{\vdots\,(*)}}{\Psi_0, \Psi, s = t, p_0 \in q_0 \vdash p_{n+1} \in q_{n+1}}}{\Theta, \Psi_0, \Psi, p_0 \in q_0 \vdash p_{n+1} \in q_{n+1}, \Sigma}}{\Gamma, \Theta, \Psi_0, \Psi \vdash p_{n+1} \in q_{n+1}, \Sigma, \Delta}$$

reduces to

$$\dfrac{\begin{array}{c}\vdots\, \tau_3 \\ \Gamma, \Theta, \Psi_0 \vdash p \in q, \Sigma, \Delta\end{array} \qquad \Psi, p \in q \vdash p_{n+1} \in q_{n+1}}{\Gamma, \Theta, \Psi_0, \Psi \vdash p_{n+1} \in q_{n+1}, \Sigma, \Delta}$$

where $(*)$ is the maximal non-critical segment with $s = t$ as one of its gates. For the case (type 5/type 3), the last inference rule of π_2 must be the comprehension. We can then use the same argument as we have just seen for (type 3/type 5). ∎

We introduce the ordering among clusters in a given proof π. Suppose that the closed cluster α is formed around the chain τ. By the definition of a cluster, none of the critical gates and non-critical end-formulas in τ are used as cut-formulas in π. However, they may appear as subformulas of formulas $s \in t$ or $s = t$ which are used in cuts within other clusters. If one of the critical gates or non-critical end-formulas of the cluster α appears as a subformula of one of the cut-formulas in the cluster β, we say that β is *lower than* α.

In order to construct a formula which contains a gate or an end-formula A as a subformula, some inference rules other than cut need to be applied to A. Since a cluster is a consecutive application of cuts, such a formula is formed outside and below the cluster itself. Therefore, it is not possible that a gate or an end-formula of the lower cluster appears as a subformula of a cut-formula in the upper cluster. Hence, the ordering thus defined among clusters does not contain a loop. Then, there are clusters which are *minima* under this ordering.

Corollary 8 *There is no proof of the empty sequent.*

Proof
By Theorem 7, it suffices to consider normal proofs. Let π be a normal proof. If π does not contain a cut, then the end sequent of π is not empty. Otherwise, choose one of the clusters, α, which is a *minimum* under the ordering. If the cluster is not closed at both ends, then one of the end-formulas remains as a subformula in the end sequent. If the cluster is closed at both ends, then the chain contains at least one critical gate, which remains as a subformula in the end sequent. ∎

5 Arithmetic in STCOM

We now sketch the development of arithmetic in STCOM. Since the formal proofs in STCOM usually become very large, we only give informal proofs.

Exponentials First of all, we may bootstrap STCOM by adding the exponential operators. For this, we extend the definition of formulas to include $!A$ and $?A$ but we do not allow comprehension over the formulas containing the exponentials. Hence, the terms of this extended system are exactly the same as before. The use of extensionality is restricted accordingly, and we have the standard inference rules for exponentials:

$$\frac{A, \Gamma \vdash \Delta}{!A, \Gamma \vdash \Delta} \quad \frac{\Gamma \vdash \Delta, A}{\Gamma \vdash \Delta, ?A} \qquad \frac{!A, !A, \Gamma \vdash \Delta}{!A, \Gamma \vdash \Delta} \quad \frac{\Gamma \vdash \Delta, ?A, ?A}{\Gamma \vdash \Delta, ?A}$$

$$\frac{!\Gamma \vdash ?\Delta, A}{!\Gamma, \vdash ?\Delta, !A} \qquad \frac{A, !\Gamma \vdash ?\Delta}{?A, !\Gamma \vdash ?\Delta}$$

where $!\Gamma$ and $?\Gamma$ are the multisets of formulas of the form $!A$ and $?A$, respectively.

The consistency of the extended system can be shown by the cut elimination argument. In this case, we first remove all the cuts with the cut-formulas $!\,A$ or $?\,A$, and then we can proceed as before [9]. Note that if we allow the exponentials inside the set terms, then the Russell's paradox will be reproduced with the term $\{x : \exists y[\,!\,(x = y) \otimes\, ?\,(y \notin y)]\}$.

The constants 1 and $\perp$ can be defined as $s = s$ and $s \neq s$, respectively. We write $!_n\, A$ for the formula

$$\underbrace{A \otimes A \otimes \ldots \otimes A}_{n \text{ times}}.$$

Furthermore, we write $A \multimap B$ for $A^{\perp} \parr B$, and $s \subset t$ for $\forall x(x \in s \multimap x \in t)$.

Fact 9 *The following sequents are provable in STCOM extended with the exponentials:*

1. $!\,s = t,\ !\,A[t/x] \vdash\ !\,A[s/x]$,
2. $!\,s = t,\ !\,A[s/x] \vdash\ !\,A[t/x]$,

and similarly for $?\,A[t/x]$ *and* $?\,A[s/x]$.

Pairs The unordered pairs $\{s, t\}$ are defined as $\{x : x = s \oplus x = t\}$. The singleton $\{s\}$ is $\{s, s\}$, which is equivalent to $\{x : x = s\}$ by the extensionality. Then, the ordered pair $\langle s, t\rangle$ can be defined in the standard way as $\{\{s\}, \{s, t\}\}$. Under this definition, we can show the standard property of the ordered pairs.

Fact 10 *The following sequents are provable in STCOM:*

1. $s = s' \otimes t = t' \vdash \langle s, t\rangle = \langle s', t'\rangle$,
2. $\langle s, t\rangle = \langle s', t'\rangle \vdash s = s'$,
3. $s = s' \otimes\, !_2\, \langle s, t\rangle = \langle s', t'\rangle \vdash t = t'$,
4. $!_2\, \langle s, t\rangle = \langle s, t'\rangle \vdash t = t'$.

Numerals and Natural Numbers The numerals $\hat{n}$ for the natural numbers n are defined in the standard way. We write $\bigcup s$ for $\{x : \exists y(x \in y \otimes y \in s)\}$ and $s \cup t$ for $\bigcup\{s, t\}$. The successor operation sc is defined as $sc(s) =_{def} s \cup \{s\}$. Then, $\hat{0} =_{def} \{x : c \neq c\}$ where c is a fixed term not containing x, and $\widehat{n+1} =_{def} sc(\hat{n})$. The set of all natural numbers has the inductive structure, for which the use of exponentials is necessary. Hence, we should not expect to obtain a term which represents such a set. Instead, we define the predicate $N(x)$ by

$$N(x) \Leftrightarrow_{def} \forall y[\,!\,(\hat{0} \in y \& \forall z(\,!\,z \in y \multimap sc(z) \in y)) \multimap x \in y].$$

Under this definition, we have the induction principle limited for the terms, and we can show that all the numerals satisfy $N(x)$.

Proposition 11 *The following are provable:*

1. $!(\hat{0} \in s \& \forall y(!y \in s \multimap sc(y) \in s)) \vdash \forall x(N(x) \multimap x \in s)$,

2. $\vdash N(\hat{n})$, *for any natural number* n.

Proof

1. This is immediate from the definition.

2. We use the (external) induction on n. For $n = 0$, assume

$$!(\hat{0} \in y \& \forall z(!z \in y \multimap sc(z) \in y)).$$

Then, $\hat{0} \in y$. Hence, $N(\hat{0})$. For $n = m + 1$, assume

$$!(\hat{0} \in y \& \forall z(!z \in y \multimap sc(z) \in y)).$$

Then, $\forall x(N(x) \multimap x \in y)$. By the inductive hypothesis, $N(\hat{m})$. Hence, $\hat{m} \in y$. So, $!\hat{m} \in y$. By the assumption again, $\forall z(!z \in y \multimap s(z) \in y)$. Hence, $\widehat{m+1} \in y$ so that $N(m \hat{+} 1)$. ∎

Fact 12 *The following are provable for any natural numbers* n*:*

1. $\vdash \widehat{n+1} \neq \hat{0}$,

2. $\vdash \hat{0} = \hat{n} \oplus \exists x(N(x) \otimes \hat{n} = s(x))$

We write $\emptyset$ for the empty multiset. This is useful for the proof by contradiction, and we can delete the contradictory cases as follows.

$$\dfrac{\dfrac{A_1 \vdash \emptyset}{A_1 \vdash B} \quad A_2 \vdash B}{A_1 \oplus A_2 \vdash B}$$

where we use weakening with the formula B.

Proposition 13 *The following are provable for any natural numbers* n *and* m*:*

1. $\hat{n} \in \hat{m} \vdash \hat{n} \subset \hat{m}$,

2. $\hat{n} \in \hat{m} \otimes \hat{m} \in \hat{p} \vdash \hat{n} \in \hat{p}$,

3. $\hat{n} \in \hat{n} \vdash \emptyset$,

4. $\hat{n} = \hat{m} \vdash \emptyset$, *for* $n \neq m$.

Proof

1. We use the induction on m. When $m = 0$, this holds vacuously. Suppose $\hat{n} \in \widehat{p+1}$. Then, $\hat{n} \in \hat{p}$ or (in the sense of $\oplus$) $\hat{n} = \hat{p}$. In the former case, $\hat{n} \subset \hat{p} \subset \widehat{p+1}$ by the inductive hypothesis. In the latter case, $\hat{n} = \hat{p} \subset \widehat{p+1}$.

2. Suppose $\hat{n} \in \hat{m} \otimes \hat{m} \in \hat{p}$. Then, $\hat{n} \in \hat{m} \otimes \hat{m} \subset \hat{p}$. Hence, $\hat{n} \in \hat{p}$.

3. We use the induction on n. When $n = 0$, this is obvious. Suppose $\widehat{p+1} \in \widehat{p+1}$. Then, $\widehat{p+1} \in \hat{p}$ or $\widehat{p+1} = \hat{p}$. In the former case, $\hat{p} \in \widehat{p+1} \subset \hat{p}$. Hence, $\hat{p} \in \hat{p}$. By the inductive hypothesis, we have then $\emptyset$. In the latter case, $\hat{p} \in \widehat{p+1} = \hat{p}$. Hence $p \in p$ so that we have again $\emptyset$.

4. Assume $\hat{n} = \hat{m}$. There are two cases. If $n < m$, then $\vdash n \in m$ so that $n \in n$. Hence, we obtain $\emptyset$. Similarly for the case $m < n$. ∎

For the proof of the next proposition, we recall that the multiplicative conjunction $\otimes$ distributes over additive disjunction $\oplus$:

$$\dfrac{\dfrac{A, C \vdash A \otimes C}{A, C \vdash (A \otimes C) \oplus (B \otimes C)} \quad \dfrac{B, C \vdash B \otimes C}{B, C \vdash (A \otimes C) \oplus (B \otimes C)}}{(A \oplus B) \otimes C \vdash (A \otimes C) \oplus (B \otimes C)}$$

$$\dfrac{\dfrac{A, C \vdash (A \oplus B) \otimes C}{A \otimes C \vdash (A \oplus B) \otimes C} \quad \dfrac{B, C \vdash (A \oplus B) \otimes C}{B \otimes C \vdash (A \oplus B) \otimes C}}{(A \otimes C) \oplus (B \otimes C) \vdash (A \oplus B) \otimes C}$$

This allows to combine cases by the multiplicative conjunction and makes the following inference admissible:

$$\dfrac{\Gamma \vdash A_1 \oplus \ldots \oplus A_m \quad \Delta \vdash B_1 \oplus \ldots \oplus B_n}{\Gamma, \Delta \vdash (A_1 \otimes B_1) \oplus \ldots \oplus (A_m \otimes Bn)}$$

where the number of cases on the right is m by n.

Proposition 14 *The following sequents are provable for any m and n:*

1. $!_2\, sc(x) = sc(\hat{m}) \vdash x = \hat{m}$,

2. $!_n\, x \in \hat{m} \vdash\, !_n\, x = \hat{0} \oplus \ldots \oplus\, !_n\, x = \widehat{m-1}$.

Proof

1. Since $\vdash x \in sc(x)$, we can derive $sc(x) = sc(\hat{m}) \vdash x \in sc(\hat{m})$. Furthermore, $x \in sc(\hat{m})$ implies $x = \hat{0} \oplus \ldots \oplus x = \hat{m}$. Then, we have

$$\begin{aligned} !_2\, sc(x) = sc(\hat{m}) \quad \vdash \quad & (x = \hat{0} \otimes sc(x) = sc(\hat{m})) \oplus \ldots \\ & \oplus (x = \widehat{m-1} \otimes sc(x) = sc(\hat{m})) \oplus (x = \hat{m}) \end{aligned}$$

and

$$!_2\, sc(x) = sc(\hat{m}) \vdash \hat{1} = sc(\hat{m}) \oplus \ldots (\hat{m} = sc(\hat{m})) \oplus (x = \hat{m}).$$

For $n \leq m$, we can derive $\hat{n} = sc(\hat{m}) \vdash \emptyset$ since $n \neq m+1$. Hence,

$$!_2\, sc(x) = sc(\hat{m}) \vdash x = \hat{m}$$

2. The proof is by induction on n. When $n = 0$, this holds obviously since $x \in \hat{m}$ implies $x = \hat{0} \oplus \ldots \oplus x = \widehat{m-1}$. For $n = p+1$, let $\alpha(\hat{r})$ be $x = \hat{r}$. Then,

$$!_p\, x \in \hat{m} \otimes x \in \hat{m} \vdash (\,!_p\, \alpha(\hat{0}) \oplus \ldots\, !_p\, \alpha(\widehat{m-1})) \otimes (\alpha(\hat{0}) \oplus \alpha(\widehat{m-1}))$$

We distribute $\otimes$ over $\oplus$. However, if $r \neq r'$, then $\alpha(\hat{r}) \otimes \alpha(\hat{r'}) \vdash \emptyset$. Hence,

$$!_{p+1}\, x \in \hat{m} \vdash\, !_{p+1}\, \alpha(\hat{0}) \oplus\, !_{p+1}\, \alpha(\widehat{m-1})$$

∎

Functions Since our terms are very much restricted, we should not expect to have terms represent numerical functions. Instead, we consider the numeralwise representation of n–ary functions f by $n+1$–ary predicates $P(\vec{x}, y)$

Definition 15 *The predicate $P(\vec{x}, y)$ numeralwise represents the function $f : \omega \to \omega$ if for any n–tuple of natutal numbers $\vec{m}$ and natural number p, we have*

$$f(\vec{m}) = p \quad \Rightarrow \quad \vdash P(\vec{\hat{m}}, \hat{p})$$
$$\vdash \forall xy\,[\,(P(\vec{\hat{m}}, x) \otimes P(\vec{\hat{m}}, y)) \multimap x = y\,]\,.$$

For example, the predicate $sc(x) = y$ represents the successor function, and $x = z \otimes y = y$ represents the first projection of pairs.

Proposition 16 *The numeralwise representable functions are closed under composition.*

Proof
Suppose that the functions f and g are represented by the predicates $F(\vec{x}, y)$ and $G(y, \vec{z}, w)$, respectively. Then, their composition is represented by

$$\exists y(F(\vec{x}, y) \otimes G(y, \vec{z}, w)).$$

To verify this, let $q = f(\vec{m})$ and $g(f(\vec{m}), \vec{n}) = p$. Then, we have $\vdash F(\vec{\hat{m}}, \hat{q}) \otimes G(\hat{q}, \vec{\hat{n}}, \hat{p})$. Hence,

$$\vdash \exists y(F(\vec{\hat{m}}, y) \otimes G(y, \vec{\hat{n}}, \hat{p}))$$

Assume $F(\vec{m}, y) \otimes G(y, \vec{n}, w)$ and $F(\vec{m}, z) \otimes G(z, \vec{n}, w')$. Since $\vdash F(\vec{m}, \hat{q})$ and $\vdash \forall xy[(F(\vec{m}, x) \otimes F(\vec{m}, y)) \multimap x = y]$, we have $y = \hat{q}$ and $z = \hat{q}$ so that $G(\hat{q}, \vec{n}, w) \otimes G(\hat{q}, \vec{n}, w')$. ¿From this, $w = w'$ follows. ∎

For primitive recursion, suppose that the functions $g(\vec{y})$ and $h(x, \vec{y}, z)$ are represented by $G(\vec{y}, w)$ and $H(x, \vec{y}, z, w)$, respectively. Let f be the function defined by

$$\begin{aligned} f(0, \vec{x}) &= g(\vec{x}) \\ f(n+1, \vec{x}) &= h(n, \vec{x}, f(n, \vec{x})). \end{aligned}$$

To represent such a function, we construct the three predicates $Fun(x, w)$, $Init(x, \vec{y}, w, G)$ and $Rec(x, \vec{y}, w, H)$ as

$$\begin{aligned} Rec(x, \vec{y}, w, H) \quad &\Leftrightarrow \quad \exists z\, [\, !_6\, x = sc(z) \\ &\qquad \otimes \exists u \exists v(\, !_2\, \langle z, u\rangle \in w \otimes H(z, \vec{y}, u, v) \otimes \langle x, v\rangle \in w)\,] \\ Init(x, \vec{y}, w, G) \quad &\Leftrightarrow \quad x = \hat{0} \otimes \exists z(\langle \hat{0}, z\rangle \in w \otimes G(\vec{y}, z)) \\ Rec(x, \vec{y}, w, H) \quad &\Leftrightarrow \\ &\exists z\, [\, !_6\, x = sc(z) \otimes \exists u \exists v(\, !_2\, \langle z, u\rangle \in w \otimes H(z, \vec{y}, u, v) \otimes \langle x, v\rangle \in w)\,] \end{aligned}$$

Then, let the predicate $F(x, \vec{y}, z)$ be as follows:

$$\begin{aligned} &\exists w\, [\, !\, \forall u (\, !_{11}\, u \in sc(x) \multimap \\ &\qquad [(Init(u, \vec{y}, w, G) \oplus Rec(u, \vec{y}, w, H)) \otimes Fun(u, \vec{y}, w)]\,) \\ &\qquad \otimes\, !_2\, \langle x, z\rangle \in w\,] \end{aligned}$$

Lemma 17 *If* $f(m, \vec{n}) = p$, *then* $\vdash F(\hat{m}, \vec{\hat{n}}, \hat{p})$.

Proof
It suffices to find a witness term s such that

1. $\vdash \langle \hat{m}, \widehat{f(m, \vec{n})}\rangle \in s$,
2. $!_7\, u \in sc(\hat{m}) \vdash Init(u, \vec{\hat{n}}, s, G) \oplus Rec(u, \vec{\hat{n}}, s, H)$,
3. $!_4\, u \in sc(\hat{m}) \vdash Fun(u, s)$.

Let $s =_{def} \{x : x = \langle \hat{0}, \widehat{f(0, \vec{n})}\rangle \oplus \ldots \oplus x = \langle \hat{m}, \widehat{f(m, \vec{n})}\rangle\}$.

1. Since $\vdash \langle \hat{m}, \widehat{f(m, \vec{n})}\rangle = \langle \hat{m}, \widehat{f(m, \vec{n})}\rangle$, we have $\vdash \langle \hat{m}, \widehat{f(m, \vec{n})}\rangle \in s$ immediately.
2. First, we have $!_7\, u \in sc(\hat{m}) \vdash\, !_7\, u = \hat{0} \oplus \ldots\, !_7\, u = \hat{m}$. Hence it suffices to show for each $r \leq m$,

$$!_7\, u = \hat{r} \vdash Init(u, \vec{\hat{n}}, s, G) \oplus Rec(u, \vec{\hat{n}}, s, H).$$

We consider two cases. Suppose $r = 0$. Let $p = f(0,\vec{n}) = g(\vec{n})$. Then, $\vdash \langle 0, \hat{p}\rangle \in s$ and $\vdash G(\hat{\vec{n}}, \hat{p})$. Hence, $\vdash \exists z(\langle 0, z\rangle \in s \otimes G(\hat{\vec{n}}, z))$. ¿From this, we have

$$u = \hat{0} \vdash u = \hat{0} \otimes \exists z(\langle 0, z\rangle \in s \otimes G(\hat{\vec{n}}, z)).$$

Suppose $r = q+1$. Then, $\vdash \hat{r} = sc(\hat{q})$. Hence, $!_7\, u = \hat{r} \vdash\, !_7\, u = sc(\hat{q})$. Let $p = f(q,\vec{n})$ and $p' = h(q,\vec{n},p) = f(r,\vec{n})$. Then,

$$\vdash \langle \hat{q}, \hat{p}\rangle \in s, \quad \vdash H(\hat{q}, \hat{\vec{n}}, \hat{p}, \hat{p'})$$

and $\vdash \langle \hat{r}, \hat{p'}\rangle \in s$. Hence, we have

$$!_7\, u = \hat{r} \vdash\, !_7\, u = sc(\hat{q}) \otimes \exists v \exists w(\,!_2\, \langle \hat{q}, v\rangle \in s \otimes H(\hat{q}, \hat{\vec{n}}, v, w) \otimes \langle \hat{r}, w\rangle \in s).$$

¿From this follows $!_7\, u = \hat{r} \vdash Rec(u, \hat{\vec{n}}, s, H)$.

3. It suffices to show for each $r \leq m$,

$$!_2\, \langle \hat{r}, v\rangle \in s \vdash v = \widehat{f(r,\vec{n})}.$$

If $q \neq r$, then $\langle \hat{r}, v\rangle = \langle \hat{q}, \widehat{f(q,\vec{n})}\rangle \vdash \emptyset$. Hence, we have $\langle \hat{r}, v\rangle \in s \vdash \langle \hat{r}, v\rangle = \langle \hat{r}, \widehat{f(r,\vec{n})}\rangle$. ¿From this follows $!_2\, \langle \hat{r}, v\rangle \in s \vdash v = \widehat{f(r,\vec{n})}$. ∎

Lemma 18 *For any natural numbers m and $\vec{n}$,*

$$F(\hat{m}, \hat{\vec{n}}, x) \otimes F(\hat{m}, \hat{\vec{n}}, y) \vdash x = y.$$

Proof

The proof is by induction on m. Let $p = f(m,\vec{n})$ and $D(u,\vec{r},w)$ be the formula

$$!_7\, u \in sc(\hat{r}) \multimap ((Init(u, \hat{\vec{n}}, w, G) \oplus Rec(u, \hat{\vec{n}}, w, H)) \otimes Fun(u, w)).$$

Then, $F(\hat{m}, \hat{\vec{n}}, x)$ is $\exists w\, [\, !\forall u D(u, \hat{m}, w) \otimes\, !_2\, \langle \hat{m}, x\rangle \in w\,]$, and it suffices to show

$$!\forall u D(u, \hat{m}, w) \otimes\, !_2\, \langle \hat{m}, x\rangle \in w \vdash x = \hat{p}.$$

$m = 0$: Since $\vdash \hat{0} \in sc(\hat{0})$, we have

$$\forall u D(u, \hat{0}, w) \vdash (Init(\hat{0}, \hat{\vec{n}}, w, G) \oplus Rec(\hat{0}, \hat{\vec{n}}, w, H)) \otimes Fun(\hat{0}, w).$$

Note that $\hat{0} = sc(z)$ implies $z \in \hat{0}$, and then $\emptyset$. Hence, $Rec(\hat{0}, \hat{\vec{n}}, w, H) \vdash \emptyset$. So,

$$\forall u D(u, \hat{0}, w) \vdash Init(\hat{0}, \hat{\vec{n}}, w, G) \otimes Fun(\hat{0}, w).$$

Since $g(\vec{n}) = f(0,\vec{n}) = p$, we have $G(\hat{\vec{n}}, z) \vdash z = \hat{p}$ and $\langle \hat{0}, z\rangle \in w \otimes G(\hat{\vec{n}}, z) \vdash \langle \hat{0}, \hat{p}\rangle \in w$. Hence, $Init(\hat{0}, \hat{\vec{n}}, w, G) \vdash \langle \hat{0}, \hat{p}\rangle \in w$. So,

$$!_2 \forall u D(u, \hat{0}, w) \vdash\, !_2\, \langle \hat{0}, \hat{p}\rangle \in w \otimes\, !_2\, Fun(\hat{0}, w).$$

However, $!_2\, \langle \hat{0}, \hat{p}\rangle \in w \otimes\, !_2\, \langle \hat{0}, x\rangle \in w \otimes Fun(\hat{0}, w) \vdash x = \hat{p}$. Hence,

$$!\forall u D(u, \hat{0}, w) \otimes\, !_2\, \langle \hat{0}, x\rangle \in w \vdash x = \hat{p}.$$

$m = q + 1$: First note that $!_{11}\, u \in sc(\hat{q})$ implies $!_{11}\, u \in sc(\hat{m})$ since $\vdash sc(\hat{q}) \in sc(\hat{m})$. Hence, $\forall u D(u, \hat{m}, w) \vdash \forall u D(u, \hat{q}, w)$ and

$$!\,\forall u D(u, \hat{m}, w) \vdash\ !\,\forall u D(u, \hat{q}, w).$$

Since $\vdash \hat{m} \in sc(\hat{m})$, we have

$$\forall u D(u, \hat{0}, w) \vdash (Init(\hat{m}, \vec{\hat{n}}, w, G) \oplus Rec(\hat{m}, \vec{\hat{n}}, w, H)) \otimes Fun(\hat{m}, w).$$

However, $\hat{m} = \hat{0}$ implies $q \in \hat{0}$, and then $\emptyset$. So, $Init(\hat{m}, \vec{\hat{n}}, w, G) \vdash \emptyset$. Hence,

$$\forall u D(u, \hat{0}, w) \vdash Rec(\hat{m}, \vec{\hat{n}}, w, H) \otimes Fun(\hat{m}, w).$$

Since $\hat{m} = sc(z)$ is identical to $sc(\hat{q}) = sc(z)$, we have $!_6\, \hat{m} = sc(z) \vdash\ !_3\, z = \hat{q}$. So,

$$\begin{aligned} &!_6\, \hat{m} = sc(z),\ !_2\, \langle z, v \rangle \in w \otimes H(z, \vec{\hat{n}}, v, v') \otimes \langle \hat{m}, v' \rangle \in w \vdash \\ &\qquad\qquad !_2\, \langle \hat{q}, v \rangle \in w \otimes H(\hat{q}, \vec{\hat{n}}, v, v') \otimes \langle \hat{m}, v' \rangle \in w. \end{aligned}$$

However, $!_2\, \langle \hat{q}, v \rangle \in w$ together with $!\,\forall u D(u, \hat{q}, w)$ implies $F(\hat{q}, \vec{\hat{n}}, v)$, and then $v = \widehat{f(q, \vec{n})}$ by the inductive hypothesis. Furthermore, $H(\hat{q}, \vec{\hat{n}}, v, v')$ together with $v = \widehat{f(q, \vec{n})}$ implies $H(\hat{q}, \vec{\hat{n}}, \widehat{f(q, \vec{n})}, v')$, and then $v' = \hat{p}$. Hence,

$$\begin{aligned} &!\,\forall u D(u, \hat{m}, w),\ !_6\, \hat{m} = sc(z),\ !_2\, \langle z, v \rangle \in w \otimes H(z, \vec{\hat{n}}, v, v') \otimes \langle \hat{m}, v' \rangle \in w \\ &\qquad\qquad \vdash \langle \hat{m}, \hat{p} \rangle \in w. \end{aligned}$$

So, we have $!\,\forall u D(u, \hat{m}, w), Rec(\hat{m}, \vec{\hat{n}}, w, H) \vdash \langle \hat{m}, \hat{p} \rangle \in w$. Hence,

$$!\,\forall u D(u, \hat{m}, w),\ !_2\, Rec(\hat{m}, \vec{\hat{n}}, w, H) \vdash\ !_2\, \langle \hat{m}, \hat{p} \rangle \in w.$$

Then,

$$!\,\forall u D(u, \hat{m}, w),\ !_2\, \forall u D(u, \hat{m}, w) \vdash\ !_2\, \langle \hat{m}, \hat{p} \rangle \in w \otimes\ !_2\, Fun(\hat{m}, w).$$

However, $!_2\, \langle \hat{m}, \hat{p} \rangle \in w \otimes\ !_2\, \langle \hat{m}, x \rangle \in w \otimes Fun(\hat{m}, w) \vdash x = \hat{p}$. Hence,

$$!\,\forall u D(u, \hat{m}, w) \otimes\ !_2\, \langle \hat{m}, x \rangle \in w \vdash x = \hat{p}.$$

∎

Therefore, we can now conclude this brief development of arithmetic with the following theorem.

Theorem 19 *The numeralwise representable functions are closed under primitive recursion.*

6 Concluding remarks

In set theory based on contraction-free logics, there seem to be two interesting issues for further study.

The first issue is the precise determination of the expressive power of such systems. Recently, Girard formulated a weak system of linear logic [3], *light linear logic,* in which the computation (normalization) through Curry-Howard style encoding halts in the polynominal time of the size of input numerals (specific form of proofs). He noted that naive set theory can be carried out in this new system for the same reason as ours (normalization by induction on the size of proofs is possible). However, it does not seem to be clear how much extra expressive power is added by the set abstraction mechanism. On the other hand, we showed that one can explicitly construct a fixpoint in a linear (affine) set theory with the non-strict comprehension and the appropriate paring, and all total recursive functions are numeralwise representable in such a system [11]. We can expect the further progress of research in this direction.

The second issue is the semantics of such systems. The standard approach to the semantics of set theoy based on a non-standard logic is to modify the Boolean-valued models with the appropriate algebra for the logic. The idea has been carried out in [9], but this construction does not yield the models for the unrestricted comprehension. The models for linear set theory with the unrestricted comprehension in terms of Kripke semantics [7] and phase semantics have been studied. However, they are not easy to construct except as the term models. It is only very recent that we find a construction of a rather satisfactory model, combining the Scott-style inverse limit construction and coherence space semantics [12]. This result seems to be a good starting point for the further investigation of semantics.

References

[1] F.B. Fitch. "A system of formal logic without an analogue of the Curry W-operator." *Journal of Symbolic Logic*, 1936.

[2] J.Y. Girard. "Linear logic." *Theoretical Computer Science*, 50, 1987, 1-102.

[3] J.Y. Girard. "Light linear logic." (manuscript).

[4] V.N. Grishin. "A nonstandard logic and its application to set theory," (Russian). *Studies in Formalized Languages and Nonclassical Logics (Russian)*, Izdat, "Nauka," Moskow. 1974, 135-171.

[5] V.N. Grishin. "Predicate and set theoretic calculi based on logic without contraction rules," (Russian). *Izvestiya Akademii Nauk SSSR*

Seriya Matematicheskaya, 45, no.1, 1981, 47-68. 239. *Math. USSR Izv.*, 18, no.1, 1982, 41-59 (English translation).

[6] L. Hallnäs. *On Normalization of Proofs in Set Theory.* Dissertiones Mathematicae 261. Polska Akademia Nauk, Instytut Matematyczny, Warszawa, 1988.

[7] Y. Komori. "Illative combinatory logic based on BCK-logic." *Math. Japonica*, 34, No. 4, 1989, 585-596.

[8] H. Ono and Y. Komori. "Logics without the contraction rule." *Journal of Symbolic Logic*, 50, 1985, 169-201.

[9] M. Shirahata. *Linear Set Theory.* Dissertation, Department of Philosophy, Stanford University, 1994.

[10] M. Shirahata. "A linear conservative extension of Zermelo-Fraenkel set theory." *Studia Logica*, 56, 1996, 361-392.

[11] M. Shirahata. "Fixpoint theorem in linear set theory." (in preparation).

[12] M. Shirahata. "A coherence space semantics for linear set theory." (in preparation).

[13] R.B. White. "A demonstrably consistent type-free extension." *Mathematica Japonica*, 32, 1987, 149-169.

[14] R.B. White. "A consistent theory of attributes in a logic without contraction." *Studia Logica*, 52, 1993, 113-142.

Department of Mathematics
Keio University, Hiyoshi Campus
sirahata@math.hc.keio.ac.jp

A Solution to a Problem of Marek and Truszczyński

Su Kaile
Inst. of Computer Science, Shantou Univ.,
Shantou 51603, PRC

Chen Huowang
Dept. of Computer Science, Changsha Inst.of Tech.
Changsha, 410073, PRC

Abstract

We first show that every normal default theory is representable in the class of free normal default theories. This gives a new feature of extensions of normal default theories. By the above result an example of default theory is obtained which is not reprsentable in the class of normal default theories but has exactly infinitely countably many extensions and those extensions are pairwisely inconsistent. It gives a solution to a problem of W. Marek and M. Truszczyński and refutes a related assertion presented by them.

1 Introduction

Reiter's default logic is one of the most prominent formalization of nonmontonic reasoning. Recently the theory of representability for default logic was introduced by W. Marek and M. Truszczyński [2]. They introduced the notions of representability and equivalence for default logic. Two default theories (D_1, W_1) and (D_2, W_2) are said to be equivalent to each other, denoted by $(D_1, W_1) \approx (D_2, W_2)$, if (D_1, W_1) and (D_2, W_2) have the same extensions. A defaut theory (D, W) is said to be representable in a class of default theories if (D, W) is equivalent to a default theory in the class.

We prove that every normal default theory is representable in the class of prerequisite-free normal default theories, and give a new essential feature distinguishing normal default theories from unrestricted ones. Using this feature we construct an example of default theory which has exactly infinitely countably many pairwisely inconsistent extensions, but is not representable in the class of normal default theories. This gives a negative answer to the question of W. Marek and M. Truszczyński ([2,p.130]) whether the following theorem holds without the assumption that the extensions are finitely generated.

Theorem A [**2, Corollary 5.10**] If a defalt theory (D, W) has at least one extension and all the extensions for (D, W) are finitely generated, then the default theory is representable in $\mathcal{N}$ (the class of normal default theories).

Moreover, the example mentioned above refutes an assertion of W. Marek and M. Truszczyński([2,p.129]) that if a default theory (D, W) has at least one and at most countably many pairwisely inconsistent extensions then the conclusion of the above theorem holds even without the assumption that extensions of (D, W) are finitely generated.

For convenience, we introduce some definitions and notations of default logic. A detailed development can be found in [3,4]. A (closed) default theory (D, W) consists of a set W of first-order formulae, and a set D of defaults. A default is an expression of the form $\frac{A:B_1,\ldots,B_k}{C}$ where A, C andB_is are called the prerequiste, consequence and justifications of the default respectively. For a default $\delta = \frac{A:B_1,\ldots,B_k}{C}$, we set $A = PRE(\delta)$,$C = CONS(\delta)$. For a set D of defaults, we set $CONS(D) = \{CONS(\delta) | \delta \in D\}$.

For an arbitrary default theory $\xi = (D, W)$, a set E of sentences is said to be an extesion of ξ if E is a fixed point of the operator Λ_ξ which is defined as follows.

For an arbitrary set S of sentences, $\Lambda_\xi(S)$ is the least set S' of formulae that satisfies:

(1)$W \subseteq W$

(2)$Th(S) = S$

(3) For every $\delta - \frac{A:B_1,\ldots,B_k}{C} \in D$, if $A \in S'$ and $\neg B_1, \ldots, \neg B_k \notin S$, then $C \in S'$.

2 Normal default theories are representable in the class of free normal default theories

In this section we shall give the representablity result of normal default theories. In order to prove the result we first give the following definition and lemmas.

Definition 2.1 For a default theory $\xi = (D, W)$ and a set S of sentences, a sequence $\{\delta_i\}(i < n)$ of defaults in D is said to be a S-default proof in ξ, if

1. All the negations of justifications of δ_is are not in S.

2. $W \cup CONS(\{\delta_j | j < i\}) \vdash PRE(\delta_i)$ for all $i < n$.

Lemma 2.1 Let $\xi = (D, W)$ be a default theory, S a set of set of sentences. If $\{\delta_i\}_{i<n}$ is an S-default proofs in ξ, then $CONS(\delta_i) \in \Lambda_\xi(S)$ for all $i < n$.

Proof Easily by induction on $i < n$.

Lemma 2.2 Let $\xi = (D, W)$ be a default theory, S a set of set of sentences. If $\{\delta_i\}_{i<n}$ and $\{\delta'_i\}_{i<n'}$ are S-default proofs in ξ, so also is the concatenation $\{\delta_i\}_{i<n} \circ \{\delta'_i\}_{i<n'}$ which is defined by $\{\delta^*_i\}_{i<n+n'}$ where $\delta^*_i = \delta_i$ for $i < n$ and $\delta^*_i = \delta'_{(i-n)}$ for $n \leq i < n + n'$.

Proof Trivially true.

Lemma 2.3 Let $\xi = (D, W)$ be a default theory, S a set of sentences, we set $TH_\xi(S) =$
$\{X | \exists$ (S-default proof $\{\delta_i\}_{i<n}$) $[W \cup CONS(\{\delta_i | i < n\}) \vdash X\}]$.
Then $\Lambda_\xi(S) = TH_\xi(S)$.

Proof $TH_\xi(S) \subseteq \Lambda_\xi(S)$ is obtained by Lemma 2.1 and $Th(\Lambda_\xi(S)) = \Lambda_\xi(S)$. $\Lambda_\xi(S) \subseteq TH_\xi(S)$ can be obtained by proving that
1 $W \subseteq TH_\xi(S)$,
2 $Th(TH_\xi(S)) \subseteq TH_\xi(S)$,
3 for every default $\frac{A:B_1,\ldots,B_k}{C} \in D$ if $A \in TH_\xi(S)$ and $\neg B_1, \ldots, \neg B_k \notin S$, then $C \in TH_\xi(S)$.

1. $W \subseteq TH_\xi(S)$ is trivially true.

2. Let $Y \in TH_\xi(S)$. There exist $Y_1, \ldots, Y_k \in TH_\xi(S)$ and S-default proofs $\{\delta^1_i\}_{i<n_1}, \ldots, \{\delta^k_i\}_{i<n_k}$ such that

   $Y_1 \wedge \ldots \wedge Y_k \vdash Y$,

   $W \cup CONS(\{\delta^1_i | i < n_1\}) \vdash Y_1, \ldots,$

   $W \cup CONS(\{\delta^k_i | i < n_k\}) \vdash Y_k$.

   Let $\{\delta^*_i\}_{i<n}$ be the concatenation of all the S-default proofs. Then $W \cup CONS(\{\delta^*_i | i < n\}) \vdash Y$, and $Y \in TH_\xi(S)$. So $Th(TH_\xi(S)) \subseteq TH_\xi(S)$ is proved.

3. For $\delta' = \frac{A:B_1,\ldots,B_k}{C} \in D$, suppose $A \in TH_\xi(S)$ and $\neg B_1, \ldots, \neg B_k \notin S$. Then there is an S-default proof $\{\delta_i\}_{i<n}$ such that

   $W \cup CONS(\{\delta_i | i < n\}) \vdash A$. It is easy to see that $\{\delta_i\}_{i<n} \circ \{\delta'\}$ is also an S-default proof in ξ.

Since $W \cup CONS(\{\delta_i | i < n\} \cup \{\delta'\}) \vdash C$, we have $C \in TH_\xi(S)$.

Our main result in this section is:

Theorem 2.1 Every normal default theory is representable in the class of free normal default theories (a default theory is called free normal if its all defaults are of the form $\frac{:B}{B}$).

Proof Let (D, W) be an arbitrary normal default theory ξ. We shall prove that ξ is equivalent to the free normal default theory $\xi' = (D', W')$, where $W' = W$, and D' is

$$\{\frac{: B_0 \wedge ... \wedge B_{(n-1)}}{B_0 \wedge ... \wedge B_{(n-1)}} \mid \exists(\{\} - d.\ p.\ \{\delta_i\}_{i<n} \in \xi)(\forall i < n)[CONS(\delta_i') = B_i]\},$$

where "{}-d. p. $\{\delta_i\}_{i<n}$" means "{}-default proof $\{\delta_i\}_{i<n}$."

We first prove that $\Lambda_{\xi'}(E) \subseteq \Lambda_\xi(E)$ for all theories E (i.e. $Th(E) = E$).

Since $\Lambda_{\xi'}(E) = Th(W \cup \{X | \frac{:X}{X} \in D'$ and $\neg X' \notin E\})$, it suffices to prove that for all $\frac{:X}{X} \in D'$ if $\neg X \notin E$, then $X \in \Lambda_\xi(E)$. Suppose $\frac{:B_0 \wedge ... \wedge B_{n-1}}{B_0 \wedge ... \wedge B_{n-1}}$ and $\{\delta_i\}_{i<n}\}$ be {}-default proof in ξ, $CONS(\delta_i) = B_i$ for all $i < n$. If $\neg(B_0 \wedge ... \wedge B_{n-1}) \notin E$, then $\neg B_0, ..., \neg B_{n-1} \notin E$, otherwise suppose that $\neg B_j \in E$ for some $j < n$, then $\neg(B_0 \wedge ... \wedge B_{n-1}) \in E$, which is impossible. Hence $\{\delta_i\}_{i<n}$ is an E-default proof, and it follows that $B_0 \wedge ... \wedge B_{n-1} \in TH_\xi(E) = \Lambda_\xi(E)$ (by Lemma 2.3). This gives the proof of $\Lambda_{\xi'}(E) \subseteq \Lambda_\xi(E)$.

Second, we prove that $\Lambda_\xi(E) \subseteq \Lambda_{\xi'}$ for all E such that $\Lambda_{\xi'} \subseteq E$ and $Th(E)$. Since $\Lambda_\xi(E) = TH_\xi(E)$, it suffices to prove that $CONS(\{\delta_i | i < n\}) \subseteq \Lambda_{\xi'}(E)$ for an arbitrary E-default proof $\{\delta_i\}_{i<n}$ in ξ. Let $\delta_i = \frac{A_i:B_i}{B_i}$ for all $i < n$. We prove, by induction on $i < n$, that $B_i \in \Lambda_{\xi'}(E)$. Assume that $i, j < n$ and $B_i \in \Lambda_{\xi'}(E)$ for all $i < j$.

Now we prove $B_j \in \Lambda_{\xi'}(E)$. By the fact that $\{\delta_i\}_{i<j+1}$ is {}-default proof in ξ we have that $\frac{:B_0 \wedge ... \wedge B_j}{B_0 \wedge ... \wedge B_j} \in D'$. It suffice to prove that $\neg(B_0 \wedge ... \wedge B_j) \notin E$. By $\Lambda_{\xi'}(E) \subseteq E = Th(E)$ and $B_i \in \Lambda_{\xi'}(E)$ for all $i < j$, we have $B_0 \wedge ... \wedge B_{j-1} \in E$. Hence $\neg(B_0 \wedge ... \wedge B_j) \notin E$, otherwise $\neg B_j \in Th(E) = E$ (for $B_0 \wedge ... \wedge B_{j-1}, \neg(B_0 \wedge ... \wedge B_j) \vdash \neg B_j$), it contradicts that $\{\delta_i\}_{i<n}$ is an E-default proof (in ξ).

Finally we prove that $\Lambda_\xi(E) = \Lambda_{\xi'}(E)$ under the condition $\Lambda_\xi(E) = E$ or $\Lambda_{\xi'}(E) = E$, and hence prove that $\Lambda_\xi(E) = E$ iff $\Lambda_{\xi'}(E) = E$, i.e. the normal default theory ξ is equivalent to the free normal default theory ξ'. Suppose $\Lambda_\xi(E) = E$ or $\Lambda_{\xi'}(E) = E$, then $Th(E) = E$. Hence $\Lambda_{\xi'}(E) \subseteq \Lambda_\xi(E)$ by the above disscussion. On the other hand, we have $\Lambda_{\xi'}(E) \subseteq E$ or $\Lambda_\xi(E) \subseteq E$, so $\Lambda_{\xi'}(E) \subseteq E$ always holds (by $\Lambda_{\xi'}(E) \subseteq \Lambda_\xi(E)$). Hence

$\Lambda_\xi(E) \subseteq \Lambda_{\xi'}(E)$. This gives that $\Lambda_\xi(E) = \Lambda_{\xi'}(E)$ under the condition that $\Lambda_\xi(E) = E$ or $\Lambda_{\xi'}(E) = E$.

3 A new feature of extensions of normal default theories and an example of default theories

It is well-known that, for every normal default theory, (1) there exists an extension and (2) two different extensions are inconsistent. W. Marek and M. Truszczyński showed that the two properties mentioned above, in several cases, are essentially the only two features distinguishing normal default theories from unrestricted ones. They gave the definition of finitely generated theory, and obtained some results such as Theorem A mentioned in the Introduction. But the question raised by W. Marek and M. Truszczyński, introduced at the beginning of this paper, remains open.

This section we give a new essential feature of normal default theories. By this feature we will construct an example of default theory which has exactly infinitely countably many pairwisely inconsistent extensions, but is not representable in the class of normal default theories. It gives a negative answer to the problem of W. Marek and M. Truzcynski and refutes the assertion of them in [2, p.129] mentioned at the beginning of this paper.

Theorem 3.1 Let (D, W) be a free normal default theory, then for all extensions E_1, E_2 of (D, W) such that $E_1 \neq E_2$ there is a sentence α such that $\alpha \in E_1$ and $E_2 \cup \{\alpha\}$ is inconsistent, and for every extension E^* of (D, W), either $\alpha \in E^*$ or $E^* \cup \{\alpha\}$ is inconsistent.

Proof Before we prove this theorem we first introduce a theorem, which can be found in [3].

Theorem([3, p.47]). For extensions E_1, E_2 there exist D_1, $D_2 \subseteq D$ such that $E_i = Th(W \cup CONS(D_i))$ ($i = 1, 2$).

Suppose that $E_1 \neq E_2$. Then $E_1 \not\subseteq E_2$, and, by the Theorem above, there is a default $\delta = \frac{:B}{B} \in D_1$ such that $B \notin E_2$. Hence $\neg B \in E_2$, otherwise $B \in \Lambda_\xi(E_2) = E_2$. So $E_2 \cup \{\alpha\}$ is inconsistent.

Moreover $B \in E_1$(by $\frac{:B}{B} \in D_1$), and for every extension E^* of (D, W), $B \in E^*$ if $\neg B \notin E^*$, i.e. we have that $B \in E^*$, or $\neg B \notin E^*$. This completes the proof of the last claim.

Corollary 3.1 Let (D, W) be a normal default theory, then for all extensions E_1, E_2 of (D, W) such that $E_1 \neq E_2$ there is a sentence α such that $\alpha \in E_1$ and $E_2 \cup \{\alpha\}$ is inconsistent, and for every extension E^* of (D, W), either $\alpha \in E^*$ or $E^* \cup \{\alpha\}$ is inconsistent.

Proof By the Theorem 2.1, every normal default theory is representable in the class of free normal default theories.

Example 3.1 The default theory (D_0, W_0) is defined as follows:

$$W_0 = \{q_i \rightarrow \neg p_i | i \neq j\},$$

$$D_0 = \{\frac{: q_i}{p_i} \mid i \in \omega\} \cup \{\frac{p_j : a_{ij}}{a_{ij}} \mid i, j \in \omega\},$$

where q_is and p_js $(i \leq j)$ and all a_{ij}s $(i < j)$ are different atoms , and $a_{ji} = \neg a_{ij}$ for $i < j$.

Theorem 3.2 The default theory (D_0, W_0) has exactly $\aleph_0$ (infinitely countably) many extensions and these extensions are pairwisely inconsistent, however, is not representable in the class of normal default theories N.

The above theorem can be proved directly by the two following lemmas and Corollary 3.1 .

Lemma 3.1 Let $E_j = Th(W_0 \cup \{p_j\} \cup \{a_{ij} | i \in \omega\})$ for all $j \in \omega$. Then E_js are exactly all extensions of (D_0, W_0) and $E_i \cup E_j$ is not consistent if $i \neq j$.

Proof It is easy to prove that E_js are extensions of (D_0, W_0) and $E_i \cup E_j$ is not consistent if $i \neq j$. We only show that for an arbitrary extension E of (D_0, W_0) there is a j such that $E = E_j$.

First, we have (by [3, Proposition 6.2.9]) $E = Th(W_0 \cup CONS(GD(E)))$, where $GD(E)$ is the set of all generating defaults for E, i.e.

$$GD(E) = \{\frac{: q_j}{p_j} | j \in \omega, \neg p_j \notin E\} \cup \{\frac{p_j : a_{ij}}{a_{ij}} | p_j \in E, \neg a_{ij} \notin E, i, j \in \omega\}.$$

Second, it is easy to see that there exists a j such that $p_j \in E$, otherwise $\frac{p_j : a_{ij}}{a_{ij}} \notin GD(E)$ for all i, j, and $CONS(\frac{:q_j}{p_j}) \notin Th(W_0 \cup CONS(GD(E)))$, hence $\frac{:q_j}{p_j} \notin GD(E)$ for all j. Thus $GD(E) = \{\}$, we have $E = Th(W_0)$, this contradicts that E is an extension.

Finally, suppose $p_{j_0} \in E$, we shall show that $E = E_{j_0}$ by proving the following facts:

Fact 1: $\frac{:q_j}{p_j} \notin GD(E)$ for $j \neq j_0$.

If $j \neq j_0$ then $q_j \rightarrow \neg p_{j_0} \in W_0 \subseteq E$. Since $Th(E) = E$, we have $\neg q_j \in E$. Thus $\frac{:q_j}{p_j} \notin GDE(E)$ is proved.

Fact 2: $p_j \notin E$ and hence $\frac{p_j:a_{ij}}{a_{ij}} \notin GD(E)$ for all $i \neq j_0$.

By Fact 1 $E = Th(W_0 \cup \{p_{j_0}\} \cup F)$ where $F \subseteq \{a_{ij} | i, j \in \omega\}$. If $p_j \in E$ for some $j \neq j_0$, then $W_0 \cup \{p_{j_0}\} \cup F \vdash p_j$. If we replace p_j by $\perp$ (false), then

$$W_0' \cup \{q_i \rightarrow \neg \perp | i \neq j\} \cup \{p_{j_0}\} \cup F \vdash \perp$$

where $W_0' \subseteq W_0$. Since $q_i \rightarrow \neg \perp$ is valid, we have $W_0' \cup \{p_{j_0}\} \cup F \vdash \perp$, this contradicts that E is consistent (since W_0 is).

Fact 3: $\frac{:q_{j_0}}{p_{j_0}} \in GD(E)$, $\frac{p_{j_0}:a_{ij_0}}{a_{ij_0}} \in GD(E)$ for all i.

By Fact 1 and Fact 2, $E \subseteq Th(W_0 \cup \{p_{j_0}\} \cup \{a_{ij_0} | i \in \omega\})$. Since

$$W_0 \cup \{p_{j_0}\} \cup \{a_{ij_0} | i \in \omega\} \not\vdash \neg q_{j_0},$$

and

$$W_0 \cup \{p_{j_0}\} \cup \{a_{ij_0} | i \in \omega\} \not\vdash \neg a_{ij_0} \ for\ all\ i,$$

we have

$$\neg q_{j_0} \notin E,$$

and

$$a_{ij_0} \notin E \ for\ all\ i.$$

Thus

$$\frac{: q_{j_0}}{p_{j_0}} \in GD(E)\ and\ \frac{p_{j_0} : a_{ij_0}}{a_{ij_0}} \in GD(E)\ for\ all\ i,$$

¿From these facts, it is easy to see that

$$GD(E) = \{\frac{: q_{j_0}}{p_{j_0}}\} \cup \{\frac{p_{j_0} : a_{ij_0}}{a_{ij_0}} | i \in \omega\}.$$

Hence

$$E = Th(W_0 \cup \{p_{j_0}\} \cup \{a_{ij_0} | i \in \omega\}) = E_{j_0}.$$

Lemma 3.2 Let E_js be as in Lemma 3.1, C an arbitrary sentence such that $C \in E_0$ and $\neg C \in E_1$. Then there is an E_i such that $C \notin E_i$ and $\neg C \notin E_i$.

Proof Suppose that C be a sentence such that $C \in E_0$ and $\neg C \in E_1$. There exists a number k such that C contains no atoms in

$$\{p_j | k \leq j\} \cup \{a_{ij} | i \leq j, k \leq j\}$$

Let j be a number greater than k. We shall prove $W_0 \cup \{p_j\} \cup \{a_{ij} | i \in \omega\} \nvdash C$. Otherwise, assume that

$$W_0 \cup \{p_j\} \cup \{a_{ij} | i \in \omega\} \vdash C,$$

i.e.

$$W_0 \cup \{p_j\} \cup \{a_{ij} | i \leq j\} \cup \{\neg a_{ji} | j < i\} \vdash C.$$

If we substitute $\top$ for $a_{ij} (i \leq j)$, and by $\bot$ for a_{ji} $(j < i)$, we have $W_0 \cup \{p_j\} \vdash C$. If substitute $\top$ for p_j, we obtain that

$W'_0 \cup \{q_i \rightarrow \top | i \neq j\} \cup \{\top\} \vdash C$ where $W'_0 \subseteq W_0$. Then $W_0 \vdash C$. This contradicts that $\neg C \in E_1$. Thus we have that $C \notin E_j$. By the same reason, $\neg C \notin E_j$ for all $j \geq k$.

Our results show that in some cases existence of extensions and inconsistency of two different extensions are not essentially only features distinguishing normal default theories from unrestricted ones.

References

[1] R. Reiter, A logic for default reasoning, *Artif. Intell.* **12**, 81-132, 1980.

[2] W. Marek and M. Truszcyński, *Nonmonotonic Logic: Context- Dependent Reasoning*, Spring-Verlag, Heiderberg, 1993.

[3] P. Besnard, *An introduction to default logic*, Springer-Verlag, Berlin, 1989.

[4] D. W. Etherington, *Reasoning with Incomplete Information, Pitman, London, 1988.*

Credulous Reasoning About Defaults*

Yao-Hua Tan† Leendert W.N. van der Torre‡

Abstract

In this paper we propose a preference-based conditional logic for credulous reasoning about defaults. A conditional default "if β then by default α" is either formalized by the strong preference "$\beta \wedge \neg\alpha$ is not preferred to or equivalent to $\beta \wedge \alpha$", or by the weak preference "the preferred β is an α". We show that these two expressions, instances of what we call the *ordering* and *minimizing* usages of preference orderings, can be considered as duals of each other. Moreover, we give a formalization of ordering and minimizing in Boutilier's modal logic CT40 and we show how to combine them in a two-phase default logic.

1 Introduction

Conditional logic is a popular framework to formalize defeasible reasoning [Del88, AM91, Alc94, Bou94a, Mor95]. The conditional sentence "if β (the antecedent or condition) then by default α (the consequent or conclusion)" is represented in this approach by the formula $\beta > \alpha$, where '$>$' is the implication of conditional logic. The popularity of this framework is based on the lack of strengthening of the antecedent of the conditional implication, which is used to formalize the specificity principle. For example, from the 'birds fly' default $b > f$ the 'penguins fly' default $(p \wedge b) > f$ cannot be derived. However, the lack of strengthening of the antecedent also has drawbacks, in particular the so-called irrelevance problem. For example, from the 'birds fly' default $b > f$ the 'red birds fly' default $(r \wedge b) > f$ cannot be derived. In the framework, the set of defaults derivable in the conditional logic is called the 'conservative core' and conditional formulas are added to this core by Delgrande's irrelevance principle [Del88] or mechanisms equivalent to system Z [Pea90] like rational closure [LM92], the minimum specificity principle [BDP92] or Boutilier's 'only knowing' construction [Bou92].

*This research was partially supported by the ESPRIT III Basic Research Project No.6156 DRUMS II and the ESPRIT III Basic Research Working Group No.8319 MODELAGE.

†Erasmus University Research Institute for Decision and Information Systems (EURIDIS). Erasmus University Rotterdam, P.O. Box 1738, 3000 DR Rotterdam, The Netherlands. E-Mail: {ytan,ltorre}@@euridis.fbk.eur.nl. Tel: (+31)10-4082601. Fax: (+31)10-4526134. Http://www.euridis.fbk.eur.nl/Euridis/welcome.html.

However, these solutions suffer from a specific instance of the irrelevance problem called the inheritance problem. For example, given that penguins are exceptional birds $b > \neg p$, from the 'birds have wings' default $b > w$ the 'penguins have wings' default $(b \wedge p) > w$ cannot be derived by system Z.

In this paper, we propose a conditional logic for credulous reasoning about defaults, i.e. a logic in which the two defaults $\top > p$ and $\top > \neg p$ can consistently exist together, where $\top$ stands for any tautology. Because it is a credulous logic, we can accept strengthening of the antecedent to solve the irrelevance and inheritance problems. Obviously, we loose the possibility to formalize specificity by accepting unrestricted strengthening of the antecedent. However, the following example shows another problem with unrestricted strengthening of the antecedent.

Example 1 (apples and pears problem) *Consider the defaults that (1) you normally buy apples or pears, and (2) you normally do not buy apples. It is counterintuitive to derive that if you buy apples, then you normally buy pears. Assume a conditional default logic that validates at least substitution of logical equivalents and the following Gentzen-style inference patterns* Strengthening of the Antecedent (SA), Weakening of the Consequent (WC) *and* Conjunction (AND).

$$\text{SA}: \frac{\beta_1 > \alpha}{(\beta_1 \wedge \beta_2) > \alpha}$$

$$\text{WC}: \frac{\beta > \alpha_1}{\beta > (\alpha_1 \vee \alpha_2)}$$

$$\text{AND}: \frac{\beta > \alpha_1, \beta > \alpha_2}{\beta > (\alpha_1 \wedge \alpha_2)}$$

Furthermore, assume as premises the defaults $\top > (a \vee p)$ and $\top > \neg a$, where a can be read as "buying apples" and p as "buying pears". The intuitive default $\top > (\neg a \wedge p)$ can be derived by AND. *¿From the latter default, the default $a > (\neg a \wedge p)$ can be derived by* SA. *Unfortunately, from this default, the counterintuitive default $a > p$ can be derived by* WC. *This default is considered to be counterintuitive, because it is not grounded in the premises. If a is true, then the first premise is fulfilled and the second one is violated. This inference can be blocked by replacing unrestricted strengthening of the antecedent by the following version of restricted strengthening of the antecedent,*[1] *in which $\Diamond$ is a modal operator and $\Diamond\phi$ is true for all consistent propositional formulas ϕ.*

$$\text{RSA}: \frac{\beta_1 > \alpha, \Diamond(\beta_1 \wedge \beta_2 \wedge \alpha)}{(\beta_1 \wedge \beta_2) > \alpha}$$

[1] This restriction in strengthening of the antecedent is uncontroversial, because the expression $p > q$ with $p \wedge q$ inconsistent is counterintuitive for logics of defeasible reasoning. However, in some other usages of conditionals and preference logics like deontic logic, such expressions do have an intuitive (and important) reading, see e.g. [TvdT96].

The default $a > (\neg a \wedge p)$ *cannot be derived from the default* $\top > (\neg a \wedge p)$ *by* RSA. *Unfortunately, the counterintuitive* $a > p$ *can still be derived in another way. ¿From the intuitive default* $\top > (\neg a \wedge p)$ *the intuitive* $\top > p$ *can be derived by* WC. *¿From this latter obligation, the counterintuitive* $a > p$ *can be derived by* RSA. *Both derivations are represented in Figure 1.*

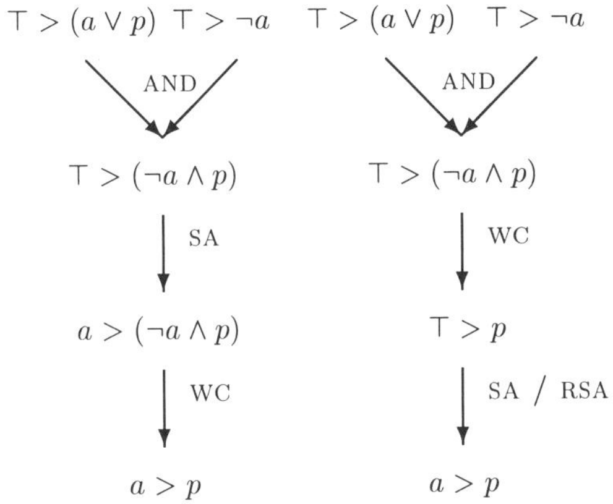

Figure 1: apples and pears problem

The apples and pears problem can be solved by a technique, which might look odd at first sight, but which turns out to work well, namely to forbid application of RSA after WC has been applied. This means that in derivations first RSA has to be applied, and only afterwards WC may be applied. We call this the *two-phase approach* in default logic. Such a sequencing in derivations is rather unnatural and cumbersome from a proof-theoretic point of view. Surprisingly, the two-phase approach can be obtained very intuitively from a semantic point of view.

In this paper we show that the two-phase approach can be obtained by combining two usages of a preference ordering in a preference-based semantics of a default logic. For the two usages we define two different types of default conditionals, which we call type-1 and type-2 defaults. The two types of defaults correspond to two different ways to evaluate formulas in a preference ordering. Type-1 defaults are formalized by strong preferences and evaluated by what we call *Ordering*, a process in which the whole ordering is used to evaluate a formula. Type-2 defaults are formalized by weak preferences and evaluated by what we call *Minimizing*, in which the ordering is used to select the minimal elements that satisfy a formula. The minimizing approach is commonly taken in preferential semantics for non-monotonic logics, see for example [Sho88, KLM90, Mak93, Bou94a]. In this paper, we consider credulous reasoning about defaults. In such reasoning, there can be several equivalence classes of preferred models. Hence, the definition of a type-2 default only considers truth in an equivalence class of preferred models instead of truth in all preferred models. In the two-phase approach, the first phase corresponds to ordering, and the second phase corresponds to minimizing. In semantic terms the two-phase approach simply means that first a preference ordering has to be constructed by ordering worlds, and subsequently the constructed ordering can be used for minimization.

In this paper we formalize type-1 and type-2 defaults in Boutilier's modal preference logic CT40 [Bou92]. The logic CT40 and the minimizing approach are well-known, but the logic of ordering and the two-phase approach to defeasible reasoning introduced in this paper are new.

This paper is organized as follows. In Section 2 we give the preference logic in which we formalize type-1 and type-2 defaults as strong and weak preferences. We show that these preferences are instances of the ordering and minimizing usages of preference orderings. In Section 3 we show how ordering and minimizing can be combined in a two-phase default logic and how this solves the problems of Example 1. Finally, in Section 4 we briefly consider the sceptical case and in Section 5 we mention some related research.

2 A logic for reasoning about defaults

Preference-based default logics are default logics of which the semantics contains a preference ordering (usually on worlds of a Kripke style possible world model). This preference ordering reflects different degrees of 'normality': a world is preferred to another world if it is, in some sense, more normal than the other world. For example, a value can be associated with each world; in such cases, the ordering is connected (for all w_1 and w_2 we have $w_1 \leq w_2$ or $w_2 \leq w_1$). However, in general the preference ordering can be any partial pre-ordering. Hence, only reflexivity and transitivity are assumed. An expression "by default p" is expressed by a preference for p, which may mean that

1. "p is preferred to $\neg p$ regardless of other things", or that
2. "p is preferred to $\neg p$ other things being equal", or
3. some intermediate reading.

Many authors (for example [TP94, Bou94a]) take the second (*ceteris paribus*) reading, because the first reading does not allow for two or more unconditional preference statements to exist consistently together, as observed by von Wright in [vW63]. For example, the preferences for p and q will quickly run into conflict when considering the worlds $p \wedge \neg q$ and $\neg p \wedge q$.

2.1 Ordering

In this paper, a strong preference for p means that "$\neg p$ is not preferred to or equivalent to p, regardless of other things". Obviously, in a connected partial pre-ordering, the expression "$\neg p$ is not preferred to or equivalent to p" is equivalent to "p is preferred to $\neg p$". Hence, for connected orderings, which are quite popular in preferential semantics, our reading has the problem described by von Wright. However, we do not restrict ourselves to connected orderings, but we allow any partial pre-ordering. With such orderings, the preferences for p and q will not run into conflict when considering the worlds $p \wedge \neg q$ and $\neg p \wedge q$: these worlds are only incomparable. Notice that the whole ordering is taken into account when a default is evaluated. That is why we call it the *ordering* approach to default logic. Similarly, a conditional default "if q then by default p" is represented by "no $\neg p \wedge q$ is preferred to or equivalent to some $p \wedge q$, regardless of other things".

The preferences are formalized in Boutilier's logic CT40, for the details and completeness results of this logic see [Bou94b]. CT40 is a bimodal propositional logic of inaccessible worlds. Boutilier notes that many applications of preference logics (like the one we describe below) do not need the complexity of inaccessible worlds, but it makes the definitions easier and the semantics clearer.

Definition 1 (Syntax of CT40) *The logic CT40 is a bimodal system with the two normal modal connectives $\Box$ and $\overleftarrow{\Box}$. The dual 'possibility' connectives are defined as usual:*

$\Diamond\alpha =_{def} \neg\Box\neg\alpha$ *and* $\overleftarrow{\Diamond}\,\alpha =_{def} \neg\,\overleftarrow{\Box}\,\neg\alpha$.

Moreover, the two following modal connectives are defined:

$\overleftrightarrow{\Box}\,\alpha =_{def} \Box\alpha\wedge\overleftarrow{\Box}\,\alpha$ *and* $\overleftrightarrow{\Diamond}\,\alpha =_{def} \Diamond\alpha\vee\overleftarrow{\Diamond}\,\alpha$.

CT40 is axiomatized by the following set of axioms and inference rules.

K	$\Box(\alpha \to \beta) \to (\Box\alpha \to \Box\beta)$
K′	$\overleftarrow{\Box}\,(\alpha \to \beta) \to (\overleftarrow{\Box}\,\alpha \to \overleftarrow{\Box}\,\beta)$
T	$\Box\alpha \to \alpha$
4	$\Box\alpha \to \Box\Box\alpha$
H	$\overleftrightarrow{\Diamond}\,(\Box\alpha\wedge\overleftarrow{\Box}\,\beta) \to \overleftrightarrow{\Box}\,(\alpha\vee\beta)$
Nec	*¿From α infer $\overleftrightarrow{\Box}\,\alpha$*
MP	*¿From $\alpha \to \beta$ and α infer β*

(Semantics of CT40) *Kripke models $M = \langle W, \leq, V\rangle$ for CT40 consist of*

W, a set of worlds, $\leq$, a binary transitive and reflexive accessibility relation, and V, a valuation of the propositions in the worlds. The modal operator $\Box$ refers to accessible worlds and the modal operator $\overleftarrow{\Box}$ to inaccessible worlds.

$$M, w \models \Box\alpha \text{ iff } \forall w' \in W \text{ if } w' \leq w, \text{ then } M, w' \models \alpha$$
$$M, w \models \overleftarrow{\Box}\, \alpha \text{ iff } \forall w' \in W \text{ if } w' \not\leq w, \text{ then } M, w' \models \alpha$$

Given this modal preference logic, we define type-1 defaults as strong preferences.

Definition 2 *Type-1 defaults "if β then by default α", written as $\beta > \alpha$, are defined as follows.*

$$\beta > \alpha =_{def} \overleftrightarrow{\Box}\, ((\beta \wedge \alpha) \rightarrow \Box(\beta \rightarrow \alpha))$$

Intuitively, a default $q > p$ expresses a strict preference of all $p \wedge q$ over $\neg p \wedge q$. The following proposition shows that this preference is represented by a negative condition: no $\neg p \wedge q$ is preferred to a $p \wedge q$,

Proposition 1 *Let$M = \langle W, \leq, V\rangle$ be a CT4O model. $M, w \models \beta > \alpha$ iff for all $w_1, w_2 \in W$ such that $M, w_1 \models \beta \wedge \neg\alpha$ and $M, w_2 \models \beta \wedge \alpha$, it is true that $w_1 \not\leq w_2$.*

Proof $\Rightarrow$ *Assume a model $M = \langle W, \leq, V\rangle$ with two worlds $w_1, w_2 \in W$ such that $M, w_1 \models \beta \wedge \neg\alpha$, $M, w_2 \models \beta \wedge \alpha$ and $w_1 \leq w_2$. We have $M, w_2 \not\models (\beta \wedge \alpha) \rightarrow \Box(\beta \rightarrow \alpha))$. Hence, $M, w \not\models \beta > \alpha$ for some world $w \in W$, because from the semantic definitions follows immediately that $M, w \models \overleftrightarrow{\Box}\, \phi$ iff for all worlds $w' \in W$, $M, w' \models \phi$.*

$\Leftarrow$ *Assume $M, w \not\models \beta > \alpha$ for some world w. Hence, there is a world $w_2 \in W$ such that $M, w_2 \not\models (\beta \wedge \alpha) \rightarrow \Box(\beta \rightarrow \alpha))$. It follows that $M, w_2 \models \beta \wedge \alpha$ and there is a world $w_1 \in W$ such that $M, w_1 \models \beta \wedge \neg\alpha$ and $w_1 \leq w_2$.*

Notice that the normality ordering is global (in the sense that the normality ordering is *not* relative to a world) and nested operators therefore do *not* have an intuitive reading, although they have a formal meaning in CT4O. The following example illustrates the definition of type-1 defaults as strong preferences.

Example 2 *Let $|\alpha|$ denote a world that satisfies α. Assume models that consist of four worlds $|p \wedge q|$, $|\neg p \wedge q|$, $|p \wedge \neg q|$ and $|\neg p \wedge \neg q|$. A model satisfies the default $\top > p$ when neither $|\neg p \wedge q|$ nor $|\neg p \wedge \neg q|$ is preferred to $|p \wedge q|$ or $|p \wedge \neg q|$. For example, the Kripke model $|p \wedge q| < |p \wedge \neg q| < |\neg p \wedge \neg q| < |\neg p \wedge q|$ in Figure 2 is a model for $\top > p$ but not for $\top > q$. Note that $\top > q$ is not true, because $|\neg p \wedge \neg q| < |\neg p \wedge q|$ and $|p \wedge \neg q| < |\neg p \wedge q|$. This shows how in the ordering approach the whole ordering is taken into account in the evaluation of a formula, and not just the most preferred $|p \wedge q|$ worlds.*

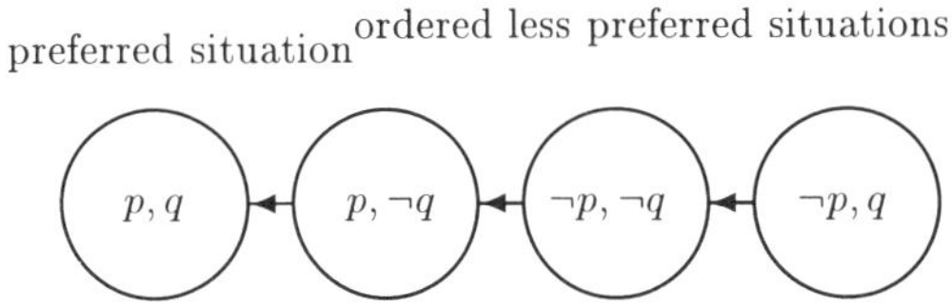

Figure 2: Preference relation with four worlds

In the beginning of this section, we mentioned von Wright's problem that absolute preferences for p and q are not mutually consistent, and the solution in our logic. This solution is illustrated in the following example.

Example 3 *Let M be a model of CT4O that satisfies the two defaults $\top > p_1$ and $\top > p_2$ and contains $|p_1 \wedge \neg p_2|$ and $|\neg p_1 \wedge p_2|$ worlds. The $|p_1 \wedge \neg p_2|$ and $|\neg p_1 \wedge p_2|$ worlds of M are incomparable, because the first is more normal than the latter with respect to default $\top > p_1$, but less normal with respect to default $\top > p_2$. An example of such a model M is represented in Figure 3.*

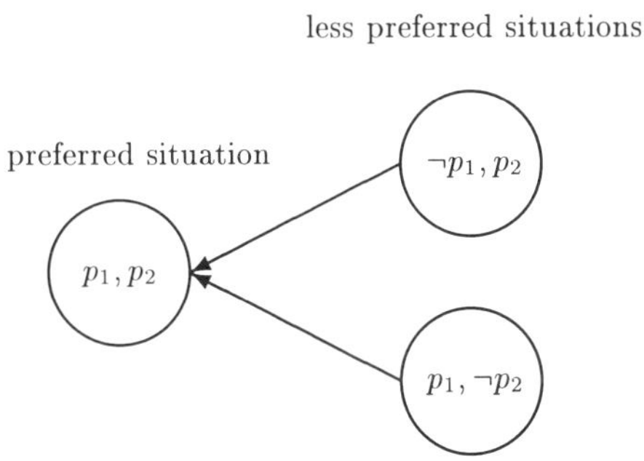

Figure 3: Preference relation with von Wright's problem

The following proposition gives several properties of the type-1 defaults.

Proposition 2 *The logic CT4O validates the following theorems of* Strengthening of the Antecedent (**SA**), Conjunction (**And**) *and* Disjunction (**Or**), *and a version of* Transitivity (**Trans′**).

SA	$\beta_1 > \alpha \rightarrow (\beta_1 \wedge \beta_2) > \alpha$
And	$(\beta > \alpha_1 \wedge \beta > \alpha_2) \rightarrow \beta > (\alpha_1 \wedge \alpha_2)$
Or	$(\beta > \alpha_1 \wedge \beta > \alpha_2) \rightarrow \beta > (\alpha_1 \vee \alpha_2)$
Trans′	$(\gamma > \beta \wedge \beta > \alpha) \rightarrow \gamma > (\alpha \wedge \beta)$

The logic CT4O does not *validate the following theorems of* Weakening of the Consequent (**WC**), Transitivity (**Trans**) *and the* sceptical axiom (**D**).

WC	$\beta > \alpha_1 \rightarrow \beta > (\alpha_1 \vee \alpha_2)$
Trans	$(\gamma > \beta \wedge \beta > \alpha) \rightarrow \gamma > \alpha$
D	$\neg(\beta > \alpha \wedge \beta > \neg\alpha)$

Proof *The (non)theorems can easily be verified by proving (un)derivability in CT4O.*

The validity of strengthening of the antecedent follows from the fact that a strong preference of p over $\neg p$ implies a preference of $p \wedge q$ over $\neg p \wedge q$. The most remarkable property of the logic is the invalidity of weakening of the consequent. Intuitively, the lack of weakening of $\top > p$ to $\top > (p \vee q)$ is the consequence of the fact that $\top > p$ expresses a preference of *all* p over $\neg p$, because from such a preference does not follow that $p \vee q$ is *always* preferred to $\neg p \wedge \neg q$. This is illustrated by the following example.

Example 4 *Reconsider the Kripke model in Figure 2. The model satisfies* $\top > p$ *but not* $\top > (p \vee q)$. *This illustrates that the type-1 defaults do not have strengthening of the antecedent.*

Transitivity, expressed by the theorem **Trans′**, is an unusual and remarkable property for default logics. The following example illustrates this property.

Example 5 *Let S be the set of conditional defaults* $\{\top > a, a > t\}$, *where a can be read as 'a certain man going to the assistance of his neighbors' and t as 'telling the neighbors that he will come'. Hence, the two defaults can be read as 'a certain man normally goes to the assistance of his neighbors', and 'if he goes, then he normally tells them he is coming'. The default* $\top > (a \wedge t)$ *can be derived from S with* **Trans′**, *which expresses that normally, the man goes to the assistance of his neighbors* and *he tells them he is coming.*

2.2 Minimizing

Type-2 defaults are defined as weak preferences in the modal preference logic. In this paper, we consider credulous reasoning about defaults. In credulous reasoning, there can be several 'extensions' or equivalence classes of preferred models. Hence, the definition of a type-2 default only considers truth in an equivalence class of preferred models instead of truth in all preferred models. For the details of this definition, see [Bou94b].[2]

[2]Boutilier makes a conditional true if the antecedent is false, i.e. he defines $\beta >_{\exists} \alpha =_{def} \overleftrightarrow{\Box} \neg\beta \vee \overleftrightarrow{\Diamond} (\beta \wedge \Box(\beta \rightarrow \alpha))$. Moreover, the definition can deal with infinite chains.

Definition 3 *Type-2 defaults "if β then by default α", written as $\beta >_\exists \alpha$, are defined as follows.*

$$\beta >_\exists \alpha =_{def} \overset{\leftrightarrow}{\Diamond} (\beta \wedge \Box(\beta \rightarrow \alpha))$$

The default $q >_\exists p$ is true in a model if p is true in an equivalence class of most preferred $|q|$ worlds of the model. Hence, the default $q >_\exists p$ refers to the preferred worlds where q is true, and $\top >_\exists p$ refers to the most preferred worlds. The following example illustrates the definition of type-2 defaults and compares it with type-1 defaults.

Example 6 *Reconsider the Kripke model in Figure 2. The model satisfies $\top >_\exists p$ and $\top >_\exists q$, whereas $\top > q$ is not true in this model. Since $\top >_\exists q$ is equivalent to $\overset{\leftrightarrow}{\Diamond} \Box q$ it is clear that q has to be true in some most preferred $|\top|$ worlds, and also that less preferred $|\top|$ worlds do not effect the truth of $\overset{\leftrightarrow}{\Diamond} \Box q$. Hence, in the evaluation of $\top >_\exists q$ only preferred elements are taken into account and not the whole ordering.*

The main properties of the logic are given by the following proposition. It illustrates that ordering and minimizing are duals, as far as we consider the properties strengthening of the antecedent and weakening of the consequent.

Proposition 3 *The logic CT40 validates the following theorem.*

WC$_\exists$ $\quad \beta >_\exists \alpha_1 \rightarrow \beta >_\exists (\alpha_1 \vee \alpha_2)$

The logic CT40 does not validate the following theorems.

SA$_\exists$ $\quad \beta_1 >_\exists \alpha \rightarrow (\beta_1 \wedge \beta_2) >_\exists \alpha$
AND$_\exists$ $\quad \beta >_\exists \alpha_1 \wedge \beta >_\exists \alpha_2 \rightarrow \beta >_\exists (\alpha_1 \wedge \alpha_2)$
DD$_\exists$ $\quad \gamma >_\exists \beta \wedge \beta >_\exists \alpha \rightarrow \gamma >_\exists \alpha$
D$_\exists$ $\quad \neg(\beta >_\exists \alpha \wedge \beta >_\exists \neg\alpha)$

Proof *The (non)theorems can easily be verified by proving (un)derivability in CT40.*

The logic validates weakening of the consequent, because the most preferred world that satisfies p_1 also satisfies $p_1 \vee p_2$. However, the logic does not have strengthening of the antecedent of $\top >_\exists p$ to $q >_\exists p$, because the preferred $|\top|$ worlds may be different from the preferred $|q|$ worlds. This property is illustrated by the following example.

Example 7 *Reconsider the Kripke model of Figure 2. The model satisfies $\top >_\exists q$ but not $\neg p >_\exists q$. For example, $\neg p >_\exists q$ is false because the preferred $|\neg p|$ worlds are the $|\neg p \wedge \neg q|$ worlds. Hence, $>_\exists$ does not have strengthening of the antecedent.*

3 Two-phase approach to default logic

3.1 Combining ordering and minimizing

The idea of combining ordering and minimizing is to combine formulas with $>$ and $>_\exists$ operators, where ordering should be strictly stronger than minimizing. However, this combination is not satisfactory in the logic CT40, because we cannot derive $\beta >_\exists \alpha$ from $\beta > \alpha$. The following proposition gives the relation between the two operators.

Proposition 4 *The logic CT40 validates the following theorem.*

$$\beta > \alpha \wedge \overleftrightarrow{\Diamond} (\beta \wedge \alpha) \rightarrow \beta >_\exists \alpha$$

Proof *The theorem can easily be verified by proving derivability in CT40. It is equivalent to the following formula:*

$$(\overleftrightarrow{\Box} ((\beta \wedge \alpha) \rightarrow \Box(\beta \rightarrow \alpha)) \wedge \overleftrightarrow{\Diamond} (\beta \wedge \alpha)) \rightarrow \overleftrightarrow{\Diamond} (\beta \wedge \Box(\beta \rightarrow \alpha))$$

Hence, when $\overleftrightarrow{\Diamond} (\alpha \wedge \beta)$ is false, then ordering is not stronger than minimizing. For example, it can easily be verified that the logic CT40 validates the theorem $\alpha > \bot$ but it does not validate $\alpha >_\exists \bot$. In the following definition, $\beta >^c \alpha$ has an additional condition which works like a 'consistency check' to test whether $\alpha \wedge \beta$ is possible.

Definition 4 *Consistent type-1 defaults "if β then by default α", written as $\beta >^c \alpha$, are defined as follows.*

$$\beta >^c \alpha =_{def} \beta > \alpha \wedge \overleftrightarrow{\Diamond} (\beta \wedge \alpha)$$

This new type of ordering is strictly stronger than minimizing, as is shown in the following proposition.

Proposition 5 *The logic CT40 validates the following theorem.*

$$\beta >^c \alpha \rightarrow \beta >_\exists \alpha$$

Proof *Follows directly from Definition 4 and Proposition 4.*

The type-1 defaults $\beta >^c \alpha$ validate weaker versions of the theorems of Proposition 2, like for example the following Restricted Strengthening of the Antecedent (**RSA**) and Restricted Conjunction (**RAnd**).

RSA $\beta_1 >^c \alpha \wedge \overleftrightarrow{\Diamond} (\beta_1 \wedge \beta_2 \wedge \alpha) \rightarrow (\beta_1 \wedge \beta_2) >^c \alpha$

RAnd $\beta >^c \alpha_1 \wedge \beta >^c \alpha_2 \wedge \overleftrightarrow{\Diamond} (\beta \wedge \alpha_1 \wedge \alpha_2) \rightarrow \beta >^c (\alpha_1 \wedge \alpha_2)$

We already saw restricted strengthening of the antecedent in Example 1 in the introduction. We elaborate on Example 1 in Section 3.2. To strengthen the theorems above, we consider only models in which all propositionally satisfiable formulas ϕ are true in *some* world. This can be 'axiomatized' with Boutilier's axiom scheme **LP**, see [Lev90, Bou94b] for a discussion. The axiom scheme **LP** states that every formula ϕ without any occurrences of modal operators, which is propositionally satisfiable, is true in some world.

Definition 5 *The logic CT40* is CT40 extended with the following axiom scheme:*
LP :$\overset{\leftrightarrow}{\Diamond}\ \phi$ for all satisfiable propositional ϕ

We write $\models$ for logical entailment in CT40.*

The following example illustrates the logic CT40* and the idea of combining ordering and minimizing.

Example 8 *Let S be the set of defaults $\{\top >^c (\neg r \wedge \neg g), g >^c r, r >^c g\}$. The intended model is given in Figure 4. We have $S \not\models \top >^c \neg r$, $S \not\models \top >^c \neg g$, $S \models \top >_\exists \neg r$ and $S \models \top >_\exists \neg g$.*

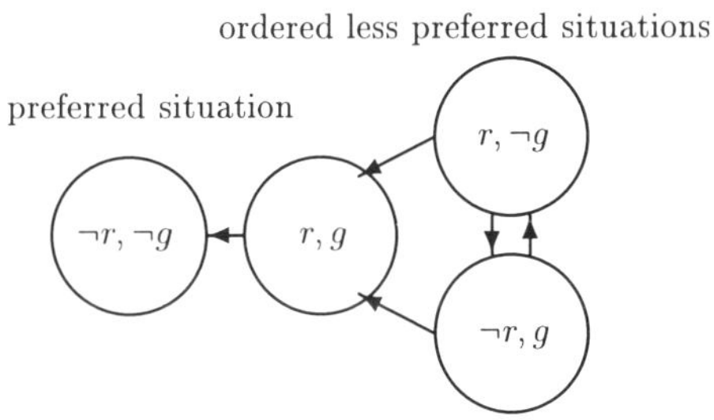

Figure 4: Preference relation

3.2 The two phases in a default logic

The two phases in a default logic correspond to the two different kinds of defaults $>^c$ and $>_\exists$. Semantically, the first phase corresponds to ordering ($>^c$) and the second phase to minimizing ($>_\exists$). ¿From a proof theoretic point of view, the first phase corresponds to applying valid inferences of $>^c$ like RSA, RAND etc, and the second phase corresponds to applying valid inferences of $>_\exists$ like WC. The basic technique of default logic as a two-phase logic is that a conclusion of the form $\beta >_\exists \alpha$ can be derived either with or without $\beta >^c \alpha$. In the first case $\beta >_\exists \alpha$ can be derived via $\beta >^c \alpha$ with

Proposition 5, which says that the latter formula implies the first one. If so, we say that $\beta >_\exists \alpha$ is derived in the first phase. In the second case we say that $\beta >_\exists \alpha$ is second phase derived. The important difference is that in the first phase we can apply RSA to $\beta >_\exists \alpha$, because of the simultaneous occurrence of $\beta >^c \alpha$. We apply RSA to $\beta >^c \alpha$ to obtain, for example, $(\beta \wedge \gamma) >^c \alpha$, and then due to Proposition 5 we also obtain $(\beta \wedge \gamma) >_\exists \alpha$. If $\beta >^c \alpha$ does not occur simultaneously with $\beta >_\exists \alpha$, then there is no way we can apply RSA to this formula. Being a minimizing formula it lacks RSA. Hence, once $\beta >_\exists \alpha$ has been derived in the second phase, we loose RSA permanently for subsequent derivations of this formula. Analogously, we can say that $\beta >_\exists \alpha$ is first phase or second phase entailed by a set of premises, depending on whether S does or does not entail $\beta >^c \alpha$.

The following example shows that the two-phase approach solves the apples and pears problem of Example 1 in the introduction.

Example 9 (apples and pears problem, continued) *Let $S = \{\top >^c (a \vee p), \top >^c \neg a\}$, where $\neg a$ does not entail the negation of p. We have $S \models \overleftrightarrow{\Diamond} (\neg a \wedge p)$, $S \models \top >^c (\neg a \wedge p)$ and $S \models \top >_\exists (\neg a \wedge p)$, $S \not\models \top >^c p$ and $S \models \top >_\exists p$. The crucial observation is that $a >_\exists p$ is not entailed by S. First of all, $a >_\exists p$ is not first phase entailed by S via $\top >_\exists p$, because $\top >_\exists p$ is not first phase entailed by S. Secondly, $a >_\exists p$ is not second phase entailed by S via $\top >_\exists p$ either, because in second phase entailment $>_\exists$ does not have strengthening of the antecedent at all. Thirdly, it is not second phase entailed by S via a first phase derivation of $a >_\exists (\neg a \wedge p)$, because $a >^c (\neg a \wedge p)$ is not entailed by $\top >^c (\neg a \wedge p)$ due to the restriction in* RSA.

4 Further research

Preference semantics are used in several different areas. In fact, the notions 'ordering' and 'minimizing' were introduced in a *deontic logic* based on preferences [TvdT96]. Although there are various subtle differences between these areas, see [Mak93] for a survey, we think that (at least) the following two discussions within deontic logic are relevant for default logic. First of all, the discussion about the validity of the deontic **D** axiom (see e.g. [TvdT96]) is relevant for sceptical reasoning about defaults. Secondly, defeasible deontic logic [vdTT95] and its multi-preference semantics [TvdT95] can be relevant to formalize specificity.

We can consider the two-phase approach with sceptical reasoning about defaults. The obvious candidate for the sceptical type-2 default is "a conditional α is true in *all* most preferred $|\beta|$ worlds", which we write as $\beta >_\forall \alpha$. The following definition of this default in CT40* is from [Bou94b].

Definition 6 *Type-2 sceptical defaults "if β then by default α", written as*

$\beta >_{\forall} \alpha$, are defined as follows.

$$\beta >_{\forall} \alpha =_{def} \overleftrightarrow{\Box} (\beta \rightarrow \Diamond(\beta \wedge \Box(\beta \rightarrow \alpha)))$$

However, to combine $>_{\forall}$ with $>^c$, the latter has to be strictly stronger than the other. The following example illustrates that this type-2 default cannot satisfactorily be combined with the type-1 default $\beta >^c \alpha$, because the ordering of worlds can be too weak.

Example 10 *Consider the default $\top >^c p$. All models that satisfy $|p| \not\leq |\neg p|$ are models of $\top >^c p$. Hence, $|p|$ worlds and $|\neg p|$ worlds are either incomparable, or $|p|$ worlds are strictly preferred to $|\neg p|$ worlds. Let M be a model in which all $|p|$ and $|\neg p|$ worlds are incomparable. M satisfies $\top >^c p$, but it does not satisfy $\top >_{\forall} p$. Hence, $\top >_{\forall} p$ is not entailed by $\top >^c p$. For minimization, we only want the models in which $|p|$ worlds are strictly preferred to $|\neg p|$.*

A solution of the previous problem is to define a preference ordering on models, which prefers models which are maximally connected with respect to the partial pre-ordering $\leq$, i.e. with the most binary relations of $\leq$. The preferred models of this ordering are the only models which are used for minimization.[3]

Definition 7 *Let $M_1 = \langle W_1, R_1, \leq_1, V_1 \rangle$ and $M_2 = \langle W_2, R_2, \leq_2, V_2 \rangle$ be two CT4O* models. M_1 is preferred to M_2 for mapping τ, written as $M_1 \sqsubseteq_\tau M_2$, iff:*

1. *τ is a one-to-one mapping of the worlds of W_2 to the worlds of W_1 such that the worlds satisfy the same propositions, and*

2. *If $w_1 \leq_2 w_2$ for $w_1, w_2 \in W_2$ then $\tau(w_1) \leq_1 \tau(w_2)$.*

We write $M_1 \sqsubset_\tau M_2$ iff $M_1 \sqsubseteq_\tau M_2$ and $M_2 \not\sqsubseteq_{\tau^{-1}} M_1$.

The ordering on models ($\sqsubseteq$) should not be confused with the ordering on worlds ($\leq$). The ordering on models is a technical trick to ensure that the worlds within a model are maximally connected, whereas the ordering on worlds expresses the normality ordering. Given the preference ordering on models, we can define a notion of preferential entailment, see [Sho88, KLM90].

[3] In [Bou92], maximally connected models related to system Z are models in which *only* the premises are known, and are formalized with Levesque's 'only knowing' (alias 'all-I-know') operator [Lev90]. However, system Z defines a unique preferred model, whereas in our case there are many distinct preferred models. Hence, we cannot simply copy this 'only knowing' concept.

Definition 8 *Let $M = \langle W, R, \leq, V\rangle$ be a model and S be a set of sentences. A world $w \in W$ of M preferentially satisfies S, written as $M, w \models_{\sqsubset} S$ iff $M, w \models S$ and there is not a model M' and a mapping τ such that $M', \tau(w) \models S$ and $M' \sqsubset_\tau M$ (M is a preferred model of S). S preferentially entails ϕ, written as $S \models_{\sqsubset} \phi$, iff for all M and w, if $M, w \models_{\sqsubset} S$ then $M, w \models \phi$.*

The following example illustrates the notion of preferential entailment.

Example 11 *Reconsider the set of defaults $S = \{\top >^c (\neg r \wedge \neg g), g >^c r, r >^c g\}$ of Example 8. Without the preference ordering on models, the $|\neg r \wedge \neg g|$ and $|r \wedge g|$ worlds could be incomparable. Such a model M would still satisfy $M \models \top >_\exists \neg r$ but it would not satisfy $M \models \top >_\forall \neg r$.*

Preferential entailment is a typical mechanism from non-monotonic reasoning. The combination of ordering and minimizing is non-monotonic, as the following example illustrates.

Example 12 *Let $S = \{\top >^c p\}$ and $S' = \{\top >^c p, \top >^c \neg p\}$. We have $S \models_{\sqsubset} \top >_\forall p$ and $S' \not\models_{\sqsubset} \top >_\forall p$. Hence, by addition of a formula we loose conclusions.*

Unfortunately, the previous example also shows that ordering is not stronger than minimizing, a condition for the two-phase approach. For this reason, we define a new type-1 default $>_D$ for the sceptical case, which trivially satisfies a counterpart of Proposition 5.

Definition 9 *Type-1 sceptical defaults "if β then by default α", written as $\beta >_D \alpha$, are defined as follows.*

$$\beta >_D \alpha =_{def} \beta > \alpha \wedge \beta >_\forall \alpha$$

The two-phase approach with $>_D$ and $>_\forall$ works exactly like the two-phase approach with $>^c$ and $>_\exists$. The following example illustrates several properties of this new type-1 default.

Example 13 *The set of defaults $\{\top >_D p, \top >_D q\}$ is consistent, but the set of defaults $\{\top >_D p, \top >_D \neg p\}$ is inconsistent. Moreover, the set of defaults $\{\top >_D p, q >_D \neg p\}$ is inconsistent too.*

The latter example shows that the specificity principle is not formalized in the proposed two-phase defeault logic. An interesting approach to this problem is the multi preference framework proposed in [TvdT95].

5 Related research

An example of a default logic with ordering and minimizing is the dynamic interpretation of defaults in Veltman's preference-based default logic [Vel91]. For example, in his logic the formula $\top[\textit{normally } p] \models \textit{presumably } (p \vee q)$ means that after ordering all worlds by preferring p to $\neg p$, $p \vee q$ is true in the preferred worlds. A distinction between the logic proposed in this paper and Veltman's logic is that in the latter, the order of the defaults influences the set of derivable formulas. Moreover, in Veltman's logic defaults are not conditionals and ordering and minimizing are not modeled in the same object language.

Two phases can be traced in several popular non-monotonic logics like Reiter's default logic, autoepistemic logic and circumscription, which we illustrate by Reiter's default logic [Rei80]. Consider default theories consisting of a set of normal default rules $\frac{\beta:\alpha}{\alpha}$, which express that α is part of an extension (a deductively closed set of formulas) if β is part of the extension and $\neg\alpha$ is not, and a factual sentence (for simplicity we assume that the facts can be represented by a single formula). For example, the 'birds fly' default rule $\frac{b:f}{f}$ expresses that f is part of an extension if b is part of the extension and $\neg f$ is not; hence, birds are assumed to fly unless there is knowledge of the contrary. Type-1 and type-2 defaults can be identified in Reiter's default logic as follows.

1. For type-1 defaults, assume a fixed factual sentence. The 'birds fly' default $\frac{b:f}{f}$ is stronger than the 'red birds fly' default $\frac{b\wedge r:f}{f}$ in the sense that if a default theory contains the first default, then the second one can be added to the default theory without changing its set of extensions.

2. For type-2 defaults, assume a fixed set of normal Reiter defaults. If the factual sentence of a default theory is b and one of the extensions contains f, then this does *not* imply that the default theory with facts $b \wedge p$ has an extension that also contains f (e.g., with default rules $\frac{\top:\neg p}{\neg p}$ and $\frac{b\wedge\neg p:f}{f}$).

In this paper we proposed a logic for credulous reasoning about both types of defaults. In the vocabulary of a logic for reasoning about type-1 defaults, the 'red birds fly' default is *derivable* from the 'birds fly' default. The example illustrates that $\frac{\beta\wedge\gamma:\alpha}{\alpha}$ can be derived from $\frac{\beta:\alpha}{\alpha}$, which means that normal Reiter defaults validate strengthening of the antecedent. Similarly, it can be shown that $\frac{\beta:\alpha\vee\gamma}{\alpha\vee\gamma}$ *cannot* be derived from $\frac{\beta:\alpha}{\alpha}$, which means that normal Reiter defaults do *not* validate weakening of the consequent. In contrast, type-2 defaults do *not* validate strengthening of the antecedent (of b to $b \wedge p$ in the example above), but they do validate weakening of the consequent (because extensions are deductively closed). In this paper, we combined type-1 and type-2 defaults by the additional condition that type-1 defaults are strictly stronger than type-2 defaults. For Reiter's default logic, this additional condition means (in the credulous case) that for every default theory with default rule $\frac{\beta:\alpha}{\alpha}$ and factual sentence β, there is an extension that contains α. Hence, the reasonable additional condition means that for every default rule $\frac{\beta:\alpha}{\alpha}$, $\beta \wedge \alpha$ is consistent.

Acknowledgement

Thanks to Patrick van der Laag for several discussions on the issues raised in this paper.

References

[Alc94] C. E. Alchourrón. Philosophical foundations of deontic logic and the logic of defeasible conditionals. In *Deontic Logic in Computer Science: Normative System Specification*, pages 43–84. John Wiley & Sons, 1994.

[AM91] N. Asher and M. Morreau. Common sense entailment: a modal theory of nonmonotonic reasoning. In *Proceedings of the Twelfth International Joint Conference on Artificial Intelligence (IJCAI'91)*, Palo Alto, 1991. Morgan Kaufman.

[BDP92] S. Benferhat, D. Dubois, and H. Prade. Representing default rules in possibilistic logic. In *Proceedings of the Second International Conference on Principles of Knowledge Representation and Reasoning (KR'92)*, pages 673–684, Cambridge, MA, 1992.

[Bou92] C. Boutilier. Conditional logics for defeault reasoning and belief revision (phd thesis). Technical Report 92-1, Department of Computer Science, University of British Colombia, 1992.

[Bou94a] C. Boutilier. Conditional logics of normality: a modal approach. *Artificial Intelligence*, 68:87–154, 1994.

[Bou94b] C. Boutilier. Unifying default reasoning and belief revision in a modal framework. *Artificial Intelligence*, 68, 1994.

[Del88] J.P. Delgrande. An approach to default reasoning based on a first-order conditional logic: revised report. *Artificial Intelligence*, 36, 1988.

[KLM90] S. Kraus, D. Lehmann, and M. Magidor. Nonmonotonic reasoning, preferential models and cumulative logics. *Artificial Intelligence*, 44:167–207, 1990.

[Lev90] H. Levesque. All I know: a study in autoepistemic logic. *Artificial Intelligence*, 42:263–309, 1990.

[LM92] D. Lehmann and M. Magidor. What does a conditional knowledge base entail? *Artificial Intelligence*, 55:1–60, 1992.

[Mak93] D. Makinson. Five faces of minimality. *Studia Logica*, 52:339–379, 1993.

[Mor95] M. Morreau. Allowed consequence. In *Proceedings of the Fourteenth International Joint Conference on Artificial Intelligence (IJCAI'95)*. Morgan Kaufman, 1995.

[Pea90] J. Pearl. System Z: a natural ordering of defaults with tractable applications to default reasoning. In *Proceedings of Theoretical Aspects of Reasoning about Knowledge (TARK)*, San Mateo, 1990. Morgan Kaufmann.

[Rei80] R. Reiter. A logic for default reasoning. *Artificial Intelligence*, 13:81–132, 1980.

[Sho88] Y. Shoham. *Reasoning About Change*. MIT Press, 1988.

[TP94] S.-W. Tan and J. Pearl. Specification and evaluation of preferences under uncertainty. In *Proceedings of the Fourth International Conference on Principles of Knowledge Representation and Reasoning (KR'94)*, pages 530–539, 1994.

[TvdT95] Y.-H. Tan and L.W.N. van der Torre. Why defeasible deontic logic needs a multi preference semantics. In *Proceedings of the ECSQARU'95. Lecture Notes in Artificial Intelligence 946*. Springer Verlag, 1995.

[TvdT96] Y.-H. Tan and L.W.N. van der Torre. How to combine ordering and minimizing in a deontic logic based on preferences. In *Proceedings of the Third Workshop on Deontic Logic in Computer Science (Δeon'96)*. Springer Verlag, 1996.

[vdTT95] L.W.N. van der Torre and Y.-H. Tan. Cancelling and overshadowing: two types of defeasibility in defeasible deontic logic. In *Proceedings of the Fourteenth International Joint Conference on Artificial Intelligence (IJCAI'95)*. Morgan Kaufman, 1995.

[Vel91] F. Veltman. Defaults in update semantics. Technical Report ITLI Prepublication Series LP-91-02, Department of Philosophy, University of Amsterdam, 1991. To appear in the *Journal of Philosophical Logic*.

[vW63] G.H. von Wright. *The logic of preference*. Edinburgh, 1963.

Computational complexity of infinite-valued Łukasiewicz propositional logic

H. Wagner
Department of Computer Science
Fernuniversität, D-58084 Hagen

Abstract

It is shown for infinite-valued Lukasiewicz propositional logic that the consequence relation is complete for coNP and that the consequence problem is Π_2-complete. Using the NP-completeness of the existential theory of real addition a simplified proof for the NP-completeness of the satisfiability problem is given.

1 Introduction

Many-valued, especially infinite-valued Łukasiewicz logic never belonged to the mainstream of logical research. Yet a lot of investigations concerning the logical foundations of infinite-valued Łukasiewicz logic has been done in the years from about 1920 to about 1965, culminating in Scarpellinis ([12]) proof of incompleteness of first order infinite-valued Łukasiewicz logic. It seems that Scarpellinis negative result breaked down the interest in these logics. However mainly caused by applications in computer science the importance of many-valued logics has grown during the last decade. So it seems natural to investigate the computational complexity of these logics. While it is quite obvious that most of the results of classical propositional logic carry over to finite-valued logics (for instance NP-completeness of the satisfiability problem, existence of a tableau calculus etc.) the situation for infinite-valued logics is much more intricated.

A starting point of investigation was Ragaz dissertation ([11]). Considering the arithmetical complexity of infinite-valued logics he especially showed the decidability of infinite-valued Lukasiewicz propositional logic. Mundici ([10]) sharpened this result by showing the NP-completeness of the satisfiability problem. In fact he only proved this result for the weak satisfiability notion, but his proof is easily adapted to the case of the strong satisfiability notion. In this paper we will show that the computational complexity of the satisfiability problem and of the consequence relation can be obtained as a trivial consequence of the NP-completeness of the set of Σ_1-formulas in the theory of real addition. Furthermore we will consider the more general consequence problem: given a recursive enumerable set M of formulas and a formula A, decide wether A semantically follows from M. This problem is interesting, because the semantical consequence relation is lacking the finiteness property. We show that this decision problem is Π_2-complete.

2 Preliminaries

We are defining the language *IPL* of infinite-valued Łukasiewicz propositional logic.

2.1 Syntax

We fix a finite alphabet $\Sigma = \{x, |, \neg, \wedge, \vee, \oplus, (,)\}$.
The set of variables and the set of formulas are defined inductively.

Definition 1 *(of the set V of propositional variables)*

- $x \in V$
- *if* $y \in V$ *then* $y| \in V$.

We also write x_m for $x|\ldots|$ (m strokes).

Definition 2 *(of the set of formulas)*

- *Every propositional variable in V is a formula.*
- *If A and B are formulas then so are $\neg A, (A \wedge B), (A \vee B)$ and $(A \oplus B)$.*

We will use $A, B, C, A_0, B_0, C_0, A_1, \ldots$ as syntactic variables for formulas. For a formula A we will also write $A(x_1, \ldots, x_n)$ if $x_1, \ldots, x_n$ are the only variables in A.

The following logical connectives for bold conjunction, implication and for equivalence are introduced as
Abbreviations:

$(A \otimes B) := \neg(\neg A \oplus \neg B)$
$(A \rightarrow B) := (\neg A \oplus B)$
$(A \leftrightarrow B) := ((A \rightarrow B) \wedge (B \rightarrow A))$
$1 := (\neg x \oplus x)$
$0 := \neg 1$

For technical reasons we use $\vee$ and $\wedge$ as basic logical operators. Of course $(A \wedge B)$ and $(A \vee B)$ could have been defined by

$$(A \vee B) := (\neg(\neg A \oplus B) \oplus B).$$

We adopt the usual conventions for omitting parentheses. So for instance we also write $A \oplus B$ instead $(A \oplus B)$ and $A_1 \wedge \ldots \wedge A_n$ instead $(\ldots((A_1 \wedge A_2) \wedge A_3) \wedge \ldots \wedge A_n)$. For every formula A in *IPL* the length $\| A \|$ of A is defined as the length of A as a string over the alphabet Σ. If each occurence of a variable p in a formula A is substituted by a formula B we denote the resulting formula as $A\,[p/B]$.

2.2 Semantics

For defining the semantics in the infinite-valued case we use the interval $[0,1]$ (this is the set of all reals r with $0 \leq r \leq 1$) as the set of logical values. The semantics of *IPL* is based on the concept of interpretations. An *interpretation* I is a mapping $I : V \to [0,1]$. Let $\mathcal{I}$ denote the set of all interpretations.

Definition 3 *The value* $\mathrm{val}\,(A, I)$ *of a formula* A *with respect to the interpretation* $I \in \mathcal{I}$ *is defined inductively:*

$\mathrm{val}\,(y, I) := I\,(y)$ *for* $y \in V$

$\mathrm{val}\,(\neg A, I) := 1 - \mathrm{val}\,(A, I)$

$\mathrm{val}\,(A \wedge B, I) := \min\,(\mathrm{val}\,(A, I)\,, \mathrm{val}\,(B, I))$

$\mathrm{val}\,(A \vee B, I) : \max\,(\mathrm{val}\,(A, I)\,, \mathrm{val}\,(B, I))$

$\mathrm{val}\,(A \oplus B, I) := \min\,(1, \mathrm{val}\,(A, I) + \mathrm{val}\,(B, I))$

A is a tautology (denoted by $\models A$) iff $\mathrm{val}\,(A, I) = 1$ *for all $I \in \mathcal{I}$. If there is $I \in \mathcal{I}$ such that* $\mathrm{val}\,(A, I) = 1$ *we say that A is (strong) satisfiable and that I satisfies A. A is weak satisfiable iff for some $I \in \mathcal{I}$* $\mathrm{val}\,(A, I) > 0$.

Let M be a set of formulas. An interpretation $I \in \mathcal{I}$ satisfies M iff for all $A \in M$ I satisfies A. M is satisfiable iff there is some $I \in \mathcal{I}$ that satisfies M. $M \models A$ iff for all $I \in \mathcal{I}$: if I satisfies M then I satisfies A. In case $M = \{B\}$ we also write $B \models A$.

Some consequences:

- $\mathrm{val}\,(A \to B, I) = \min\,(1, 1 - \mathrm{val}\,(A, I) + \mathrm{val}\,(B, I))$
- $\mathrm{val}\,(A \otimes B, I) = \max\,(0, \mathrm{val}\,(A, I) + \mathrm{val}\,(B, I) - 1)$
- $\mathrm{val}\,(A \leftrightarrow B, I) = 1 - |\mathrm{val}\,(A, I) - \mathrm{val}\,(B, I)|$

For later we will introduce some further abbreviations:

Definition 4 *For all natural numbers n and formulas $A \in IPL$:*

$$\sum_{i=1}^{0} A := 0 \quad , \quad \sum_{i=1}^{n+1} A := A \oplus \sum_{i=1}^{n} A$$

$$\theta\,(n, A, x) := \sum_{i=1}^{n+1} A \wedge \left(\sum_{i=1}^{n} A \to x \right)$$

We state without proof:

Lemma 1 *For all natural numbers n, all formulas $A \in IPL$ and all interpretations I:*

1.

$$val\left(\sum_{i=1}^{n} A, I\right) = \min(1, n \cdot val(A, I))$$

2.

$$val(\theta(n, A, x), I) = 1 \Rightarrow I(x) \geq \frac{n}{n+1}$$

Definition 5 *To each formula A of IPL we define a formula A^* and a formula F_A by:*
$(y)^* := c_y \leftrightarrow y$ *for each propositional variable y*
$(\neg A)^* := (c_{\neg A} \leftrightarrow \neg c_A) \wedge A^*$
$(A \wedge B)^* := (c_{A \wedge B} \leftrightarrow c_A \wedge c_B) \wedge A^* \wedge B^*$
$(A \vee B)^* := (c_{A \vee B} \leftrightarrow c_A \vee c_B) \wedge A^* \wedge B^*$
$(A \oplus B)^* := (c_{A \oplus B} \leftrightarrow (c_A \oplus c_B)) \wedge A^* \wedge B^*$
where the c_B's for subformulas B of A are new propositional variables of V.
$F_A := c_A \wedge A^*$.

Obviously the length of F_A is linear in the length of A.

Lemma 2 *Let A be an arbitrary formula of IPL. Let further I be an interpretation with $I(c_B) = \mathrm{val}(B, I)$ for all subformula B of A.*
Then $\mathrm{val}(A, I) = \mathrm{val}(F_A, I)$.

Proof:
We prove by induction on the definition of A, that $\mathrm{val}(A^*, I) = 1$.
Our claim is trivial in case $A = y$.
Case $A = \neg B$:
By induction hypothesis we have $\mathrm{val}(B^*, I) = 1$. By our assumption on I we further have $\mathrm{val}(c_{\neg B}, I) = \mathrm{val}(\neg B, I) = 1 - \mathrm{val}(B, I) = 1 - \mathrm{val}(c_B, I) = \mathrm{val}(\neg c_B, I)$, so $\mathrm{val}(c_{\neg B} \leftrightarrow \neg c_B, I) = 1$, therefore $\mathrm{val}(A^*, I) = 1$.
Case $A = B \oplus D$:
By induction hypothesis we have $\mathrm{val}(B^*, I) = \mathrm{val}(D^*, I) = 1$. Further we have

$$\begin{aligned} \mathrm{val}(c_{B \oplus D}, I) &= \mathrm{val}(B \oplus D, I) = \min(1, \mathrm{val}(B, I) + \mathrm{val}(D, I)) \\ &= \min(1, \mathrm{val}(c_B, I) + \mathrm{val}(c_D, I)) = \mathrm{val}(c_B \oplus c_D, I). \end{aligned}$$

Again we get $\mathrm{val}(A^*, I) = 1$.

The remaining cases are proved in a similar way. So our induction is finished.
But now we get $\text{val}(F_A, I) = \text{val}(c_A \wedge A^*, I) = \text{val}(c_A, I) = \text{val}(A, I)$.
□

Corollary 1 *A formula A of IPL is (strong resp. weak) satisfiable if and only if its translation F_A is (strong resp. weak) satisfiable.*

3 Complexity of the satisfiability problem and the consequence relation

While it is well known by Cook's theorem ([2]) that the satisfiability problem for classical 2-valued propositional logic is NP-complete the complexity of the corresponding problem for the infinite-valued propositional logic of Lukasiewicz wasn't determined for a long time. In 1987 Mundici ([10]) proved the NP-completeness for this problem too, using the weak satisfiability notion (but his proof can easily be adapted for the case of the ordinary satisfiability notion). The difficult part of his proof was showing containment in NP. Hähnle ([5]) and independently the author ([13]) then gave simplified proofs for the NP-completeness of the satisfiability problem by reduction to some linear programming problems. Developing a tableau calculus for many-valued logics Hähnle was led to a polynomial time reduction of the (strong) satisfiability problem to a mixed integer programming problem, while in [13] a nondeterministic Turing reduction to (a special case of) linear programming was obtained. As mixed integer programming is NP-complete and linear programming has been proven to be P-complete by Khachian ([7]) and Kamarkar ([6]), both methods proved containment in NP. However, as we will see now, the classical proof of decidability of IPL, which belonged to the folklore of the logical community and which e.g. is given in [11] and in [4], contains nearly all necessary details for an even more trivial proof of NP-completeness. Furthermore this approach has the advantage that for many decision problems of IPL the computational complexity can be determined in an easy way.

Let SAT_{IPL} denote the set of satisfiable formulas of IPL,

$$SAT_{IPL} := \{A \in IPL \mid \exists I \in \mathcal{I}.val(A, I) = 1\}$$

and

$$TAUT_{ITL} := \{A \in IPL \mid \ \models A\}$$

the set of all valid formulas. *Satisfiability problem* and *tautology problem* mean the corresponding decision problems for these sets.

Theorem 1 *SAT_{IPL} is NP-complete.*

Proof:

"SAT_{IPL} is NP-hard:"

Let SAT denote the satisfiability problem in the classical two-valued case. We give a polynomial time many-one reduction of SAT to SAT_{IPL}. To any instance $A(x_1, \ldots, x_n)$ to SAT we assign $F(A) :\equiv A(x_1, \ldots, x_n) \wedge (x_1 \vee \neg x_1) \wedge \ldots \wedge (x_n \vee \neg x_n)$ as an instance to SAT_{IPL}. It is easy to see that $A(x_1, \ldots, x_n) \in SAT$ iff $F(A) \in SAT_{IPL}$ and that $F(A)$ is computable from $A(x_1, \ldots, x_n)$ in polynomial time. So it follows that SAT_{IPL} is NP-hard.

"SAT_{IPL} is in NP:"

We define a translation, which assigns to each formula A of IPL a Σ_1-formula $T(A)$ in the theory TH($\mathcal{R}, +, \leq, 0, 1$), the theory of real addition, in such a way, that this translation is computable by a deterministic Turing machine in polynomial time. For the language of TH($\mathcal{R}, +, \leq, 0, 1$) we are using the same symbols $+, \leq, 0$ and 1, for the individual variables we are using the same symbols as for the propositional variables in IPL. Further we introduce the following formulas as abbreviations in the theory of real addition:

$x = \max(y, z) :\Leftrightarrow (x = y \wedge z \leq y) \vee (x = z \wedge y \leq z)$

$x = \min(y, z) :\Leftrightarrow (x = y \wedge y \leq z) \vee (x = z \wedge z \leq y)$

$x = y :\Leftrightarrow x \leq y \ \wedge \ y \leq x$

$x < y :\Leftrightarrow \neg (y \leq x)$

$x = 1 - y :\Leftrightarrow x + y = 1.$

Here the introduced max and min function symbols have the usual meanings of the maximum and minimum functions.

By Lemma 2 it suffices to define such a translation for the formulas F_A ($= c_A \wedge A^*$), where A^* is a conjunction $A^* = \bigwedge_{1 \leq i \leq n} A_i$ of formulas A_i of the form $(c_y \leftrightarrow y)$, $(c_{\neg B} \leftrightarrow \neg c_B)$, $(c_{B \wedge D} \leftrightarrow c_B \wedge c_D)$, $(c_{B \vee D} \leftrightarrow c_B \vee c_D)$, and $(c_{B \oplus D} \leftrightarrow (c_B \oplus c_D))$. We use here $\equiv$ as symbol for the equality in the metalanguage of the theory of real addition. Let

$T'(c_y \leftrightarrow y) :\equiv c_y = y \wedge 0 \leq y \wedge y \leq 1$

$T'(c_{\neg B} \leftrightarrow \neg c_B) :\equiv c_{\neg B} = 1 - c_B \wedge 0 \leq c_B \wedge c_B \leq 1$

$T'(c_{B \wedge D} \leftrightarrow c_B \wedge c_D) :\equiv c_{B \wedge D} = \min(c_B, c_D) \wedge 0 \leq c_B \wedge c_B \leq 1 \wedge 0 \leq c_D \wedge c_D \leq 1$

$T'(c_{B \vee D} \leftrightarrow c_B \vee c_D) :\equiv c_{B \vee D} = \max(c_B, c_D) \wedge 0 \leq c_B \wedge c_B \leq 1 \wedge 0 \leq c_D \wedge c_D \leq 1$

$T'(c_{B \oplus D} \leftrightarrow (c_B \oplus c_D)) :\equiv c_{B \oplus D} = \min(1, c_B + c_D) \wedge 0 \leq c_B \wedge c_B \leq 1 \wedge 0 \leq c_D \wedge c_D \leq 1$

Essentially $T'(A_i)$ corresponds to the definition of the semantics of the formula A_i. Now let $T'(A^*) :\equiv \bigwedge_{1 \leq i \leq n} T'(A_i)$. Let $\mathbf{x}$ be the sequence of the individual variables occuring in $T'(A^*)$. Then $T(A) :\equiv T(F_A) :\equiv$

$\exists \mathbf{x}(c_A = 1 \wedge 0 \leq c_A \wedge c_A \leq 1 \wedge T'(A^*))$. Obviously this translation is polynomial time computable, and F_A is satisfiable (in IPL) iff $T(F_A)$ is valid in the theory of real addition.However, by a well known result of von zur Gathen and Sieveking ([3]), the set of Σ_1-formulas in the theory of real addition is NP-complete. We therefore get $SAT_{IPL} \in NP$.

□

Corollary 2 *1. The weak satisfiablity problem is NP-complete.*

2. $TAUT_{IPL}$ is complete for coNP.

3. The consequence relation of IPL, i.e. the set

$$\{(A, B) \mid A, B \in IPL \text{ and } A \models B\}$$

is complete for coNP.

Proof: In each case the proof of hardness for the corresponding complexity class is trivial, so we prove only containment.

1. Just replace in the definition of $T(F_A)$ the equality $c_A = 1$ by $c_A > 0$. We get: F_A is weak satisfiable (in IPL) iff $T(F_A)$ is valid in the theory of real addition.
2. Immediate consequence of (1).
3. Let
   $T(A, B) :\equiv \exists \mathbf{x}(c_A = 1 \wedge 0 \leq c_A \wedge c_A \leq 1 \wedge T'(A^*) \wedge c_B < 1 \wedge 0 \leq c_B \wedge c_B \leq 1 \wedge T'(B^*))$, where $\mathbf{x}$ is the sequence of individual variables occuring in $T'(A^*)$ or in $T(B^*)$. (Observe, c_A and c_B are contained in $\mathbf{x}$). We get $A \not\models B$ iff $T(A, B)$ is valid in the theory of real addition. So containment in coNP follows for the consequence relation.

   □

4 Complexity of the consequence problem

In classical two-valued propositional logic we have the finiteness theorem: if a formula A semantically follows from an infinite set M of formulas, there is already a finite subset M' of M such that A already follows from M'. This property fails for the infinite-valued Lukasiewicz propositional logic. So it seems interesting to consider the following *consequence problem*: Given a recursive (enumerable) set M of formulas of *IPL* and a formula A of *IPL*, decide wether $M \models A$. The corresponding problem for classical two-valued propositional logic is Σ_1-complete (we assume here some knowledge of Kleene's arithmetical hierarchy).

Theorem 2 *The consequence problem for infinite-valued Lukasiewicz propositional logic is complete for Π_2 (with respect to many-one reduction).*

Proof:

"The consequence problem is in Π_2:"

This follows already from the corresponding result of Ragaz ([11]) for inifinite-valued first order logic.

"The consequence problem is hard for Π_2:"

We will show this by reducing a Π_2-complete problem to the consequence problem. We will use special two-counter machines introduced by ([8], p.61). Such a special two-counter machine is an automaton with finite-state control and exactly two counters, which are memory units capable of storing any natural numbers. The finite control may modify its counters as it changes state. Formally we present special two-counter machines as a sequence $\langle P_1, \ldots, P_k \rangle$ of instructions P_i, $1 \leq i \leq k$, where each instruction P_i is one of the following three kinds:

H	("halt")
$add\,(j)$	("add 1 to the number in the second counter and go to instruction j")
$sub\,(j,l)$	("if the number in the first counter is not zero, then subtract 1 from it and go to instruction j; otherwise interchange the counters and go to instruction l")

We stipulate that P_k, and only P_k, is H, and that for each P_i, $1 \leq i < k$, if P_i is $add\,(j)$ or $sub\,(j,l)$ then $1 \leq j, l \leq k$. With input "n" we mean that the special two-counter machine starts its instructions with the natural number n in both counters. A configuration for a special two-counter machine is given by a triple (i, u, v), where i is the index of the instruction to be executed next, u is the number stored in the first counter and v is the number stored in the second counter. $(1, r, r)$ is the start configuration for input r. If $i = k$ we say that (i, u, v) is an end configuration.

The halting problem is then given by the decision problem for the set $H_1 := \{(m,n) \mid$ the special two-counter machine with code m halts for input $n\}$. (We assume some standard numeration of the special two-counter machines.)

Lewis had shown in [8] that the inputless version (this means input 0 for machine m) of the halting problem is Σ_1-complete. A simple reduction proof shows that also H_1 is Σ_1-complete. Then it easily follows that the decision problem for the following set

$$H_\infty := \{m \mid \text{there exist infinite many } n \text{ with } (m,n) \in H_1\}$$

is Π_2-complete. We reduce H_∞ to the consequence problem.

Given a code m of a special two-counter machine we assign to this machine m a pair (Γ_m, B_m) where Γ_m is an infinite recursive set of formulas of IPL and $B_m \in IPL$ in such a way that machine m halts for infinitely many inputs n iff $\Gamma_m \models B_m$.

Assume machine m codes a sequence $\langle P_1, \ldots, P_k \rangle$ of k instructions. We introduce an abbreviation for some of the variables. Let $\langle\,\rangle$ be a recursive bijective mapping from $\mathcal{N}^4$ in $\mathcal{N}$. We write $\langle a, b, c, d \rangle$ instead $\langle\,\rangle\,(a, b, c, d)$. Define for all natural numbers e, r, n

$$
\begin{aligned}
&q_r :\equiv x_{\langle r,0,0,1\rangle}\\
&c^n_{r,e} :\equiv x_{\langle r,n,e,2\rangle}\\
&d^n_{r,e} :\equiv x_{\langle r,n,e,3\rangle}\\
&p^n_{1,r} :\equiv x_{\langle r,n,0,4\rangle}\\
&\qquad\qquad\vdots\\
&p^n_{k,r} :\equiv x_{\langle r,n,0,k+3\rangle}\;.
\end{aligned}
$$

We use the variable $c^n_{r,e}$ to express that machine m after doing n steps for input r has stored the number e in its first counter: Similarly $d^n_{r,e}$ expresses this property for the second counter. The variables $p^n_{i,r}$, $1 \leq i \leq k$, determine which instruction is next to be executed when machine m for input r has done n steps. Further we define $\theta_r := \theta\,(r, q_r, x)$ for all $r \in \mathcal{N}$.

Γ_m contains for all natural numbers r, n, e, f the following formulas:

$$
\begin{aligned}
&& c^0_{r,r}\\
&& d^0_{r,r}\\
&& p^0_{1,r}\\
p^n_{i,r} \wedge c^n_{r,f} \wedge d^n_{r,e} & \;\rightarrow\; & p^{n+1}_{j,r} \wedge c^{n+1}_{r,f} \wedge d^{n+1}_{r,e+1}
\end{aligned}
$$

for all i and j with $1 \leq i < k$ and $1 \leq j \leq k$, such that P_i is $add\,(j)$

$$
\begin{aligned}
p^n_{i,r} \wedge c^n_{r,0} \wedge d^n_{r,e} & \;\rightarrow\; & p^{n+1}_{l,r} \wedge c^{n+1}_{r,e} \wedge d^{n+1}_{r,0}\\
p^n_{i,r} \wedge c^n_{r,f+1} \wedge d^n_{r,e} & \;\rightarrow\; & p^{n+1}_{j,r} \wedge c^{n+1}_{r,f} \wedge d^{n+1}_{r,e}
\end{aligned}
$$

for all i, j and l with $1 \leq i < k$ and $1 \leq j, l \leq k$, such that P_i is $sub\,(j, l)$

$$p^n_{k,r} \rightarrow \theta_r$$

If m isn't the code of a machine let $\Gamma_m := \emptyset$.
B_m is just the formula x.

We show: $m \in H_\infty$ iff $\Gamma_m \models B_m$.
"only if":
By induction on n it is easily proved that for all natural numbers n and r:
if machine m for input r after doing n steps without stopping reaches the configuration (i, u, v) then $\Gamma_m \models p^n_{i,r} \wedge c^n_{r,u} \wedge d^n_{r,v}$.

Now assume $m \in H_\infty$. Then there exist infinitely many r with $(m,r) \in H_1$. So for every interpretation I, if I satisfies Γ_m the there exist infinitely many r, such that for some n I satisfies $p^n_{k,r}$. Therefore for infinitely many r I satisfies θ_r. We get by by lemma 1, that I satisfies x and therefore $\Gamma_m \models x$.

"if":

Assume $m \notin H_\infty$ and m is the code of a special two-counter machine. So there exists $r' \in \mathcal{N}$ such that for all $r \geq r'$ $(m,r) \notin H_1$. Therefore for all inputs $r \geq r'$ machine m runs through an infinite sequence

$$(i_{0,r}, u_{0,r}, v_{0,r}), (i_{1,r}, u_{1,r}, v_{1,r}), \ldots$$

of configurations with $(i_{0,r}, u_{0,r}, v_{0,r}) = (1, r, r)$. Especially we have $i_{n,r} \neq k$ for all $n \in \mathcal{N}$. For all $r \in \mathcal{N}$ we define

$$D_{r,0} := \{p^0_{1,r}, c^0_{r,r}, d^0_{r,r}\}$$

$$D_{r,n+1} := \begin{cases} \left\{p^{n+1}_{j,r}, c^{n+1}_{r,f}, d^{n+1}_{r,e+1}\right\} & \text{if } D_{r,n} = \left\{p^n_{i,r}, c^n_{r,f}, d^n_{r,e}\right\} \text{ and} \\ & P_i = add(j) \\ \left\{p^{n+1}_{l,r}, c^{n+1}_{r,e}, d^{n+1}_{r,0}\right\} & \text{if } D_{r,n} = \{p^n_{i,r}, c^n_{r,0}, d^n_{r,e}\} \text{ and} \\ & P_i = sub(j,l) \\ \left\{p^{n+1}_{j,r}, c^{n+1}_{r,f}, d^{n+1}_{r,e}\right\} & \text{if } D_{r,n} = \left\{p^n_{i,r}, c^n_{r,f+1}, d^n_{r,e}\right\} \text{ and} \\ & P_i = sub(j,l) \\ \emptyset & \text{if } D_{r,n} = \emptyset \text{ or } p^n_{k,r} \in D_{r,n} \end{cases}$$

$$D_r := \bigcup_{n \in \mathcal{N}} D_{r,n}$$

$$D := \bigcup_{r \in \mathcal{N}} D_r$$

An interpretation I is defined by

$$I(p) := \begin{cases} 1 & \text{if } p \in D \\ \frac{1}{r+1} & \text{if } p = q_r \text{ and } \exists n \in \mathcal{N}. \\ & p^n_{k,r} \in D_r \\ \sup\left\{\frac{r}{r+1} \mid r \in \mathcal{N}, \exists n. p^n_{k,r} \in D_r\right\} & \text{if } p = x \\ 0 & \text{else} \end{cases}$$

It is easy to prove that this interpretation I satisfies Γ_m and that I satisfies $p^n_{i,r} \wedge c^n_{r,u} \wedge d^n_{r,v}$ iff machine m for input r is in configuration (i,u,v) after executing n steps.

¿From the definition of D_r we get that for all $r \geq r'$ there is no n such that $p_{k,r}^n \in D_r$. Further

$$I(x) = \sup\left\{\frac{r}{r+1} \mid r \in \mathcal{N}, \exists n.p_{k,r}^n \in D_r\right\} \leq \frac{r'}{r'+1} < 1.$$

As I satisfies Γ_m it follows that $\Gamma_m \not\models x$, i.e. $\Gamma_m \not\models B_m$.

It is not difficult to prove that Γ_m is a recursive set and that (an index of) Γ_m can be recursively computed from m. (We won't go into details here as we have omitted coding details.) So we get a recursive many-one reduction of H_∞ to the consequence problem of *IPL*. Therefore the consequence problem is Π_2-hard.

□

References

[1] I.M. Bomze, Optimierung - Theorie und Algorithmen, BI Wissenschaftsverlag, 1993

[2] S.A. Cook, The complexity of theorem-proving procedures, in Proc. 3rd Ann. ACM Symp. on Theory of Computing (1971) 151-158

[3] J. von zur Gathen, M. Sieveking, Weitere zum Erfüllungsproblem polynomial äquivalente kombinatorische Aufgaben, in: E. Specker, V. Strassen, Komplexität von Entscheidungsproblemen, LNCS 43, Springer Verlag 1976, 49-71

[4] S. Gottwald, Mehrwertige Logik Eine Einführung in Theorie und Anwendungen, Akademie Verlag 1989

[5] R. Hähnle, Automated deduction in multiple-valued logics, Oxford University Press 1993

[6] N. Karmarkar, A new polynomial-time algorithm for linear programming, Combin. 4 (1984) 373-395

[7] G. Khachian, A polynomial algorithm in linear programming, Soviet Mathematics Doklady 20 (1979) 191-194

[8] H.R. Lewis, Unsolvable cases of quantificational formulas, Addison-Wesley Publ. Comp., Inc., 1979

[9] R. McNaughton, A theorem about infinite-valued sentential logic, The Journal of Symbolic Logic 16 (1951) 1-13

[10] D. Mundici, Satisfiability in many-valued sentential logic is NP-complete, Theoret. Comp. Sci. 52 (1987) 145-153

[11] M.E. Ragaz, Arithmetische Klassifikation von Formelmengen der unendlichwertigen Logik, Dissertation ETH Zürich 1981

[12] B. Scarpellini, Die Nichtaxiomatisierbarkeit des unendlichwertigen Prädikatenkalküls von Łukasiewicz, The Journal of Symbolic Logic 27 (1962) 159-170

[13] H. Wagner, Computational complexity of infinite-valued Łukasiewicz propositional logic, Technical Report 184 - 5/1995, Fernuniversität Hagen

NDK and Natural Reasoning

Mariko Yasugi *   Masahiro Nakata

Abstract

A revised version of NDK, a system of first order predicate calculus formulated in the natural deductions where the rule of inference can be applied to any *disjunctive component*. NDK thus does not have axioms. NDK-deductions are simpler than NK-ones in many cases. Various features of classical proofs are investigated by taking the advantage of the formulation of NDK. The major result is a conversion algorithm of NDK-deductions to NK-type-deductions, which are regarded as more natural, preserving the logical structures.

Introduction

In [7], we proposed a new formal system, NDK, of first order classical predicate calculus. It is an extension of NJ, natural deduction system of constructive logic, where the rule of inference can be applied to a *disjunctive component* of a formula.

Since an axiom of the excluded middle is a theorem of NDK, NDK does not need any axioms.

The system NDK in this article is a revised version of the system given in [7]. For this reason as well for the reader's convenience, we will present definitions and statements in some detail.

The system NDK is reviewed first (Section 1). Equivalence between NDK and NK is established in Section 2, giving explicit translations of deductions. Proofs (deductions) are compared for LK,NK and NDK, especially by proving Peirce axiom (Section 3). For many cases, it is easier proving in NDK than in NK. (See [2], for example, for the systems NJ, NK and LK.)

In Section 4, Wang's algorithm for the propositional part of NDK was dealt with in 4 of [7]. A generalized version of it is stated here. For the reader's convenience, we prove it in detail.

Let us break the introduction temporarily and explain the reasons why we came up with the system NDK. They are listed below.

*This work has been supported in part by Grant-in-Aid for Scientific Researches (No. 07864012) and by Program of Okawa Foundations (No. 95-12).

(1) The first author has worked on the termination problems of reductions in sequent-type calculi, by way of ordinal notations. The accessibility of ordinal notations would guarantee the termination of reductions of proofs. The accessibility proofs of ordinal notations were then carried out in systems of constructive arithmetic, which were in turn mapped to certain term systems. The last ones would act as tools of termination mechanism of the reductions of proofs. (The main references are [3], [4], [5], [6].)

This is too an indirect method. We expect that, similarly to the case of NJ-type systems, there be term systems inherent in NDK-type systems. This is yet a theme to be worked out. (The normalization of NDK is defined in 3 of [7].)

(2) It is well known that there is a neat corelation between LJ and NJ, but the mutual relations between LK and NK do not yield fruitful results. On the other hand, there is a beautiful relation between LK and NDK (2 of [7]).

(3) Interesting properties which hold for sequent calculi such as the mid-sequent property and Wang's algorithm also hold for NDK (See Theorem 11 and 4 in [7] as well as Section 4 of this article).

(4) In many cases, NDK-deductions are simpler than, for example, NK-deductions (Section 3), and hence more convenient in automatically constructing deductions. We plan to elaborate on this theme later on.

Our major concern in this article starts at Section 5, where various notions concerning NDK-deductions are given. Among them are *critical inferences* and *negative dual pairs*. These definitions prelude subsequent three sections. In Section 6, NDK-inferences are classified into natural ones and *artificial* ones, and then the *Oshiba's project* is explained. Oshiba in [1] proposed a conversion algorithm to transform LK-proofs to LJ-type proofs allowing the axioms of the excluded middle. We will carry out a similar task concerning NDK-deductions

We regard *classical* $\supset I, \neg I$ and $\forall I$ as artificial, and hence try to remove them, by adding *the excluded middle type axioms*, preserving thereby the logical structures of original deductions. The transformation algorithm is described in Sections 7 and 8, and some typical examples are given in Section 9.

At the end, in Section 10, we remark on the characteristics of the critical $\supset I$ as an annihilator.

The authors would like to express their gratitude to T. Oshiba and N. Saneto for their interest in our work.

1 System NDK

We introduce the system NDK, which is an extension of constructive logic NJ to classical logic. NDK has no axiom. Inference rules of NDK can apply to an arbitrary component of the disjunctive form of the premise. Here we give some notions and notations as well as the inference rules of NDK.

Remark 1 The system NDK was first introduced in 1 of [7]. We are presenting a revised version of NDK here. In particular, the symbol $\bot$ designating falsefood is not adopted. Instead, the empty conclusion is admitted. This streamlines the description of inference rules as well as of natural conversions between NDK and LK. Formulas of NDK are those of the first order predicate calculus, and hence no details will be given.

Formulas are denoted by letters $A, B, \cdots$, and we use the letters $\Gamma, \Delta, \cdots$ to denote a formula or the emptiness. $\Gamma \vee A$ will stand for honest $\Gamma \vee A$ if Γ is a formula; it will denote A if Γ is empty.

Definition 1.1 The notion of *disjunctive components* is defined recursively as follows.

(1) If A is atomic, then the *disjunctive component* of A is A itself.

(2) If A is of the form $B \wedge C, B \supset C, \neg B, \forall x B$ and $\exists x B$, then the *disjunctive component* of A is A itself.

(3) If A is of the form $B \vee C$, then *disjunctive components* of A are *disjunctive components* of B, *disjunctive components* of C and A itself.

For example, *disjunctive components* of $X \vee (\neg X \vee Y)$ are $X \vee (\neg X \vee Y), X, \neg X \vee Y, \neg X$ and Y. ($X \vee \neg X$ is *not* a disjunctive component.)

Remark 2 We can paraphrase this definition as follows. An occurrence of an expression in a formula A is said to be *surface* if it is *not* in the scope of any connective other than $\vee$ within A. A subformula of A, say B, is called a *disjunctive component* of A if it is surface in A. (Notice that A is itself a disjunctive component.)

Definition 1.2 An NDK-deduction is defined in the same manner as that of NJ, with generalized rules of inference.

The rules of inference are listed below. They are *Introduction* and *Elimination* (abbreviated to I and E respectively) of logical connectives except for the rule of *contradiction*, abbreviated to Con. The explicitly written $\vee$'s are *surface*, so that the auxiliary formulas and the principal formulas of inferences are *disjunctive components*. An assumption with *label a* will be denoted by $[A : a]$, though the label will be omitted most of the time. (An assumption is never empty. That is, $[\ : n]$ is *not* admitted.) The labels of discharged assumptions are indicated at the inferences in parentheses.

$$\frac{\begin{array}{c}\vdots\\ \Gamma_1 \vee A \vee \Gamma_2\end{array} \quad \begin{array}{c}\vdots\\ \Gamma_1 \vee B \vee \Gamma_2\end{array}}{\Gamma_1 \vee (A \wedge B) \vee \Gamma_2} \wedge I \qquad \frac{\begin{array}{c}\vdots\\ \Gamma_1 \vee (A_1 \wedge A_2) \vee \Gamma_2\end{array}}{\Gamma_1 \vee A_i \vee \Gamma_2} \wedge E \quad (i = 1, 2)$$

$$\frac{\begin{array}{c}\vdots\\ \Gamma_1 \vee A_i \vee \Gamma_2\end{array}}{\Gamma_1 \vee (A_1 \vee A_2) \vee \Gamma_2} \vee I \quad (i = 1, 2)$$

$$\frac{\begin{array}{c}\vdots\\ \Gamma_1 \vee (A \vee B) \vee \Gamma_2\end{array} \quad \begin{array}{c}[A:a]\\ \vdots\\ D\end{array} \quad \begin{array}{c}[B:b]\\ \vdots\\ D\end{array}}{\Gamma_1 \vee D \vee \Gamma_2} (a, b) - \vee E$$

$$\frac{\begin{array}{c}[A:a]\\ \vdots\\ \Gamma_1 \vee B \vee \Gamma_2\end{array}}{\Gamma_1 \vee (A \supset B) \vee \Gamma_2} (a)- \supset I \qquad \frac{\begin{array}{c}\vdots\\ \Delta_1 \vee A \vee \Delta_2\end{array} \quad \begin{array}{c}\vdots\\ \Gamma_1 \vee (A \supset B) \vee \Gamma_2\end{array}}{\Delta_1 \vee \Gamma_1 \vee B \vee \Delta_2 \vee \Gamma_2} \supset E$$

$$\frac{\begin{array}{c}[A:a]\\ \vdots\\ \Gamma_1 \vee \Gamma_2\end{array}}{\Gamma_1 \vee \neg A \vee \Gamma_2} (a) - \neg I \qquad \frac{\begin{array}{c}\vdots\\ \Delta_1 \vee A \vee \Delta_2\end{array} \quad \begin{array}{c}\vdots\\ \Gamma_1 \vee \neg A \vee \Gamma_2\end{array}}{\Delta_1 \vee \Gamma_1 \vee \Delta_2 \vee \Gamma_2} \neg E$$

$$\frac{\begin{array}{c}\vdots\\ \Gamma_1 \vee F \vee \Gamma_2\end{array}}{\Gamma_1 \vee \forall x F \vee \Gamma_2} \forall I \qquad \frac{\begin{array}{c}\vdots\\ \Gamma_1 \vee \forall x F \vee \Gamma_2\end{array}}{\Gamma_1 \vee F[t/x] \vee \Gamma_2} \forall E$$

(eigenvariable condition)

$$\frac{\begin{array}{c}\vdots\\ \Gamma_1 \vee F[t] \vee \Gamma_2\end{array}}{\Gamma_1 \vee \exists x F \vee \Gamma_2} \exists I \qquad \frac{\begin{array}{c}\vdots\\ \Gamma_1 \vee \exists x F \vee \Gamma_2\end{array} \quad \begin{array}{c}[F:a]\\ \vdots\\ D\end{array}}{\Gamma_1 \vee D \vee \Gamma_2} (a) - \exists E$$

(eigenvariable condition)

$$\frac{\begin{array}{c}\vdots\\ \overline{}\end{array}}{A} \; Con$$

The inferences $\forall I$ and $\exists E$ satisfy the *eigenvariable condition*, that is, $\forall I$ has no free occurence of x in the assumptions, Γ_1 and Γ_2, and $\exists E$ has no free occurence of x in D and in the assumptions excepting $[F : a]$.

An inference is called *classical* when one of the $\Gamma_1, \Gamma_2, \Delta_1$ and Δ_2 as above is *not* empty.

Remark 3 (1) The empty premise in *Con* is introduced by the application of $\neg E$ that is not classical.

(2) In $\supset E, \neg I$ and $\neg E$, the associations of components in the conclusion are arbitrary. For example, in $\supset E$, there are several possibilities. $((\Delta_1 \vee \Gamma_1) \vee B) \vee (\Delta_2 \vee \Gamma_2)$ and $(\Delta_1 \vee \Gamma_1) \vee (B \vee (\Delta_2 \vee \Gamma_2))$ are some of the examples. This much of freedom makes the deduction forming easier.

According to Proposition 2 of [7], we can claim some elementary properties with regards to NDK. For example, $A \vee \neg A$ is a theorem of NDK; NDK is equivalent with NK; from $\Gamma_1 \vee (D \vee D) \vee \Gamma_2$, can one derive $\Gamma_1 \vee D \vee \Gamma_2$, where the only elimination used will be the $\vee$-elimination applied to $D \vee D$.

The last fact suggests that we can introduce the following as a derived rule (named *Contraction*).

$$\frac{\Gamma_1 \vee (D \vee D) \vee \Gamma_2}{\Gamma_1 \vee D \vee \Gamma_2}\ Contraction$$

In 2 of [7], natural mappings between NDK-deductions and LK-proofs are defined. We can easily modify them to the present version of NDK.

Reductions and normal forms of NDK-deductions are defined in 3 of [7] similarly to those of NJ. Proposition 9 there claims that a cut-free proof of LK is transformed to a normal deduction of NDK, and hence it holds that every NDK-theorem has a normal NDK-proof.

2 Equivalence between NDK and NK

In this section, we give a revised proof of the fact that NDK is equivalent to NK. For this purpose, we show the following lemma as the first step.

Lemma 1 Let $X \vee B$ and $\neg X$ be deducible from respectively Γ and Δ in NK. Then B is deducible from $\Gamma \cup \Delta$ in NK without adding the axiom of the excluded middle. (Here Γ and Δ denote finite sets of assumptions)

Proof.

$$
\dfrac{\begin{array}{c}\Gamma\\ \vdots\\ X \vee B\end{array} \qquad \dfrac{\dfrac{[X:1] \quad \begin{array}{c}\Delta\\ \vdots\\ \neg X\end{array}}{}}{B} \qquad [B:2]}{B}\ (1,2)
$$

Remark 4 The thorem below has been shown in [7]. The proof that NDK is a subsystem of NK given here is an improved one, so that it renders a clear insight of where the axiom of the excluded middle is necessary.

Theorem 1 NDK is equivalent to NK.

Proof. Inference of NK are those of constructive NDK inferences and the axiom of the excluded middle is deducible in NDK;

$$
\dfrac{[A:1]}{A \vee \neg A}\ (1)
$$

Therefore we obtain that NK is a subsystem of NDK.

Next, in order to establish that NDK is a subsystem of NK, we show that any NDK-deduction Π can be translated into an NK-deduction, by induction on the construction of Π. Let I be the last inference in Π.

If I is not classical, then we can immediately apply the induction hypotheses. Let us thus assume that *I is classical.*

If I is $\wedge I$, then Π is of the form

$$
\dfrac{\begin{array}{c}\vdots\\ \Gamma_1 \vee A \vee \Gamma_2\end{array} \quad \begin{array}{c}\vdots\\ \Gamma_1 \vee B \vee \Gamma_2\end{array}}{\Gamma_1 \vee (A \wedge B) \vee \Gamma_2}
$$

Let Π' be the NK-deduction

$$
\dfrac{\begin{array}{c}\vdots\\ \Gamma_1 \vee B \vee \Gamma_2\\ \vdots\\ \Gamma_1 \vee \Gamma_2 \vee B\end{array} \quad \begin{array}{c}[\Gamma_1 \vee \Gamma_2 : 2]\\ \vdots\\ \Gamma_1 \vee (A \wedge B) \vee \Gamma_2\end{array} \quad \begin{array}{c}\dfrac{[A:3] \quad [B:4]}{A \wedge B}\\ \vdots\\ \Gamma_1 \vee (A \wedge B) \vee \Gamma_2\end{array}}{\Gamma_1 \vee (A \wedge B) \vee \Gamma_2}\ (2,4)
$$

Π can be transformed into the following deduction in NK.

$$\dfrac{\begin{matrix} \vdots \\ \Gamma_1 \vee A \vee \Gamma_2 \\ \vdots \\ \Gamma_1 \vee \Gamma_2 \vee A \end{matrix} \quad \begin{matrix} [\Gamma_1 \vee \Gamma_2 : 1] \\ \vdots \\ \Gamma_1 \vee (A \wedge B) \vee \Gamma_2 \end{matrix} \quad \Pi'}{\Gamma_1 \vee (A \wedge B) \vee \Gamma_2}\ (1,3)$$

If I is $\wedge E$, then Π is of the form

$$\dfrac{\begin{matrix} \vdots \\ \Gamma_1 \vee (A_1 \wedge A_2) \vee \Gamma_2 \end{matrix}}{\Gamma_1 \vee A_i \vee \Gamma_2}\quad (i = 1, 2)$$

Π can be transformed into the following deduction in NK.

$$\dfrac{\begin{matrix} \vdots \\ \Gamma_1 \vee (A_1 \wedge A_2) \vee \Gamma_2 \\ \vdots \\ \Gamma_1 \vee \Gamma_2 \vee (A_1 \wedge A_2) \end{matrix} \quad \begin{matrix} [\Gamma_1 \vee \Gamma_2 : 1] \\ \vdots \\ \Gamma_1 \vee A_i \vee \Gamma_2 \end{matrix} \quad \begin{matrix} \dfrac{[A_1 \wedge A_2 : 2]}{A_i} \\ \vdots \\ \Gamma_1 \vee A_i \vee \Gamma_2 \end{matrix}}{\Gamma_1 \vee A_i \vee \Gamma_2}\ (1,2)$$

If I is $\vee E$, then Π is of the form

$$\dfrac{\Gamma_1 \vee (A \vee B) \vee \Gamma_2 \quad \begin{matrix} [A : 1] \\ \vdots \\ D \end{matrix} \quad \begin{matrix} [B : 2] \\ \vdots \\ D \end{matrix}}{\Gamma_1 \vee D \vee \Gamma_2}\ (1,2)$$

Let Π' be the deduction

$$\dfrac{[A \vee B : 4] \quad \begin{matrix} [A : 1] \\ \vdots \\ D \\ \vdots \\ \Gamma_1 \vee D \vee \Gamma_2 \end{matrix} \quad \begin{matrix} [B : 2] \\ \vdots \\ D \\ \vdots \\ \Gamma_1 \vee D \vee \Gamma_2 \end{matrix}}{\Gamma_1 \vee D \vee \Gamma_2}\ (1,2)$$

Π can be transformed into the following deduction in NK.

$$\dfrac{\begin{matrix} \vdots \\ \Gamma_1 \vee (A \vee B) \vee \Gamma_2 \\ \vdots \\ \Gamma_1 \vee \Gamma_2 \vee (A \vee B) \end{matrix} \quad \begin{matrix} [\Gamma_1 \vee \Gamma_2 : 3] \\ \vdots \\ \Gamma_1 \vee D \vee \Gamma_2 \end{matrix} \quad \Pi'}{\Gamma_1 \vee D \vee \Gamma_2}\ (3,4)$$

Other cases except for $\supset I, \neg I$ and $\forall I$ can be dealt with similarly.

If I is $\neg I$, then Π is of the form

$$\dfrac{\begin{matrix}[A:1]\\ \vdots \\ \Gamma_1 \vee \Gamma_2\end{matrix}}{\Gamma_1 \vee \neg A \vee \Gamma_2}\ (1)$$

Π can be translated into the following deduction in NK with the axiom of the excluded middle.

$$\dfrac{A \vee \neg A \quad \begin{matrix}[A:1]\\ \vdots \\ \Gamma_1 \vee \Gamma_2 \\ \vdots \\ \Gamma_1 \vee \neg A \vee \Gamma_2\end{matrix} \quad \begin{matrix}[\neg A:2]\\ \vdots \\ \Gamma_1 \vee \neg A \vee \Gamma_2\end{matrix}}{\Gamma_1 \vee \neg A \vee \Gamma_2}\ (1,2)$$

If I is $\supset I$, then Π is of the form

$$\dfrac{\begin{matrix}[A:1]\\ \vdots \\ \Gamma_1 \vee B \vee \Gamma_2\end{matrix}}{\Gamma_1 \vee (A \supset B) \vee \Gamma_2}\ (1)$$

Π can be translated into the following deduction in NK with the axiom of the excluded middle. Let Π' be the deduction below.

$$\begin{matrix}\begin{matrix}\begin{matrix}[A:1]\\ \vdots \\ \Gamma_1 \vee B \vee \Gamma_2\end{matrix} \quad [\neg\Gamma_1:5] \\ \vdots \\ B \vee \Gamma_2\end{matrix} \qquad [\neg\Gamma_2:3] \\ \vdots \\ \dfrac{B}{A \supset B}\ (1) \\ \vdots \\ \Gamma_1 \vee (A \supset B) \vee \Gamma_2\end{matrix}$$

By virtue of Lemma 1, this is an NK-deduction. Using Π', we obtain the following.

$$\dfrac{\Gamma_2 \vee \neg\Gamma_2 \quad \begin{matrix}[\Gamma_2 : 2] \\ \vdots \\ \Gamma_1 \vee (A \supset B) \vee \Gamma_2\end{matrix} \quad \dfrac{\Gamma_1 \vee \neg\Gamma_1 \quad \begin{matrix}[\Gamma_1 : 4] \\ \vdots \\ \Gamma_1 \vee (A \supset B) \vee \Gamma_2\end{matrix} \quad \Pi'}{\Gamma_1 \vee (A \supset B) \vee \Gamma_2}\,(4,5)}{\Gamma_1 \vee (A \supset B) \vee \Gamma_2}\,(2,3)$$

Notice that the newly added axioms of the excluded middle are $\Gamma_1 \vee \neg\Gamma_1$ and $\Gamma_2 \vee \neg\Gamma_2$.

If I is $\forall I$, then Π is of the form

$$\frac{\Gamma_1 \vee F \vee \Gamma_2}{\Gamma_1 \vee \forall x F \vee \Gamma_2}$$

Let Π' be the deduction below.

$$\begin{matrix} \vdots & & \\ \Gamma_1 \vee F \vee \Gamma_2 \quad [\neg\Gamma_1 : 4] & & \\ \vdots & & \\ F \vee \Gamma_2 & & [\neg\Gamma_2 : 2] \\ & \vdots & \\ & \dfrac{F}{\forall x F} & \\ & \vdots & \\ & \Gamma_1 \vee \forall x F \vee \Gamma_2 & \end{matrix}$$

By virtue of Lemma 1, this is an NK-deduction. Π can then be transformed as follows

$$\dfrac{\Gamma_2 \vee \neg\Gamma_2 \quad \begin{matrix}[\Gamma_2 : 1] \\ \vdots \\ \Gamma_1 \vee \forall x F \vee \Gamma_2\end{matrix} \quad \dfrac{\Gamma_1 \vee \neg\Gamma_1 \quad \begin{matrix}[\Gamma_1 : 3] \\ \vdots \\ \Gamma_1 \vee \forall x F \vee \Gamma_2\end{matrix} \quad \Pi'}{\Gamma_1 \vee \forall x F \vee \Gamma_2}\,(3,4)}{\Gamma_1 \vee \forall x F \vee \Gamma_2}\,(1,2)$$

This completes the proof of Theorem 1.

Remark 5 In the proof above, the use of the axioms of the excluded middle is essential in cases of classical $\supset I, \neg I$ and $\forall I$, because they deduce the axiom of the excluded middle (or propositions which are equivalent to it). In the case of $\neg I$, we have

$$\frac{[A : 1]}{A \vee \neg A}\,(1)$$

In the case of $\supset I$, we have

$$\cfrac{\cfrac{[A:1]}{A \vee (X \wedge \neg X)}}{A \vee (A \supset X \wedge \neg X)}\,(1)$$

In the case of $\forall I$ (assuming that A has no free occurrence of x), we have

$$\cfrac{\cfrac{[\forall x(A \vee F(x))]}{A \vee F(x)}}{A \vee \forall x F(x)}$$

All other inferences can be transformed into NJ-inferences.

3 Comparison of proofs in LK, NK and NDK

In many cases, NDK-deductions are simpler than NK-deductions and LK-proofs.

We will see this through proofs of Peirce Axiom in respective systems. Peirce Axiom is of the form $((A \supset B) \supset A) \supset A$, and the system obtained from NJ by adding it is the classical logic.

NDK:

$$\cfrac{\cfrac{\cfrac{[A:1]}{A \vee B}}{A \vee (A \supset B)}\,(1) \qquad [(A \supset B) \supset A:2]}{\cfrac{\cfrac{A \vee A}{A}\,\mathit{Contraction}}{((A \supset B) \supset A) \supset A}\,(2)}$$

NK: Let Π be the following deduction.

$$\cfrac{A \vee \neg A \qquad \cfrac{\cfrac{\cfrac{\cfrac{[A:4] \quad [\neg A:5]}{\ }}{B}}{A \supset B}\,(4)}{A \vee (A \supset B)} \qquad \cfrac{\cfrac{[A:1]}{A \vee B} \quad \cfrac{[A:2]}{A \vee (A \supset B)} \quad \cfrac{\cfrac{[B:3]}{A \supset B}}{A \vee (A \supset B)}}{A \vee (A \supset B)}\,(2,3)}{A \vee (A \supset B)}\,(5,1)$$

Then the deduction as required is the following.

$$\cfrac{\cfrac{\Pi \quad [A:7] \quad \cfrac{[A \supset B:6] \quad [(A \supset B) \supset A:8]}{A}}{A}\,(7,6)}{((A \supset B) \supset A) \supset A)}\,(8)$$

LK:

$$\cfrac{\cfrac{\cfrac{\cfrac{A \Rightarrow A}{A \Rightarrow A, B}}{\Rightarrow A, A \supset B} \quad \begin{matrix}\vdots \\ (A \supset B) \supset A, A \supset B \Rightarrow A\end{matrix}}{(A \supset B) \supset A \Rightarrow A, A}\ cut(A \supset B)}{\cfrac{(A \supset B) \supset A \Rightarrow A}{\Rightarrow ((A \supset B) \supset A) \supset A}}$$

cut-free

$$\cfrac{\cfrac{\cfrac{\cfrac{\cfrac{A \Rightarrow A}{A \Rightarrow A, B}}{\Rightarrow A, A \supset B} \quad A \Rightarrow A}{(A \supset B) \supset A \Rightarrow A, A}}{(A \supset B) \supset A \Rightarrow A}}{\Rightarrow ((A \supset B) \supset A) \supset A}$$

4 Wang's algorithm

We consider Wang's algorithm, which is an algorithm deciding the validity of a given formula in the propositional part of NDK. The result in 4 of [7] regarding Wang's algorithm can be generalized to the question whether a formula is valid relative to given formulas, with a minor modification.

Definition 4.1 Let Σ be a finite sequence of formulas $B_1, \cdots, B_n (n \geq 0)$, and let F be any formula. Then the figure S

$$\frac{[B_1] \cdots [B_n]}{F}$$

is called a *sequent* (of F from Σ). $[B_1], \cdots, [B_n]$ are said to be the assumptions and F is said to be the conclusion of S. If F is true whenever $B_1, \cdots, B_n$ are true, then the sequent S is said to be valid. One can also say that F is valid relative to Σ. We assume that assumptions are labelled.

We can define the resolution rules, which generate new sequents from given ones. $\Longrightarrow_{(*a)}$ ($\Longrightarrow_{(*c)}$) means the resolution of an assumption (the conclusion) whose outermost logical connective is $*$.

$$\frac{\cdots [A \wedge B] \cdots}{\Gamma} \Longrightarrow_{(\wedge a)} \frac{\cdots [A][B] \cdots}{\Gamma}$$

$$\frac{\cdots}{\Gamma_1 \vee (A \wedge B) \vee \Gamma_2} \Longrightarrow_{(\wedge c)} \frac{\cdots}{\Gamma_1 \vee A \vee \Gamma_2} \; ; \; \frac{\cdots}{\Gamma_1 \vee B \vee \Gamma_2}$$

$$\frac{\cdots [A \vee B] \cdots}{\Gamma} \Longrightarrow_{(\vee a)} \frac{\cdots [A] \cdots}{\Gamma} \; ; \; \frac{\cdots [B] \cdots}{\Gamma}$$

$$\frac{\cdots [A \supset B] \cdots}{\Gamma} \Longrightarrow_{(\supset a)} \frac{\cdots [B] \cdots}{\Gamma} \; ; \; \frac{\cdots}{\Gamma \vee A}$$

$$\frac{\cdots}{\Gamma_1 \vee (A \supset B) \vee \Gamma_2} \Longrightarrow_{(\supset c)} \frac{\cdots [A] \cdots}{\Gamma_1 \vee B \vee \Gamma_2}$$

$$\frac{\cdots [\neg A] \cdots}{\Gamma} \Longrightarrow_{(\neg a)} \frac{\cdots}{\Gamma \vee A}$$

$$\frac{\cdots}{\Gamma_1 \vee \neg A \vee \Gamma_2} \Longrightarrow_{(\neg c)} \frac{\cdots [A] \cdots}{\Gamma_1 \vee \Gamma_2}$$

Starting with a sequent S, the resolution process should continue until all assumptions become atomic and the minimal components of the conclusion become atomic. Such a sequent will be called terminal. The collection of sequents thus resolved, a tree begining with a given sequent S, is called a *resolution tree* of S and is denoted by T_S or $T_{\Sigma;F}$.

Theorem 2 (1) All resolution trees are finite.

(2) A sequent S is valid if and only if, for each terminal sequent S' of T_S, there is a same atomic formula in an assumption and in the conclusion of S'.

(3) If a sequent of F from Σ is valid, then $T_{\Sigma;F}$ induces an NDK-deduction of F from Σ without live assumptions except Σ.

Proof. (1) and (2) are easily examined.

(3) When the condition holds, we can construct an NDK-deduction of F from Σ from the resolution tree of S. It is defined as follows by induction on the construction of the tree T_S. A terminal sequent is of the form

$$\frac{\cdots [X] \cdots}{\Gamma_1 \vee X \vee \Gamma_2}$$

where X is atomic. The conclusion can be obtained from $[X]$ by some $\vee I$'s. Let $(*a)$ be the first resolution rule applied to S. (See Definition 4.1.) Let us assume the induction hypotheses, and construct the last step of a deduction.

$(\wedge a)$

$$\dfrac{[A \wedge B]}{A}\wedge E \quad \dfrac{[A \wedge B]}{B}\wedge E \quad \begin{array}{c}\vdots\\ \Gamma\end{array}$$

$(\wedge c)$

$$\dfrac{\Gamma_1 \vee A \vee \Gamma_2 \quad \Gamma_1 \vee B \vee \Gamma_2}{\Gamma_1 \vee (A \wedge B) \vee \Gamma_2}\wedge I$$

$(\vee a)$

$$\dfrac{[A \vee B] \quad \begin{array}{c}[A:a]\\ \vdots\\ \Gamma\end{array} \quad \begin{array}{c}[B:b]\\ \vdots\\ \Gamma\end{array}}{\Gamma}(a,b) - \vee E$$

$(\supset a)$

$$\dfrac{\dfrac{[A \supset B] \quad \begin{array}{c}\vdots\\ \Gamma \vee A\end{array}}{\Gamma \vee B}\supset E \\ \vdots \\ \Gamma \vee \Gamma}{\Gamma}\ Contraction$$

$(\supset c)$

$$\dfrac{\begin{array}{c}[A:a]\\ \vdots\\ \Gamma_1 \vee B \vee \Gamma_2\end{array}}{\Gamma_1 \vee (A \supset B) \vee \Gamma_2}(a) - \supset I$$

$(\neg a)$

$$\dfrac{\begin{array}{c}\vdots\\ \Gamma \vee A\end{array} \quad [\neg A]}{\Gamma}\neg E$$

$(\neg c)$

$$\dfrac{\begin{array}{c}[A:a]\\ \vdots\\ \Gamma_1 \vee \Gamma_2\end{array}}{\Gamma_1 \vee \neg A \vee \Gamma_2}(a) - \neg I$$

This completes the proof of Theorem 2.

Example. Let T_S be the following.

$$\dfrac{[\neg(\neg A \wedge \neg B) : 1]}{A \vee B} \Longrightarrow_{(\neg a)} \overline{(A \vee B) \vee (\neg A \wedge \neg B)}$$

$$\Longrightarrow_{(\wedge c)} \overline{(A \vee B) \vee \neg A} \; ; \; \overline{(A \vee B) \vee \neg B}$$

$$\Downarrow_{(\neg c)} \qquad\qquad \Downarrow_{(\neg c)}$$

$$\frac{[A:2]}{A \vee B} \qquad\qquad \frac{[B:3]}{A \vee B}$$

The coresponding NDK-deduction is constructed below.

$$\cfrac{\cfrac{\cfrac{\cfrac{[A:2]}{A \vee B}}{(A \vee B) \vee \neg A}\,(2) \quad \cfrac{\cfrac{[B:3]}{A \vee B}}{(A \vee B) \vee \neg B}\,(3)}{(A \vee B) \vee (\neg A \wedge \neg B)} \qquad [\neg(\neg A \wedge \neg B):1]}{A \vee B}$$

5 Notions on deductions

We will consider an NDK-deduction Π, and define some concepts regarding formulas and inferences occurring in Π. When we speak of a formula A, we thus mean an occurrence of A in Π.

Definition 5.1 1) Consider two *disjunctive components* in Π, say A and B. A will be called a predecessor of B (in Π) with respect to inference I, if one of the following holds. In the figures below, $C(X)$ will denote a disjunctive component containing X as a disjunctive compoment.

(1) I is either an introduction or an elimination with one premise:

$$\frac{\Gamma \vee C(X) \vee \Delta}{\Gamma \vee C(Y) \vee \Delta}\, I$$

Y is obtained from X by an application of I. A is $C(X)$ and B is $C(Y)$, or A is a component of Γ (respectively Δ) in the premise and B is the corresponding component in the conclusion (with several exceptions). Exceptional cases are the following.

(1.1) I is $\vee I$ which derives $X \vee Z$ from X, hence Y is of the form $(X \vee Z)$, and B is the Z in Y. In this case, B has *no* predecessor.

(1.2) I is $\vee I$ which derives $X \vee Z$ from X, A is X and B is X. In this case, A is the predecessor of B.

(1.3) I is $\neg I$, and B is Y. In this case, X is empty, Y is of the form $\neg Z$ and B has *no* predecessor.

(2) I is $\wedge I$:

$$\frac{\Gamma \vee C(X) \vee \Delta \quad \Gamma \vee C(Y) \vee \Delta}{\Gamma \vee C(X \wedge Y) \vee \Delta}\, I$$

A is $C(X)$ or $C(Y)$ and B is $C(X \wedge Y)$, or A and B are the corresponding components in Γ or Δ.

(3) I is $\supset E$:

$$\frac{\Gamma_1 \vee X \vee \Delta_1 \quad \Gamma_2 \vee X \supset Y \vee \Delta_2}{\Gamma_1 \vee \Gamma_2 \vee Y \vee \Delta_1 \vee \Delta_2}\ I$$

A is $X \supset Y$ and B is Y, or A and B are the corresponding formulas in Γ's or Δ's. (Precise definition would be more complicated. For example, if B is $\Gamma_1 \vee \Gamma_2$, then A can be either Γ_1 or Γ_2. If B is the entire formula of the conclusion, then A can be either of the premises.)

(4) I is $\exists E$:

$$\frac{\begin{array}{c}\vdots \\ \Gamma \vee \exists x F \vee \Delta\end{array} \quad \begin{array}{c}[F] \\ \vdots \\ D\end{array}}{\Gamma \vee D \vee \Delta}\ I$$

There are three cases.

(4.1) A and B are the corresponding components in D, Γ or Δ.

(4.2) A is the $\exists xF$ in the major premise and B is the assumption F

(4.3) Situations as were noted in (3). For example, if B is $\Gamma \vee D$, then A is either Γ or D in the premises.

(5) I is $\vee E$:

$$\frac{\begin{array}{c}\vdots \\ \Gamma \vee (X \vee Y) \vee \Delta\end{array} \quad \begin{array}{c}[X] \\ \vdots \\ D\end{array} \quad \begin{array}{c}[Y] \\ \vdots \\ D\end{array}}{\Gamma \vee D \vee \Delta}\ I$$

There are three cases.

(5.1) A and B are the corresponding components in D, Γ or Δ.

(5.2) B is the X in the assumption and A is the X in the major premise. Similarly with Y.

(5.3) Similarly to (4.3).

(6) I is $\neg E$:

$$\frac{\begin{array}{c}\vdots \\ \Gamma_1 \vee X \vee \Delta_1\end{array} \quad \begin{array}{c}\vdots \\ \Gamma_2 \vee \neg X \vee \Delta_2\end{array}}{\Gamma_1 \vee \Gamma_2 \vee \Delta_1 \vee \Delta_2}\ I$$

A and B are the corresponding components in Γ's or Δ's, or the cases similar to (4.3) apply.

2) A *disjunctive component* A is said to be *initial* if A has no predecessor. This happens in one of the following cases.

(2.1) A is an assumption which is not discharged by $\exists E$ or $\vee E$.

(2.2) A is introduced by *Con*.

(2.3) A is the principal formula of a $\neg I$.

(2.4) A is the newly introduced component of a $\vee I$.

3) A *string* of (occurrences of) formulas in P, say $C_1, C_2, \ldots, C_n$, is a sequence of formulas in P with the property that, for each i, $1 \leq i \leq n$, C_i is a disjunctive component in P and C_{i+1} is a predecessor of C_i.

4) A string as above is called a *path* of C_1 if C_n is initial.

5) Let C and D be occurrences of disjunctive components in Π. D is said to be an *ancestor* of C (C is a *descendant* of D) if there is a string of C in Π containing D.

6) A *disjunctive component* C is said to be *irrelevant* if there is a path of C whose initial formula is the newly introduced component of a $\vee I$.

7) A *path* of a deduction Π is a *path* of its conclusion.

Definition 5.2 A *classical* $\supset I$ or $\forall I$ is said to be *critical* if it is applied to an *irrelevant* (occurrence of a) formula. A classical $\neg I$ is regarded as *critical.*

Definition 5.3 *Excluded middle type axiom* (cited from (6) of 3 in [1])

A pair of formulas E and F is called a *negatively dual pair* if it satisfies one of the following. (In this case F is called a negative dual of E, and vice versa.)

(1) $E \equiv A$ and $F \equiv \neg A$ for any formula A.

(2) $E \equiv \neg A$ and $F \equiv A$ for any formula A.

(3) Suppose (E_i, F_i) is a *negatively dual pair*, $i = 1, 2$.

(3.1) $(E, F) \equiv (E_1 \wedge E_2, F_1 \vee F_2)$

(3.2) $(E, F) \equiv (E_1 \vee E_2, F_1 \wedge F_2)$

(4) Supposse (E', F') is a *negatively dual pair.*

(4.1) $(E, F) \equiv (\forall x E', \exists x F')$

(4.2) $(E, F) \equiv (\exists x E', \forall x F')$

When (E, F) is a *negatively dual pair*, $E \vee F$ is called an *excluded middle typc axiom*. It is obvious that an *excluded middle type axiom* is a theorem of NK.

Definition 5.4 According to 3 of [1], we deal with the *excluded middle type axiom* instead of the pure excluded middle. (See Definition 5.3.)

(1) The system obtained from NDK with *excluded middle type axioms* added will be called $NDKE$.

(2) The system obtained from NK with *excluded middle type axioms* added will be called NKE.

(3) The notions defined in Definitions 5.1 and 5.2 can be extended to $NDKE$.

In particular, a disjunctive component of an axiom does not have any predecessor, and hence it is initial. The notion of a critical inference will remain invariant under addition of axioms. It is obvious that NK, NKE, NDK and $NDKE$ are all equivalent.

Definition 5.5 Let Π be an $NDKE$-deduction. The *rank* of Π, $rank(\Pi)$, will be defined as follows.

Let β be a *path* of Π. r_β is the number of *critical inferences* along β. (See 7) of Definition 5.1 and Definition 5.2.) Then

$$rank(\Pi) := \sum_{\beta:path} r_\beta$$

Definition 5.6 A *disjunctive component* in a deduction is said to be essential if it is not irrelevant.

6 Natural reasoning

According to the comment made at the end of Section 1, it holds that, given a set of assumptions Γ and a conclusion A, if A be NDK-deducible from Γ, then one can construct a normal NDK-deduction from Γ to A. This fact could be made use of for automated theorem proving. Our present interest lies, however, elsewhere.

In 5 of [7], we briefly explained Oshiba's project in [1], and outlined how to execute his idea in NDK. The purpose of the remaining sections is to give an accurate treatment to it.

Oshiba speculates that automated theorem proving (of first order classical predicate calculus) in terms of cut-free proofs is, though efficient, far from natural human thinking. According to him, employment of excluded middle type axioms appeal to human thinking better. He first proves a sharpened version of the interpolation theorem for LK, and then applies it to automatical transformation of an LK-proof into an LJ-proof with excluded middle type axioms added.

Dealing with the same subject in NDK reveals to us some features of classical deductions.

We first speculate on the *natural* and *artificial* reasonings with regards to the classical inferences of NDK. Consider, as an instance, $\wedge I$:

$$\frac{\Gamma \vee A \vee \Delta \quad \Gamma \vee B \vee \Delta}{\Gamma \vee (A \wedge B) \vee \Delta}$$

One can be easily convinced that this is a valid reasoning. That is, if Γ or Δ is true, then so is the conclusion. If A and B are true, then (regardless of the truth values of Γ and Δ) so is $A \wedge B$, and hence so is the conclusion.

On the other hand, $\supset I$ of a certain kind is hard to be intuitively comprehended. Consider the following.

$$\dfrac{\dfrac{\dfrac{\dfrac{[A:1]}{\exists x A}}{\exists x A \lor C} \quad \dfrac{[D]}{\exists x A \lor D}}{\exists x A \lor (C \land D)}}{\exists x A \lor (A \supset C \land D)}\ (1)-\supset I$$

The assumption $[A : 1]$ is discharged to imply $C \land D$, leaving $\exists x A$ alone. In order to justify this inference, one has to reason as follows. Assume A. If A is true, then so is $\exists x A$, and hence the conclusion is true. Otherwise, A has to be false. In this case, $A \supset C \land D$ is true, regardless of C and D. This flow of reasoning can be formalized as follows.

$$\dfrac{A \lor \neg A \quad \dfrac{\dfrac{[A:2]}{\exists x A}}{\exists x A \lor (A \supset C \land D)} \quad \dfrac{\dfrac{\dfrac{\dfrac{\dfrac{[A:3] \quad [\neg A:4]}{\quad}}{C} \quad [D]}{C \land D}}{A \supset C \land D}\ (3)}{\exists x A \lor (A \supset C \land D)}}{\exists x A \lor (A \supset C \land D)}\ (2,4)$$

The assumptions $[A : 2]$ and $[A : 3]$ correspond to the assumption $[A : 1]$ above.

Various experiments in thoughts have thus convinced us that, all the inference rules of NDK except classical $\supset I, \neg I$ and $\forall I$ can be accepted as natural, while the latters are somewhat artificial.

As was explained in the Remark 5 in Section 2, these two groups of inference rules are distinguished so that the *natural* ones can be converted to NJ-deductions (without the axiom of the excluded middle), while the *artificial* ones essentially need the axiom of the excluded middle in such a conversion.

We are thus led to the

Definition 6.1 Let $NDKE^*$ be the system $NDKE$ in which there are no *classical* inferences of $\supset I, \neg I$ and $\forall I$ admitted.

Our aim is then to define a conversion algorithm of NDK-deductions into $NDKE^*$-deductions, preserving the structure of the original deduction as much as possible. What this last condition means will become clear later.

Note A uniform transformation procedure of classical NDK-inferences has been given in Section 2. Here we aim at a more delicate analysis of deductions.

The transformation algorithms will be successively given in the subsequent sections.

Let us state our objective of the next two sections explicitly.

Objective Define a conversion algorithm of any NDK-deduction to an $NDKE^*$-deduction, preserving the original logical structure.

7 Preparatory transformations

We first classify the *artificial* inferences, that is, the classical $\supset I$, $\neg I$ and $\forall I$, into four types. Let the indicated inference in each figure below be called J. J is assumed to satisfy each of the conditions (1) through (4) respectively.

Type I:

$$\begin{array}{c}[A:n]\\ \vdots\\ \dfrac{\Gamma\vee C\vee\Delta}{\Gamma\vee(A\supset C)\vee\Delta}\,(n)-\supset I\\ \vdots\end{array}$$

(1) C is essential.

Type II:

$$\begin{array}{c}\vdots\\ \dfrac{\Gamma\vee F\vee\Delta}{\Gamma\vee\forall xF\vee\Delta}\,\forall I\\ \vdots\end{array}$$

(2) F is essential.

Type III: $\supset I$ where
(3) C is irrelevant.

Type IV: $\forall I$ where
(4) F is irrelevant.

We first eliminate Type I and Type II by the transformations defined in Section 2.

Preparatory transformations. ¿From any $NDKE$-deduction, eliminate Type I inferences and Type II inferences with the transformation procedures in the proof of Theorem 1 in Section 2.

We thus have only to eliminate inferences of Type III and Type IV. We will henceforth deal with an $NDKE$-deduction Π which has no inference of Type I or Type II

8 Conversion algorithm

What is left for us to do is to define an algorithm to eliminate critical inferences, that is, inferences of Type III and Type IV. For this purpose, we need some lemmas.

Lemma 2 Let Π be an $NDKE$-deduction with the conclusion $\Gamma\vee G\vee\Delta$. Suppose G is irrelevant. Then one can trnsform Π to obtain a deduction Π'

of $\Gamma \vee \Delta$ satisfying the following conditions.

(i) The assumptions and the axioms of Π' are those of Π as well as a new assumption E, which is characterized as follows. E is of the form $\forall x_m \cdots \forall x_1 E'$, where E' is the conjunction of the assumptions which are discharged by $\supset I$ or $\neg I$ applied to the ancestors of G. (Such inferences $\supset I$ and $\neg I$ are critical.) $x_1, \cdots, x_m$ are the eigenvariables in Π occurring free in E'.

(ii) Π' is obtained from Π by eliminating the inferences which are applied to the ancestors of G and by adding E as well as some $\forall E'$s.

In case a critical inference indeed applies to an ancestor of G in Π,

$$rank(\Pi') < rank(\Pi)$$

Π' can be constructed by induction on the number of the inferences in Π below the initial components of paths of G.

Since the construction is a routine work, we will explain it with an example so that the reader can get an insight.

Π:

$$\dfrac{\dfrac{\dfrac{\dfrac{\dfrac{\dfrac{\dfrac{\dfrac{\dfrac{[A(x):1]}{\exists x A(x)} \qquad \dfrac{[B(y):2]}{\exists x A(x) \supset B(y)}}{B(y)}}{B(y) \vee C}}{B(y) \vee (B(y) \supset C)}\,(2)\,I_1}{B(y) \vee \exists y(B(y) \supset C)}}{\forall y B(y) \vee \exists y(B(y) \supset C)}}{\forall y B(y) \vee (A(x) \supset \exists y(B(y) \supset C))}\,(1)\,I_2}{\forall y B(y) \vee \forall x(A(x) \supset \exists y(B(y) \supset C))}\,I_3}{}$$

Π':

$$\dfrac{\dfrac{\dfrac{\dfrac{\dfrac{[\forall y(A(x) \wedge B(y)):3]}{A(x) \wedge B(y)}}{A(x)}}{\exists x A(x)} \qquad \dfrac{\dfrac{\dfrac{[\forall y(A(x) \wedge B(y)):3]}{A(x) \wedge B(y)}}{B(y)}}{\exists x A(x) \supset B(y)}}{B(y)}}{\forall y B(y)}$$

G is $\forall x(A(x) \supset \exists y(B(y) \supset C))$. The critical inferences I_1, I_2 and I_3 have been eliminated and the conjunction of $A(x)$ and $B(y)$, each discharged in

Π, is formed. Since y becomes an eigenvariable, $\forall y$ is prefixed to $A(x) \wedge B(y)$. We thus obtain $\forall y(A(x) \wedge B(y))$ as E.

Lemma 3 Under the same situation as in Lemma 2, we can construct an $NDKE$-deduction of G, say Π'', satisfying the conditions below.

(i) The assumption of Π'' is the negative dual F of the E obtained in Lemma 2.

(ii) Π'' is obtained from Π by copying the inferences which derive G. In Π'', G is not irrelevant and there are no critical inferences.

Let formula P be $\neg A(x) \vee \neg B(y)$ and Λ be the deduction below.

$$\dfrac{[P:8] \qquad \dfrac{\dfrac{\dfrac{[A(x):4] \quad [\neg A(x):5]}{\quad}}{\exists y(B(y) \supset C)}}{A(x) \supset \exists y(B(y) \supset C)}\,(4) \qquad \begin{array}{c} \dfrac{\dfrac{\dfrac{[B(y):6] \quad [\neg B(y):7]}{\quad}}{C}}{B(y) \supset C}\,(6) \\ \vdots \\ A(x) \supset \exists y(B(y) \supset C) \end{array}}{A(x) \supset \exists y(B(y) \supset C)}\,(5,7)$$

Π'' for the Π as above looks like this:

$$\dfrac{\dfrac{[\exists y(\neg A(x) \vee \neg B(y)):9] \quad \Lambda}{A(x) \supset \exists y(B(y) \supset C)}\,(8)}{\forall x(A(x) \supset \exists y(B(y) \supset C))}$$

Let us now consider a deduction Π with conclusion H. Let $\Gamma \vee G \vee \Delta$ denote a formula in Π, where G is irrelevant. Let Ψ be the subdeduction of Π whose conclusion is $\Gamma \vee G \vee \Delta$. Among those, take a lowermost $\Gamma \vee G \vee \Delta$ such that $rank(\Psi) > 0$. By virtue of Lemma 2, we have a deduction

$$\begin{array}{c} [E] \\ \vdots \\ \Gamma \vee \Delta \end{array}$$

satisfying the conditions (i) and (ii). Introducing G by $\vee I$ or Con, and copying Π, we obtain the deduction Π_1:

$$\begin{array}{c} [E] \\ \vdots \\ \dfrac{\Gamma \vee \Delta}{\Gamma \vee G \vee \Delta} \\ \vdots \\ H \end{array}$$

Notice that, since $\supset I$ does not apply to a descendant of G below $\Gamma \vee G \vee \Delta$ in Π, there is no critical inference there. So $rank(\Pi_1) < rank(\Pi)$, and hence Π_1 can be transformed into an $NDKE^*$-deduction Π_1^*.

Next, Lemma 3 yields a figure

$$\begin{array}{c} [F] \\ \vdots \\ G \end{array}$$

In Π, G is eliminated immediately, due to the condition on $\Gamma \vee G \vee \Delta$. Simulate Π below G, and introduce some components of H at the end if necessary. One then obtains a figure Π_2:

$$\begin{array}{c} [F] \\ \vdots \\ G \\ \vdots \\ H \end{array}$$

$rank(\Pi_2) < rank(\Pi)$, and hence Π_2 can be transformed into an $NDKE^*$-deduction Π_2^*. Finally, we obtain Π^*:

$$\dfrac{E \vee F \quad \begin{array}{c} [E] \\ \vdots\ \Pi_1^* \\ H \end{array} \quad \begin{array}{c} [F] \\ \vdots\ \Pi_2^* \\ H \end{array}}{H}$$

One can see that Π^* preserves the structure of Π. We can thus conclude that Π^* is the desired deduction. It especially holds that no new Type I or Type II inference sneaks in.

Remark 6 The conversion algorithm described above for Type III and Type IV inferences splits the original deduction Π into two parts: the essensial part (Π_1^*) and the irrelevant part (Π_2^*). Each part simulates Π and, at the end, merges with the other in terms of $\vee E$ with an excluded middle type axiom as the major premise. We have thus attained our objective.

9 Examples of conversion

(1) Exampe of [1]

Π:

$$\dfrac{\dfrac{\dfrac{\dfrac{\dfrac{\dfrac{\dfrac{\dfrac{[P(b):1]}{P(b)\lor P(c)}\,\lor I}{P(b)\lor(P(b)\supset P(c))}\,(1)}{(P(a)\supset P(b))\lor(P(b)\supset P(c))}}{(P(a)\supset P(b))\lor\forall y(P(b)\supset P(y))}}{(P(a)\supset P(b))\lor\exists x\forall y(P(x)\supset P(y))}}{\forall y(P(a)\supset P(y))\lor\exists x\forall y(P(x)\supset P(y))}}{\exists x\forall y(P(x)\supset P(y))\lor\exists x\forall y(P(x)\supset P(y))}}{\exists x\forall y(P(x)\supset P(y))}\,\mathit{Contraction}$$

$\Pi_1 \equiv \Pi_1^*$:

$$\begin{array}{c}\dfrac{\dfrac{[\forall z P(z):1]}{P(b)}}{P(a)\supset P(b)}\\ \vdots \\ \exists x\forall y(P(x)\supset P(y))\end{array}$$

$\Pi_2 \equiv \Pi_2^*$:

$$\dfrac{[\exists z\neg P(z):4] \qquad \begin{array}{c}\dfrac{\dfrac{\dfrac{[\neg P(b):2]\quad [P(b):3]}{}}{P(c)}}{P(b)\supset P(c)}\,(3)\\ \vdots \\ \exists x\forall y(P(x)\supset P(y))\end{array}}{\exists x\forall y(P(x)\supset P(y))}\,(2)$$

Π^*:

$$\dfrac{\forall z P(z)\lor\exists z\neg P(z) \qquad \begin{array}{c}[\forall z P(z):1]\\ \vdots\,\Pi_1^*\\ \exists x\forall y(P(x)\supset P(y))\end{array} \qquad \begin{array}{c}[\exists z\neg P(z):4]\\ \vdots\,\Pi_2^*\\ \exists x\forall y(P(x)\supset P(y))\end{array}}{\exists x\forall y(P(x)\supset P(y))}\,(1,4)$$

(2) Π:

$$\dfrac{\dfrac{[\forall x(F(x)\lor C)]}{F(x)\lor C}}{\forall x F(x)\lor C}$$

Π^*:

$$\dfrac{C\vee\neg C \qquad \dfrac{[C:2]}{\forall xF(x)\vee C} \qquad \dfrac{\dfrac{\begin{array}{c}\dfrac{[\forall x(F(x)\vee C)]}{F(x)\vee C} \quad [\neg C:1] \\ \vdots \\ F(x)\end{array}}{\forall xF(x)}}{\forall xF(x)\vee C}}{\forall xF(x)\vee C}\ (2,1)$$

10 Critical inference as an annihilator

At the end, we speculate on the algorithmic characterization of the critical $\supset I$. Consider

$$\dfrac{\dfrac{\begin{array}{c}[A:1]\\ \vdots \\ B\end{array}}{B\vee C}}{B\vee(A\supset C)}\ (1)-\supset I$$

This can be explained as follows. "If A is true, then so is B: otherwise (A is false and hence) $A\supset C$ is true regardless of C."

$A\supset C$ thus has an active role only when A is not true, but then the validity of $A\supset C$ is independent of C. We can thus take $X\wedge\neg X$, a contradiction, as C. This means that a critical $\supset I$ annihilates the content of C. It will then be acceptable to call such an inference an *annihilator*.

References

[1] T. Oshiba, *An Approach to Human Reasoning in Automatic Theorem Proving*, Jouhou Shori Gakkaishi(in Japanese), vol.35(1994), 222-231.

[2] M. E. Szabo, *The Collected Papers of Gerhard Gentzen*, North-Holland, Amsterdam-London, 1969.

[3] M. Yasugi, *Construction principle and transfinite induction up to* ε_0, J. Austral. Math. Soc., vol.31(1982), 24-47.

[4] M. Yasugi, *Hyper-principle and the functional structure of ordinal diagrams*, Comment. Math. Univ. St. Pauli, vol.34, no.2(1985), 227-263, the opening part; vol.35, no.1(1986), 1-38, the concluding part.

[5] M. Yasugi, *The machinery of consistency proofs*, Annals of Pure and Applied Logic, vol.44(1989), 139-152.

[6] M. Yasugi and S. Hayashi, *A functional system with transfninitely defined types*, Logic, Languages and Computation, Lecture Notes in Computer Science 792(1994), 31-60 (edited by N. Jones, M. Sato and M. Hagiya), Springer-Verlag.

[7] M. Yasugi and K. Ryu, *NDK, A New Classical System,* The Bull. of the Inst. of Computer Science of Kyoto Sangyo Univ., vol.11(1994), 1-25.

Faculty of Science, Kyoto Sangyo University
Kita-ku, Kyoto 603, Japan
e-mail: yasugi@cc.kyoto-su.ac.jp.

Graduate School of Science, Kyoto Sangyo University
Kita-ku, Kyoto 603, Japan
e-mail: nakata@cc.kyoto-su.ac.jp.

DEFAULT LOGIC AND IT'S VARIANTS: A SEMANTICAL VIEW *

ZHANG MINGYI

Applied Mathematics Institute,
Guizhou Academy of Sciences
Guiyang, Guizhou, 550001, P.R.China
Department of Computer Science,
Guizhou University
Guiyang, Guizhou, 550025, P.R.China

Abstract

In view of importance of well-understood semantics for knowledge representation systems, various semantical views for default logic and it's variants have been presented. But they are different each other in form. This paper provides a semantical framework, which supports a uniform model theoretic semantics for Reiter's default logic, Lukaszewicz and Brewka's variants of default logic.

1. Introduction

Reiter's default logic (DL)[1980][1] is currently one of the most popular and widely used formalization of default reasoning, and has attracted attention as a formal system for nonmonotonic reasoning. People have tended to be satisfied with intuitive characterization of the extensions of default logic, together with the Tarskian semantics of individual extension. Even though default logic has intuitively been well understood it took several years until a model theoretic semantic was given. In order to get a semantic characterization of default as a whole, Etherington [1986,1987] [2],[3] viewed defaults as extending the first order knowledge about incompletely specified world and selected restricted subsets of the models of the underlying first order theory. Lukaszewicz [1985][6] formalized this idea for a restricted class of default theories. This work was generalized to cover the entire class of default

* Supported by National Advanced Research and Development Project of China.

Key Words: Semantical Framework, Compatible Frame, Partial Order $\geq_D$, First Order Interpretation, Model.

theories by Etherington [1986, 1987] [3],[4]. Lukaszewicz [1988][7], Guerreira and Casanova [1990][8] presented alternative semantics respectively.

In order to overcome drawbacks of Reiter's logic (they are mainly lacking cumulativity and joint consistency with the generated extension, as has been discussed in various papers, e.g. in [9,10,11]), Brewka [1991] [11], introduced a slight variant of default logic, called cumulative default logic(CDL). Since those semantics proposed for DL [Etherington 1986,1987; Guerreiro and Casanova,1990 etc.] are not applicable any more, Schaub [1991][12] provided a semantic characterization of CDL. Considering how to solve the floating conclusion problem, Brewka [13] further extended the idea underlying CDL and presented a framework which is based on a fundamental shift of the role of defaults within the formalism, i.e. defaults are not considered as inference rules but elements of the language. This framework lacks a general model theoretic semantic. Zhang Mingyi [1992,1993][14],[15],[16] gave characterizations of the extensions for Reiter's DL and Brewka's CDL respectively, which are based on a set of facts and a set of defaults. This motivates us to make the research reported in this paper.

In what follows, we present a semantic framework to cover Reiter's DL, Lukaszewicz and Brewka's variants. We begin by providing a brief introduction to DL and it's variants. Afterwards a uniform semantical framework and notions of compatibility are introduced. Then we show how Brewka's L_{DL} and L_1 extensions are characterized by the presented semantical framework. Finally we argue that our framework supports the design of a model theoretic semantics of variants default logics by cope with Reiter's DL and Brewka's CDL and Lukaszewicz' modified DL.

2. Primary

Let L be a first order language. A default is any expression of the form $\alpha : \beta_1, ..., \beta_n/\gamma$, where $\alpha, \beta_i (1 \leq i \leq n)$ and γ are closed first order formulas. A default theory, DT, is an order pair (D, W), where D is a set of defaults and W a set of first order formulas.

Definition 2.1 (DL extension)[1]

Let (D, W) be a DT, S a set of formulas. Let $\Gamma(S)$ be the smallest set satisfying the following conditions:

1. $W \subseteq \Gamma(S)$
2. $Th_L(\Gamma(S)) = \Gamma(S)$
3. if $\alpha : \beta_1, ..., \beta_n/\gamma \in D, \alpha \in \Gamma(S)$, $\neg\beta_i \notin S (1 \leq i \leq n)$, then $\gamma \in \Gamma(S)$.

A set E of closed wffs is an extension of (D,W) iff $E = \Gamma(E)$, i.e. E is a fixed point of the operator Γ.

Definition 2.2 [14] Let D be a set of defaults, we define

$$PRE(D) = \{\alpha | (\alpha : \beta_1, ..., \beta_n/\gamma) \in D\}$$

$$CON(D) = \{\gamma | (\alpha : \beta_1, ..., \beta_n / \gamma) \in D\}$$

$$CCS(D) = \{\beta_i | (\alpha : \beta_1, ..., \beta_n / \gamma) \in D, 1 \leq i \leq n\}$$

$$KERN(D) = \{d \in D | d = true : true/\alpha\}$$

$$ASS(D) = \{d \in D | d = true : \beta_1, ..., \beta_n / \gamma\}$$

The logic CDL is based on a shift from simple proposition to more complicated structure called assertions. An assertion is a pair (p, X) consisting of a first order formula p and a (finite) set of formulas X, the support of p. Let K be an arbitrary set of assertions, we use the following notations:

FORM(K), the set of formulas of K, is the set $\{p | (p, \{r_1, ..., r_n\}) \in K\}$

SUPP(K), the support of K, is the set $\{r_i | (p, \{r_1, ..., r_n\}) \in K, 1 \leq i \leq n\}$

Given an assertional default theory, ADT, (D, W), where D is a set of defaults and W a set of assertions, a CDL extension is defined using a Reiter-like fixed point construction as follows:

Definition 2.3 (CDL extension)[11] An extension of an AST (D, W) is a fixed point of the operator Γ which, given a set of assertions S, produces the smallest set of assertions S' satisfying the following conditions:

1. $W \subseteq S'$
2. $Th_S(S') = S'$
3. if $\alpha : \beta/\gamma \in D$, $(\alpha, \{r_1, ..., r_n\}) \in S'$ and $\{\beta, \gamma\} \cup FORM(S) \cup SUPP(S)$ is consistent then $(\gamma, \{r_1, ..., r_n, \beta, \gamma\}) \in S'$.

Here $Th_S(A)$ is the smallest set such that $A \subseteq Th_S(A)$ and if

$$(\alpha_1, X_1), ..., (\alpha_k, X_k) \in Th_S(A)$$

and $\alpha_1, ..., \alpha_k \vdash \alpha$ ($\vdash$ stands for classical provability) then $(\alpha, X_1 \cup ... \cup X_k) \in Th_S(A)$.

As in the framework for CDL gave by Brewka [13], the language DEF is the set of all defaults. A default theory, DT_B, T is a set of defaults, i.e. $T \subseteq DEF$. There is no distinct treatment of facts and defaults as in DL: facts can be represented as a special type of defaults of the form $true : true/\alpha$. Defaults of the form $true : \beta_1, ..., \beta_n/\gamma$ are called assumable. We will often use α as shorthand for the facts $true : true/\alpha$ as well as $: \beta_1, ..., \beta_n/\gamma$ for the assumable $true : \beta_1, ..., \beta_n/\gamma$. Extensions will be generated in a stepwise manner. Let R be a set of monotonic inference rules and T a set of defaults. The closure of T under R is denoted $C_R(T)$ and called potential belief set.

Definition 2.4[13] (default logic) A default logic is a pair (R, Ext), where

1.R is a set of (monotonic) inference rules on defaults.

2. $Ext : P(DEF) \mapsto P(P(DEF))$ is a function mapping a set of defaults D to a set of subsets of D, the extensions of D.

We usually are only interested in the value of Ext for sets of defaults which are closed under R. Extensions of the closure $C_R(T)$ will also simply be called extensions of T. It is worth to point out that Brewka defined skeptical provability. This makes trivial failure of cumulativity avoided in the limiting case, where no extension exit. That is, every instance of Brewka's framework satisfies cumulativity. To reconstruct Reiter's DL in this framework Brewka gave a logic $L_{DL} = (R_{DL}, Ext_{DL})$ in the following way.

Definition 2.5[13] R_{DL} is the following set of rules:

DL_0

$$\Rightarrow true : true/true$$

DL_1

$$true : \beta_1, ..., \beta_n/\gamma, \gamma \vdash \sigma \Rightarrow true : \beta_1, ..., \beta_n/\sigma$$

DL_2

$$true : \beta_1, ..., \beta_n/\gamma, true : \beta_{n+1}, ..., \beta_m/\sigma \Rightarrow true : \beta_1, ..., \beta_m/\gamma \wedge \sigma$$

DL_3

$$true : \beta_1, ..., \beta_n/\gamma, \gamma : \beta_{n+1}, ..., \beta_m/\sigma \Rightarrow true : \beta_1, ..., \beta_m/\sigma$$

Definition 2.6[13] Let T be a set of defaults, C the closure of T under R_{DL}, E is an L_{DL} extensions of T, $E \in Ext_{DL}(C)$ iff

1. $E \subseteq ASS(C)$

2. for each assumable $d = true : \beta_1, ..., \beta_n/\gamma \in ASS(C)$: $d \in E$ iff $\{\beta_i\} \cup CON(E)$ is consistent for all i $(1 \leq i \leq n)$.

Brewka also gave an exact correspondence between DL extensions and consequents of L_{DL} extensions [13].

Theorem 2.1 [13] Let (D,W) be a Reiter DT, and $T = D \cup \{true : true/p | p \in W\}$. The function CON is a bijective mapping from the set of L_{DL} extensions of T to the set of DL extensions of (D,W).

To avoid some of the drawbacks of DL he introduced the following non-monotonic logic, which is quite well-behaved for semi-normal default theories.

Definition 2.7[13] R_{L_1} consists of the following rules:

R_0 tautologies:

$$\Rightarrow true : true/true$$

R_1 weakening:

$$true : \beta/\gamma, \gamma \vdash \sigma \Rightarrow true : \beta/\sigma$$

R_2 combination:

$$true : \beta_1/\gamma_1, true : \beta_2/\gamma_2 \Rightarrow true : \beta_1 \wedge \beta_2/\gamma_1 \wedge \gamma_2$$

R_3 chaining:

$$true : \beta/\gamma, \gamma : \tau/\delta \Rightarrow true : \beta \wedge \tau/\delta$$

R_4 equivalence:

$$\alpha : \beta/\gamma, \alpha \equiv \eta, \beta \equiv \tau \Rightarrow \eta : \tau/\gamma$$

R_5 disjoining:

$$true : \beta/\gamma, true : \tau/\gamma \Rightarrow true : \beta \vee \tau/\gamma$$

R_6 cases:

$$\alpha : \beta/\gamma, \eta : \tau/\delta \Rightarrow \alpha \vee \eta : \beta \wedge \tau/\gamma \vee \delta$$

Definition 2.8[13] Let T be a DT_B, C it's closure under R_{L_1}. E is an L_1 extension of T ,i.e. $E \in Ext_{L_1}(C)$ iff E is a maximal subset of C such that

1. $E \subseteq ASS(C)$, that is E consists of assumable in C.
2. $KERN(C) \subseteq E$
3. $CCS(E) \cup CON(E)$ is consistent.

Remark. If $E = \emptyset$ is an L_{DL} extension of a default theory ,then there is some $\beta_\delta \in CCS(\{\delta\})$ such that $\{\beta_\delta\}$ is inconsistent for all $\delta \in ASS(C)$. Without lose of generality we always suppose that an L_{DL} extension E of T is not empty in this paper.

Theorem 2.2 [13] Let T be a semi-normal DT. T has an extension iff CON(KERN(T)) is consistent.

Theorem 2.3 [13] (semi-monotony) Let T be a semi-normal DT, $d = \alpha : \beta/\gamma$ a default such that $\beta \vdash \gamma$ and β is not eguivalent to true. For every extension E of T there is an extension E' of $T' = T \cup \{d\}$ such that $E \subseteq E'$.

3. The Semantics Framework

In this section a uniform semantics framework and notions of compatibility of a frame are introduced, and properties of frames are studied. Let Φ be a class of first order interpretations for the first order language L. Like those approaches introduced by Etherington and Schaub, we consider a pair of the form $(\Psi, \hat{\Psi})$,where Ψ is a set of models and $\hat{\Psi}$ a class of sets of models. For any set F of closed wffs we denote the set of all models of F by MOD(F), i.e. $MOD(F) = \{\psi \in \Phi | \psi \models F\}$. If $F = \emptyset$ we define $MOD(\emptyset) = \Phi$. Here $\emptyset$ stands for empty set. In particular, $MOD(\{false\}) = \emptyset$, $MOD(\{true\}) = \Phi$.

Definition 3.1 Call a pair $(\Psi, \hat{\Psi})$ a frame ,where $\Psi \neq \emptyset$, $\Psi \in P(\Phi)$, $\hat{\Psi} \in P(P(\Phi))$. A frame $(\Psi, \hat{\Psi})$ is distributed compatible iff $\Psi \cap \Omega \neq \emptyset$ for all $\Omega \in \hat{\Psi}$. A frame $(\Psi, \hat{\Psi})$ is joint compatible iff $\Psi \cap \cap\{\Omega | \Omega \in \hat{\Psi}\} \neq \emptyset$. For any frames $(\Psi, \hat{\Psi})$ and $(\Omega, \hat{\Omega})$, $(\Psi, \hat{\Psi}) = (\Omega, \hat{\Omega})$ iff $\Psi = \Omega, \hat{\Psi} = \hat{\Omega}$.

Definition 3.2 Let $\delta = \alpha : \beta_1, ..., \beta_n/\gamma$ be a default. We define a relation $\geq_\delta$ on frames as follows: for any frames $(\Psi_1, \hat{\Psi}_1)$ and $\Psi_2, \hat{\Psi}_2)$, $(\Psi_2, \hat{\Psi}_2) \geq_\delta (\Psi_1, \hat{\Psi}_1)$ holds iff

1. $\Psi_1 \models \alpha$, i.e. for any $\psi \in \Psi_1, \psi \models \alpha$
2. $\Psi_2 = \{\psi \in \Psi_1 | \psi \models \gamma\}$
3. $\hat{\Psi}_2 = \hat{\Psi}_1 \cup \cup_{1 \leq i \leq n}\{MOD(\{\beta_i\})\}$

We can extend this relation to a sequence of defaults and a set of defaults respectively.

Definition 3.3 Let $(\Psi_1, \hat{\Psi}_1)$ and $(\Psi_2, \hat{\Psi}_2)$ be any two frames and (δ_i) a α-type (α is an ordinal number) sequence of defaults. The induced relation $\geq_{(\delta_i)}$ is defined as follows:

$(\Psi_2, \hat{\Psi}_2) \geq_{(\delta_i)} (\Psi_1, \hat{\Psi}_1)$ holds iff there is a sequence $((\Omega_i, \hat{\Omega}_i))$ of frames such that

1.$(\Omega_0, \hat{\Omega}_0) = (\Psi_1, \hat{\Psi}_1)$
2. for a successor ordinal i: $(\Omega_i, \hat{\Omega}_i) \geq_{\delta_{i-1}} (\Omega_{i-1}, \hat{\Omega}_{i-1})$
3. for a limit ordinal i: $\Omega_i = \cap_{j<i}\Omega_j, \hat{\Omega}_i = \cup_{j<i}\Omega_j$
4. $(\Psi_2, \hat{\Psi}_2) = (\Omega_\alpha, \hat{\Omega}_\alpha)$

For $\{\}$, the empty sequence of defaults, we define $(\Psi, \hat{\Psi}) \geq_{\{\}} (\Psi, \hat{\Psi})$ for any frame $(\Psi, \hat{\Psi})$.

Definition 3.4 Let D be a set of defaults . The part order $\geq_D$ is defined as follows: $(\Psi_2, \hat{\Psi}_2) \geq_D (\Psi_1, \hat{\Psi}_1)$ holds iff there is a sequence (δ_i) of defaults in D (possibly, the empty sequence of defaults) such that $(\Psi_2, \hat{\Psi}_2) \geq_{(\delta_i)}$ $(\Psi_1, \hat{\Psi}_1)$

Lemma 3.1 Let D_1, D_2 be sets of defaults and $D_1 \subseteq D_2$. If $(\Psi, \hat{\Psi}) \geq_{D_1}$ $(\Omega, \hat{\Omega})$ then $(\Psi, \hat{\Psi}) \geq_{D_2} (\Omega, \hat{\Omega})$. In particular $(\Psi, \hat{\Psi}) \geq_D (\Psi, \hat{\Psi})$ for any set of defaults D and any frame $(\Psi, \hat{\Psi})$.

Lemma 3.2 Let δ be a default and D a set of defaults. The relation $\geq_\delta$ is anti-symmetric and transitive.(In particular, if $(\Psi_1, \hat{\Psi}_1) \geq_\delta (\Psi_2, \hat{\Psi}_2) \geq_\delta$ $(\Psi_3, \hat{\Psi}_3)$ then $(\Psi_1, \hat{\Psi}_1) = (\Psi_2, \hat{\Psi}_2)$). The relation $\geq_D$ is reflexive, anti-symmetric and transitive, i.e. $\geq_D$ is a partial order.

Proof. Reflexivity and transitivity are obvious. For the anti-symmetry, it is clear that $\Psi_2 \subseteq \Psi_1$ and $\hat{\Psi}_1 \subseteq \hat{\Psi}_2$ if $(\Psi_2, \hat{\Psi}_2) \geq_D (\Psi_1, \hat{\Psi}_1)$. So, anti-symmetry is true.

Remark It is worth noting that the definition of $\geq_D$ here is differed from that in [4] and [12]. As in [4], $\Gamma_1 \geq_D \Gamma_2$ iff there is $\delta \in D$ such that $\Gamma_1 \geq_\delta \Gamma_2$, where Γ is a set of models and D a set of defaults and $\Gamma_1, \Gamma_2 \in$ $P(\Gamma)$. We point out that the definition is not applied to the soundness and completeness results for a default theory. In fact, if $D = (: A/B, B : C/F)$,

$W = \emptyset$ then $E = Th(\{B, F\})$ is the extension of (D, W). But there is no any subset D' of D such that $\Gamma \geq_{D'} MOD(W)$, since there is no any default $\delta \in D$ such that $\Gamma \geq_\delta MOD(W)$, where $\Gamma = \{\phi \in \Phi | \phi \models F\}$. Similarly the definition of $\geq_D$ given by Schaub [12] is incorrect for an assertional theory.

Definition 3.5 Let D be a set of defaults. A frame $(\Psi, \hat{\Psi})$ is $\geq_D$-maximal iff there is no any frame $(\Omega, \hat{\Omega})$ such that $(\Omega, \hat{\Omega}) >_D (\Psi, \hat{\Psi})$. (i.e. $(\Omega, \hat{\Omega}) \geq_D (\Psi, \hat{\Psi})$ and $(\Omega, \hat{\Omega}) \neq (\Psi, \hat{\Psi})$). $(\Psi, \hat{\Psi})$ is $\geq_D$-maximally distributed compatible iff there is no any distributed compatible frame (DCF) $(\Omega, \hat{\Omega})$ such that $(\Omega, \hat{\Omega}) >_D (\Psi, \hat{\Psi})$. $(\Psi, \hat{\Psi})$ is $\geq_D$-maximally joint compatible iff there is no any joint compatible frame (JCF) $(\Omega, \hat{\Omega})$ such that $(\Omega, \hat{\Omega}) >_D (\Psi, \hat{\Psi})$.

Definition 3.6 A set of defaults D is consistent iff $CON(D)$ is consistent.

Lemma 3.3 Let T be a DT_B, C it's closure under R_{DL} and $T^* = T \cap C$. T^* is consistent iff so is C. In particular, if T is consistent then so is C.

Proof. "If Part". Trivial.

"Only If Part". Since T^* is consistent there is $\psi \in \Psi$ such that $\psi \models CON(T^*)$. We inductively show that $\Psi \models CON(\{\delta\})$ for any $\delta \in C$.

Base. Clear.

Step. 1. If $\delta = true : true/true$ then $\psi \models true$.

2. If $\psi \models \sigma$ for some $true : \beta_1, ..., \beta_n/\sigma \in C$ and $\sigma \vdash \gamma$ then $\psi \models \gamma$

3. If $\psi \models CON(\{true : \beta_1, ..., \beta_n/\gamma, true : \beta_{n+1}, ..., \beta_m/\sigma\})$ then $\psi \models \gamma \wedge \sigma$.

4. If $\psi \models \sigma$ for $true : \beta_1, ..., \beta_n/\sigma$ and $\sigma : \beta_{n+1}, ..., \beta_m/\gamma \in C$ then $\psi \models \gamma$ since $\sigma : \beta_{n+1}, ..., \beta_m/\gamma \in T$.

Definition 3.7 Let D be a set of defaults. A frame $(\Psi, \hat{\Psi})$ is the model of D, denoted $(\Psi, \hat{\Psi}) \models D$, iff $\Psi = MOD(PRE(D) \cup CON(D))$ and $\hat{\Psi} = \{MOD(\{\beta\}) | \beta \in CCS(D)\}$.

Clearly, $(\Phi, \{\Phi\})$ is the model of $\{true : true/true\}$.

Corollary 3.4 If $(\Psi, \hat{\Psi}) \geq_{(\delta_i)} (\Phi, \{\Phi\})$ then $(\Psi, \hat{\Psi})$ is the model of (δ_i).

Lemma 3.5 Let T be a DT_B, C it's closure under R_{DL}. Assume that T is consistent. A frame $(\Psi, \hat{\Psi})$ is the model of ASS(C) iff $(\Psi, \hat{\Psi})$ is $\geq_C$-maximal such that $(\Psi, \hat{\Psi}) \geq_C (\Phi, \{\Phi\})$.

Proof.

"If Part". For any $d = true : \beta_1, ..., \beta_n/\gamma \in ASS(C)$ it is clear that $(\{\psi \in \Psi | \psi \models \gamma\}, \hat{\Psi} \cup \{MOD(\{\beta_i\}) | 1 \leq i \leq n\}) \geq_C (\Psi, \hat{\Psi})$ since T is consistent. By maximality of $(\Psi, \hat{\Psi})$ we have $\{\psi \in \Psi | \psi \models \gamma\} = \Psi$, $\hat{\Psi} \cup \{MOD(\{\beta_i\}) | 1 \leq i \leq n\} = \hat{\Psi}$. So, $\Psi \models \gamma, MOD(\{\beta_i\}) \in \hat{\Psi}$ for all $i, 1 \leq i \leq n$.

"Only If Part". Using the well ordered set $ASS(C) = \{d_1, d_2, ...\}$ we can define a sequence of frames $((\Psi_i, \hat{\Psi}_i))$ as follows

1. $(\Psi_0, \hat{\Psi}_0) = (\Phi, \{\Phi\})$
2. for a successor ordinal i:

(1) if $\delta_{i-1} = true : \beta_1, ..., \beta_n/\gamma \in T$, then $(\Psi_i, \hat{\Psi}_i) = (\{\psi \in \Psi_{i-1} | \psi \models \gamma\}, \hat{\Psi}_{i-1} \cup \{MOD(\{\beta_j\}) | 1 \leq j \leq n\})$

(2) if

I. $\delta_{i-1} = true : true/true$; or

II. $\delta_{i-1} = true : \beta_1, ..., \beta_n/\sigma,\ true : \beta_1, ..., \beta_n/\gamma \in \{\delta_j | j < (i-1)\}$ and $MOD(\{\sigma\}) \subseteq MOD(\{\gamma\})$; or

III. $\delta_{i-1} = true : \beta_1, ..., \beta_m/\gamma \wedge \sigma,\ true : \beta_1, ..., \beta_n/\gamma \in \{\delta_j | j < (i-1)\}$, and $true : \beta_{n+1}, ..., \beta_m/\sigma \in \{\delta_j | j < (i-1)\}$

then $(\Psi_i, \hat{\Psi}_i) = (\Psi_{i-1}, \hat{\Psi}_{i-1})$

(3) if $\delta_{i-1} = true : \beta_1, ..., \beta_m/\sigma,\ true : \beta_1, ..., \beta_n/\gamma \in \{\delta_j | j < (i-1)\}$, and $\gamma : \beta_{n+1}, ..., \beta_m/\sigma \in C$, then $(\Psi_i, \hat{\Psi}_i) = (\{\psi \in \Psi_{i-1} | \psi \models \sigma\}, \hat{\Psi}_{i-1})$

3. for a limit ordinal i: $(\Psi_i, \hat{\Psi}_i) = (\cap_{j<i}\Psi_j, \cup_{j<i}\hat{\Psi}_j)$

where for each $i \geq 0$ we defined δ_i to be the default $\delta_i = d_k = \alpha : \beta_1, ..., \beta_n/\gamma \in C$ with $k = min\{j | d_j \notin \{\delta_m | m \leq i\}\}$ and one of the above cases in term 2 holds for d_j.

Clearly, the sequence $((\Psi_i, \hat{\Psi}_i))$ is well defined and each of the defaults in $ASS(C)$ is chosen only once. So $(\cap_{i \geq 0}\Psi_i, \cup_{i \geq 0}\hat{\Psi}_i) = (\Psi, \hat{\Psi})$ and $(\Psi, \hat{\Psi})$ is the model of $ASS(C)$.

Finally we prove that $(\Psi, \hat{\Psi})$ is $\geq_C$-maximal such that

$$(\Psi, \hat{\Psi}) \geq_C (\Phi, \{\Phi\}).$$

Clearly, $(\Psi, \hat{\Psi}) \geq_C (\Phi, \{\Phi\})$. If there is $(\Psi^*, \hat{\Psi}^*)$ such that $(\Psi^*, \hat{\Psi}^*) >_\delta (\Psi, \hat{\Psi})$ for some $\delta = \alpha : \beta_1, ..., \beta_n/\gamma \in C$ then $\Psi \models \alpha$. So there are defaults $d_1, ..., d_m \in ASS(C)$ such that $CON(\{d_1, ..., d_m\}) \vdash \alpha$ by compactness of first order logic. Repeatedly using rules DL_2 and DL_3 we have $d = true : \beta_1, ..., \beta_k/\gamma \in ASS(C)$, where $\{\beta_{n+1}, ..., \beta_k\} = CCS(\{d_1, ..., d_m\})$. As above we know that each default in ASS(C) is in (δ_i), which implies that d is in (δ_i). Hence $\Psi \models \gamma$ and $\{MOD(\{\beta_i\}) | 1 \leq i \leq n\} \subseteq \hat{\Psi}$. This shows that $(\Psi^*, \hat{\Psi}^*) = (\Psi, \hat{\Psi})$, a contradiction.

Lemma 3.6 Suppose that the sequences $((\Psi_i, \hat{\Psi}_i))$ and (δ_i) are defined as that in the proof of Lemma 3.5. Then the model $(\Psi, \hat{\Psi})$ of ASS(C) is $\geq_T$-maximal such that $(\Psi, \hat{\Psi}) \geq_T (\Phi, \{\Phi\})$.

Proof. we inductively show that for each $j \geq 0$: if $\delta_j \notin T$ then $(\Psi, \hat{\Psi}) \geq_{(\delta_i^{**})} (\Psi^*, \hat{\Psi}^*) \geq_{(\delta_i^*)} (\Phi, \{\hat{\Phi}\})$, where $(\delta_i^{**}) = (\delta_i) - \{\delta_1, ...\delta_j\}$ and $(\delta_i^*) = T \cap \{\delta_1, ..., \delta_j\}$.

Base. $j = 0$, then $\delta_0 = true : true/true$. Clearly, the induction conclusion holds and $(\delta_i^*) = \emptyset$.

Step. Suppose that the induction proposition holds for all $k < j$, i.e. if $\delta_k \notin T$, then $(\Psi, \hat{\Psi}) \geq_{(\delta_i^{**})} (\Psi^*, \hat{\Psi}^*) \geq_{(\delta_i^*)} (\Phi, \{\Phi\})$, where $(\delta_i^{**}) = (\delta_i) - \{\delta_k | k < j\}, (\delta_i^*) = T \cap \{\delta_k | k < j\}$. Now assuming $\delta_j \notin T$, it is easy to prove the induction proposition is true by induction on structure of δ_j.

Base. Trivial.

Step. It is needed to consider the case where $\delta_j = true : \beta_1, ..., \beta_m/\sigma$, $true : \beta_1, ..., \beta_n/\gamma \in (\delta_i^*)$ and $\gamma : \beta_{n+1}, ..., \beta_m/\sigma \in C$. Let $V = (\delta_i^*)$. Clearly, $\Psi^* \models \gamma$ and $CON(V) \vdash \gamma$. By compactness of first order logic and Rules DL_1, DL_2 and DL_3 we have $true : \beta_1, ..., \beta_m, \beta_{m+1}, ..., \beta_p/\sigma \in ASS(C)$ where $\{\beta_{m+1}, ..., \beta_p\} = CCS(V)$. According to the definition of $((\Psi_i, \hat{\Psi}_i))$ there is some $q(q \leq j)$ such that $\delta_q = true : \beta_1, ..., \beta_p/\sigma$. Hence $(\Psi_{j+1}, \hat{\Psi}_{j+1}) = (\Psi^*, \hat{\Psi}^*)$.

Therefore there is a subsequence (δ_i') of (δ_i) such that $(\delta_i') \subseteq T$ and $(\Psi, \hat{\Psi}) \geq_{(\delta_i')} (\Phi, \{\Phi\})$. Since $(\Psi, \hat{\Psi})$ is $\geq_C$-maximal then it is $\geq_T$-maximal by Lemma 3.5.

Theorem 3.7 Let T be a DT_B, C it's closure under R_{DL}. $(\Psi, \hat{\Psi})$ is a $\geq_T$-maximal element such that $(\Psi, \hat{\Psi}) \geq_T (\Phi, \{\Phi\})$ iff $(\Psi, \hat{\Psi})$ is $\geq_C$-maximal one such that $(\Psi, \hat{\Psi}) \geq_C (\Phi, \{\Phi\})$.

Proof.

"Only If Part" We inductively prove that if $(\Psi^*, \hat{\Psi}^*) \geq_\delta (\Psi, \hat{\Psi})$ then $(\Psi^*, \hat{\Psi}^*) = (\Psi, \hat{\Psi})$ for any $\delta \in C$. First, We note that for any $\delta = \alpha : \beta_1, ..., \beta_n/\gamma \in C$ and $(\Psi, \hat{\Psi})$ if $\Psi \models \alpha$ then $\{\psi \in \Psi | \psi \vdash \gamma\} \neq \emptyset$ by consistency of T and Lemma 3.3. So, $(\{\psi \in \Psi | \psi \models \gamma\}, \hat{\Psi} \cup \{MOD(\{\beta_i\}) | 1 \leq i \leq n\}) \geq_\delta (\Psi, \hat{\Psi})$.

Base. Clear.

Step. 1. If $\delta = true : true/true$ then it is obvious.

2. If $\delta = true : \beta_1, ..., \beta_n/\gamma$ and $d = true : \beta_1, ..., \beta_n/\sigma \in T$, $\sigma \vdash \gamma$, then $(\{\psi \in \Psi^* | \psi \models \sigma\}, \hat{\Psi}^*) \geq_d (\Psi^*, \hat{\Psi}^*) \geq_\delta (\Psi, \hat{\Psi})$. So, $(\{\psi \in \Psi | \psi \models \sigma\}, \hat{\Psi}^*) \geq_d (\Psi, \hat{\Psi})$. By the induction hypothesis we get $(\{\psi \in \Psi | \psi \models \sigma\}, \hat{\Psi}^*) = (\Psi, \hat{\Psi})$. But $\{\psi \in \Psi | \psi \models \sigma\} = \{\psi \in \Psi^* | \psi \models \sigma\}$, so $(\Psi^*, \hat{\Psi}^*) = (\Psi, \hat{\Psi})$.

The other two cases are similar.

"If Part". It is immediate by Lemmas 3.5 and 3.6.

Corollary 3.8 Let T be a DT_B, C it's closure under R_{DL}. $(\Psi, \hat{\Psi})$ is a $\geq_T$-maximally (distributed or joint) compatible frame such that $(\Psi, \hat{\Psi}) \geq_T (\Phi, \{\Phi\})$ iff it is a $\geq_C$- maximally (distributed or joint) compatible frame such that $(\Psi, \hat{\Psi}) \geq_C (\Phi, \{\Phi\})$.

4. Semantics for L_{DL} and L_1 extensions

In this section we give semantical characterizations of Brewka's L_{DL} and L_1 extensions.

Definition 4.1 Let T be a DT_B, C it's closure under R_{DL} and E it's L_{DL} extension. Then T_E is the generating set of E iff

1. $T_E \subseteq T$

2. for each $d \in T$:$d \in T_E$ iff $d \in E$ or there are defaults $true : \beta_1, ..., \beta_m/\gamma$, $true : \beta_1, ..., \beta_n/\sigma \in E$ such that $d = \sigma : \beta_{n+1}, ..., \beta_m/\gamma$.

By the above definition we have

Corollary 4.1 T_E is consistent.

Lemma 4.2 Let E be an L_{DL} extension of a DT_B T, T_E the generating set of E. If F is the closure of T_E under R_{DL} then $E = ASS(F)$.

Proof First we inductively show that $d \in E$ for all $d \in ASS(F)$.

Base. If $d \in T_E$ then $d \in E$.

Step. 1. If $d = true : true/true$ then $d \in E$

2. If $d = true : \beta_1, ..., \beta_n/\sigma$ then there is $d' = true : \beta_1, ..., \beta_n/\gamma \in F$ with $\gamma \vdash \sigma$. By the hypothesis we obtain $d' \in E$. So $d \in E$.

3. If $d = true : \beta_1, ..., \beta_m/\gamma \wedge \sigma$ then there are defaults $d_1 = true : \beta_1, ..., \beta_n/\gamma$, $d_2 = true : \beta_{n+1}, ..., \beta_m/\sigma \in F$. By the hypothesis $d_1, d_2 \in E$. So $d \in E$.

4. If $d = true : \beta_1, ..., \beta_m/\sigma$ then there are defaults $d_1 = true : \beta_1, ..., \beta_n/\gamma$, $d_2 = \gamma : \beta_{n+1}, ..., \beta_m/\sigma \in F$. By the hypothesis $d_1 \in E$. Since $d_2 \in T_E$ we have $d \in E$.

Hence $ASS(F) \subseteq E$. Vice versa, we can prove $E \subseteq ASS(F)$ similarly. Assume that d is any default in E.

Base. Clear.

Step. 1. $d = true : true/true$. it is obvious.

2. If $d = true : \beta_1, ..., \beta_n/\sigma$ then there is a default

$$d_1 = true : \beta_1, ..., \beta_n/\gamma \in C$$

with $\gamma \vdash \sigma$. Since $\{\beta_i\} \cup CON(E)$ is consistent for all $1 \leq i \leq n$ then $d_1 \in E$. By the hypothesis $d_1 \in ASS(F)$. So $d \in ASS(F)$.

3. If $d = true : \beta_1, ..., \beta_m/\gamma \wedge \sigma$ then there are defaults $d_1 = true : \beta_1, ..., \beta_n/\gamma$, $d_2 = true : \beta_{n+1}, ..., \beta_m/\sigma \in C$. Since $\{\beta_i\} \cup CON(E)$ is consistent for $1 \leq i \leq m$ then $d_1, d_2 \in E$. By the hypothesis $d_1, d_2 \in ASS(F)$. So $d \in ASS(F)$.

4. If $d = true : \beta_1, ..., \beta_m/\sigma$ then there are defaults $d_1 = true : \beta_1, ..., \beta_n/\gamma$, $d_2 = \gamma : \beta_{n+1}, ..., \beta_m/\sigma \in C$. Similarly, $d_1 \in E$ and $d_2 \in T$. By the hypothesis $d_1 \in ASS(F)$. So $d \in ASS(F)$.

Definition 4.2 Let $(\Psi, \hat{\Psi})$ be a distributed compatible frame and $\delta = \alpha : \beta_1, ..., \beta_n/\gamma$ a default. $(\Psi, \hat{\Psi})$ is stable w.r.t. δ iff if $\Psi \models \alpha$ and there is $\psi_i \in \Psi$ such that $\psi_i \models \gamma, \beta_i$ for each $i : 1 \leq i \leq n$ then $(\{\psi \in \Psi | \psi \models \gamma\}, \hat{\Psi} \cup \{MOD(\{\beta_i\}) | 1 \leq i \leq n\})$ is distributed compatible.

Lemma 4.3 Let E be an L_{DL} extension of a DT_B T, T_E the generating set of E. There is a $\geq_T$-maximally distributed compatible frame $(\Psi, \hat{\Psi})$ such that

1. $(\Psi, \hat{\Psi}) \geq_T (\Phi, \{\Phi\})$
2. $(\Psi, \hat{\Psi})$ is the model of E.
3. $(\Psi, \hat{\Psi})$ is stable w.r.t. δ for all $\delta \in T$.

Proof. By Corollary 4.1 and Lemma 3.5 there is a $\geq_T$-maximal frame $(\Psi, \hat{\Psi})$ such that $(\Psi, \hat{\Psi}) \geq_{T_E} (\Phi, \{\Phi\})$ and that $(\Psi, \hat{\Psi})$ is the model of

$ASS(F)$, where F is the closure of T_E under R_{DL}. So $(\Psi, \hat{\Psi})$ is the model of E by Lemma 4.2. Clearly, $(\Psi, \hat{\Psi})$ is distributed compatible. Now we show that $(\Psi, \hat{\Psi})$ is stable w.r.t. δ for all $\delta \in T$. Assume that $\delta = \alpha : \beta_1, ..., \beta_n/\gamma \in T$. If $\Psi \models \alpha$, i.e. $CON(E) \vdash \alpha$, then there are defaults $\delta_1, ..., \delta_m \in T_E$ with $CON(\{\delta_1, ..., \delta_m\}) \vdash \alpha$ by compactness of first order logic. Repeatedly using rules DL_1, DL_2 and DL_3 we obtain

$true : \beta_{n+1}, ..., \beta_p/\alpha \in ASS(F) = E$

$\delta' = true : \beta_1, ..., \beta_n, \beta_{n+1}, ..., \beta_p/\gamma \in ASS(C)$

where $\{\beta_{n+1}, ..., \beta_p\} = CCS(\{\delta_1, ..., \delta_m\}$ and C is the closure of T under R_{DL}. Note that $CON(E) \cup \{\beta_i\}$ is consistent for all $i : n+1 \leq i \leq p$. If there is $\psi_i \in \Psi$ such that $\psi_i \models \beta_i, \gamma$ then $CON(E) \cup \{\beta_i\}$ is consistent for all $i : 1 \leq i \leq p$. So $\delta' \in E$. Hence $(\{\psi \in \Psi | \psi \models \gamma\}, \hat{\Psi} \cup \{MOD(\{\beta_i\}) | 1 \geq i \geq n\} = (\Psi, \hat{\Psi})$, i.e. it is distributed compatible.

From Lemma 4.3 it is easy to get the following proposition.

Theorem 4.4 (Correctness) If E is an L_{DL} extension of a DT_B T then the model $(\Psi, \hat{\Psi})$ of E is a $\geq_T$-maximally distributed compatible frame such that

1. $(\Psi, \hat{\Psi}) \geq_T (\Phi, \{\Phi\})$
2. $(\Psi, \hat{\Psi})$ is stable w.r.t. δ for all $\delta \in T$.

Theorem 4.5 (Completeness) Let T be a DT_B. If $(\Psi, \hat{\Psi})$ is a $\geq_T$-maximally distributed compatible frame such that

1. $(\Psi, \hat{\Psi}) \geq_T (\Phi, \{\Phi\})$
2. $(\Psi, \hat{\Psi})$ is stable w.r.t. δ for all $\delta \in T$

then there is an L_{DL} extension E of T such that $(\Psi, \hat{\Psi})$ is the model of E.

Proof. Let (δ_i) be a sequence of defaults in T with $(\Psi, \hat{\Psi}) \geq_{(\delta_i)} (\Phi, \{\Phi\})$. Suppose F is the closure of (δ_i) under R_{DL} and $E = ASS(F)$. Clearly, $(\Psi, \hat{\Psi})$ is the model of E by Lemmas 3.5 and 3.7. Now we show that E is an L_{DL} extension of T. Assume that $\delta = true : \beta_1, ..., \beta_n/\gamma \in ASS(C)$, where C is the closure of T under R_{DL}. If $\delta \in E$ then $CON(E \cup \{\beta_i\})$ is consistent for all $i : 1 \leq i \leq n$. Vice versa, if $CON(E) \cup \{\beta_i\}$ is consistent for all $i : 1 \leq i \leq n$ then $(\{\psi \in \Psi | \psi \models \gamma\}, \hat{\Psi} \cup \{MOD(\{\beta_i\}) | 1 \leq i \leq n\})$ is distributed compatible by stability of $(\Psi, \hat{\Psi})$ w.r.t. δ. On the other hand, $(\{\psi \in \Psi | \psi \models \gamma\}, \hat{\Psi} \cup \{MOD(\{\beta_i\}) | 1 \leq i \leq n\}) \geq_\delta (\Psi, \hat{\Psi})$, and $(\Psi, \hat{\Psi})$ is $\geq_C$-maximally distributed compatible by Corollary 3.8. So, $\Psi \models \gamma$, $\{MOD(\{\beta_i\}) | 1 \leq i \leq n\} \subseteq \hat{\Psi}$, which implies that $\delta \in E$.

In a similar method we can get a semantic characterization of L_1 extensions. Comparing the definition of L_1 extensions with that of L_{DL} and CDL extensions, we observe two basic differences:

1. The set of generating rules of closure increase two rules: disjoining and cases.
2. The joint consistent condition is applied to a set of assumable de-

faults such that it is possible that applied default does not belong to the corresponding extension.

In addition restricting ourselves to defaults with single justification makes the recorded justifications more complex.

These are sketched by following examples:

Example 4.1 The default theory $(\{A : C/D, B : E/F\}, \{A \vee B\})$ (ADT $(\{A : C/D, B : E/F\}, \{(A \vee B, \emptyset)\}))$ has one DL extension $Th_L(\{A \vee B\})$ (one CDL extension $Th_S(\{(A \vee B, \emptyset)\})$. The corresponding DT_B has one L_1 extension which contains $A \vee B$ and $: C \wedge D/D \vee F$. Clearly, the applied defaults $A : C/D$ and $B : E/F$ are not contained in L_1 extension.

Example 4.2 $(\{: A/C \wedge D, : B/\neg(C \wedge D)\}, \emptyset)$ has no DL extension. As a ADT, it has two CDL extensions $Th_S(\{(C \wedge D, \{A\})\})$ and $Th_S(\{(\neg(C \wedge D), \{B\})\})$. The corresponding DT_B has two extensions, one of which contains :A/C, another contains :A/D. And neither L_1 extension contains the applied default $: A/C \wedge D$ by R_3.

Example 4.3 $(\{: A/B, B : C/D\}, \emptyset)$ has no DL extension, but it has one CDL extension $Th_S(\{(B, \{A\})\}$. The corresponding DT_B has two extensions, which contain $: A/B$ and $: A \wedge C/\neg B$ respectively.

This added our difficulties. In a sense, we can only get a weaker semantics characterization of L_1 extensions. But this just suggests useness of our semantics framework. As that we will see in section 5, the framework is also applied to CDL extensions. And it is not easy to reconstruct CDL extensions in Brewka's framework.

Theorem 4.6 (Correctness) Let T be a DT_B, C its closure under R_{L_1}. If E is an L_1 extension of T, then there is a $\geq_C$-maximally joint compatible frame $(\Psi, \hat{\Psi})$ such that

1. $(\Psi, \hat{\Psi}) \geq_C (\Phi, \{\Phi\})$;
2. $\Psi = MOD(CON(E))$;
3. $MOD(\{\beta\}) \in \hat{\Psi}$ for any $\beta \in CCS(E)$.

Proof (outline) In a method similar to that in Lemma 3.5, we can define a frame $(\Psi, \hat{\Psi^*})$ such that $\Psi = MOD(CON(E)), \hat{\Psi^*} = \{MOD(\{\beta\}|\beta \in CCS(E)\}$ (using the well ordered set of E). Here we simply apply each default in order of E to constrict the sequence $((\Psi_i, \hat{\Psi}_i))$. Since $CON(E) \cup CCS(E)$ is consistent, it is easy to see that $(\Psi, \hat{\Psi^*})$ is a $\geq_E$-maximally joint consistent frame such that $(\Psi, \hat{\Psi^*}) \geq_E (\Phi, \{\Phi\})$. Let $\hat{\Psi} = \hat{\Psi^*} \cup \{MOD(\{\beta\})|\alpha : \beta/\gamma \in C - E, \Psi \models \alpha, \{\beta, \gamma\} \cup CON(E) \cup CCS(E)$ is consistent $\}$. It is easy to show that $(\Psi, \hat{\Psi})$ is $\geq_C$- maximally joint compatible such that $(\Psi, \hat{\Psi}) \geq_C (\Phi, \{\Phi\})$. In fact, if there exists a default $\delta = \alpha : \beta/\gamma \in C - E$ such that $(\{\psi \in \Psi|\psi \models \gamma\}, \hat{\Psi} \cup \{MOD(\{\beta\})\}) \geq_\delta (\Psi, \hat{\Psi})$ and that the left of the inequality is joint compatible, then $\Psi \models \alpha, \{\beta, \gamma\} \cup CON(E) \cup CCS(E)$ is consistent. Hence $MOD(\{\beta\}) \in \hat{\Psi}$. By compactness there are defaults in E : $\beta_i/\gamma_i (1 \leq i \leq n)$ such that $\gamma_1 \wedge ... \wedge \gamma_n \vdash \alpha$. Therefore $: \beta_1 \wedge ... \wedge \beta_n/\alpha \in E$. This implies that

$: \beta_1 \wedge ... \wedge \beta_n \wedge \beta/\gamma \in E$. So, $\{\psi \in \Psi | \psi \models \gamma\} = \Psi$.

Theorem 4.7 (Completeness) Let T be a DT_B, C its closure under R_{L_1} and KERN(T) consistent. If $(\Psi, \hat{\Psi})$ is a $\geq_C$-maximally joint compatible frame such that $(\Psi, \hat{\Psi}) \geq_C (\Phi, \{\Phi\})$, then there is an L_1 extension E such that $\Psi = MOD(CON(E))$ and that $MOD(\{\beta\}) \in \hat{\Psi}$ for each $\beta \in CCS(E)$.

Proof (outline) Let $E = \{: \beta/\gamma \in C | \Psi \models \gamma, MOD(\{\beta\}) \in \hat{\Psi}\}$. It is clear that $E \subseteq ASS(C), KERN(C) \subseteq E$. We easy show that E is an L_1 extension of T by $\geq_C$-maximally joint compatibility.

5. Application to DL and it's variants

In this section we illustrate the expressiveness of our semantics frame by applying it to Reiter's DL, Lukaszewicz' modified extension and Brewka's CDL. So it seems interesting that this frame supports a model theoretic semantic for the design of a particular logic.

A semantics of a Reiter DT, though it can be made similarly to Theorems 4.4 and 4.5 by the characterization of extensions for DL (Theorem 2.11 of [14]), is easy got by Theorem 2.1 in section 2. In fact it is sufficient to note that $(\Psi, \hat{\Psi}) \geq_D (\Phi, \{\Phi\})$ iff $(\Phi, \hat{\Psi}) \geq_T (MOD(W), \{\Phi\})$, where (D,W) is a DT and $T = D \cup \{true : true/p | p \in W\}$. So we have

Theorem 5.1 (Correctness) If E is a DL extension of (D,W) then there is a $\geq_D$-maximally distributed compatible frame $(\Psi, \hat{\Psi})$ satisfying the following conditions:

1. $(\Psi, \hat{\Psi}) \geq_D (MOD(W), \{\Phi\})$
2. $(\Psi, \hat{\Psi})$ is stable w.r.t. δ for all $\delta \in D$
3. $\Psi = MOD(E)$ and $\hat{\Psi} = \{MOD(\{\beta\}) | \beta \in CCS(GD(E))\} \cup \{\Phi\}$, where GD(E) is the generating set of defaults for E.

Theorem 5.2 (Completeness) Let (D,W) be a default theory. If $(\Psi, \hat{\Psi})$ is a frame satisfying the following conditions:

1. $(\Psi, \hat{\Psi}) \geq_D (MOD(W), \{\Phi\})$
2. $(\Psi, \hat{\Psi})$ is $\geq_D$-maximally distributed compatible and is stable w.r.t. δ for all $\delta \in D$

then there is a DL extension E of (D,W) such that $\Psi = MOD(E)$ and $\hat{\Psi} = \{MOD(\{\beta\}) | \beta \in CCS(GD(E))\} \cup \{\Phi\}$ where GD(E) is the generating set of defaults for E.

In what follows, we give a semantics characterization of CDL extensions. To do this, the following notations and propositions in [16] are needed, and we always suppose that any ADT (D,W) is well-based, i.e. $FORM(W) \cup SUPP(W)$ is consistent.

Definition 5.1[16] Let $\Delta = (D, W)$ be an ADT. For any subset D' of D, $\Lambda(D', \Delta) = \cup_{i \geq 0} D'_i(\Delta)$ where

$D'_0(\Delta) = \{\alpha : \beta/\gamma \in D' | FORM(W) \vdash \alpha\}$

$D'_{i+1}(\Delta) = \{\alpha : \beta/\gamma \in D' | FORM(W) \cup CON(D'_i(\Delta)) \vdash \alpha\}$ for $i : i \geq 0$

Whenever $\Delta = (D, W)$ is clear from the context we may write $\Lambda(D')$ instead of $\Lambda(D', \Delta)$.

Definition 5.2 Let E be a CDL extension of $\Delta = (D, W)$. Then the generating set of defaults of E, $GD(E, \Delta)$, is the set of all $\alpha : \beta/\gamma \in D$ such that

$$\alpha \in FORM(E), \{\beta, \gamma\} \cup FORM(E) \cup SUPP(E)$$

is consistent.

Definition 5.3[16] Let $\Delta = (D, W)$ be an ADT and D' a subset of D. We say that D' is Δ-robust (or just robust, if Δ is understood), if the following conditions are satisfied:

1. D' is joint compatible w.r.t. Δ
2. $\Lambda(D') = D'$
3. D' is a maximal subset of D w.r.t. the properties 1 and 2, i.e. there is no $D'' \subseteq D$ such that $D' \subseteq D''$, $D' \neq D''$ and that D'' is joint compatible and $\Lambda(D'') = D''$.

Theorem 5.3[16] Let $\Delta = (D, W)$ be an ADT. If E is a CDL extension of Δ then $GD(E, \Delta)$ is robust. Vice versa, if $D' \subseteq D$ is a robust set then there is a CDL extension E of (D,W) with $GD(E, \Delta) = D'$.

Suppose that $\Delta = (D, W)$ be an ADT. Let

$$T = D \cup \{true : r_1, ..., r_n/p | (p, \{r_1, ..., r_n\}) \in W)$$

and T a DT_B. The following properties are easy obtained by Corollary 3.8, Definitions 5.2, 5.3 and Theorem 5.3.

Lemma 5.4 If D' is a robust set w.r.t.(D,W) then there is a $\geq_D$-maximally joint compatible frame $(\Psi, \hat{\Psi})$ such that

$$\Psi = MOD(FORM(W) \cup CON(D')),$$

$$\hat{\Psi} = \{MOD(\{q\}) | q \in SUPP(W) \cup CCS(D') \cup \{true\}\}$$

and that
$(\Psi, \hat{\Psi}) \geq_D (MOD(FORM(W)), \{MOD(\{q\})| \ q \in SUPP(W)\} \ or \ q = true)$.

Proof. (outline) In a way similar to that in Lemma 3.5 we use the well ordered set $D' = \{d_1, d_2, ...\}$ and get sequences (δ_i) and $((\Psi_i, \hat{\Psi}_i))$ such that

1. (δ_i) consists of all defaults in D',
2. $(\Psi, \hat{\Psi}) \geq_{(\delta_i)} (MOD(FORM(W)), \{MOD(\{q\}) | q \in SUPP(W) \ or \ q = true\})$.
3. $(\Psi, \hat{\Psi}) = (\cap_{i \geq 0} \Psi_i, \cup_{i \geq 0} \hat{\Psi}_i)$ is $\geq_D$-maximally joint compatible.

Lemma 5.5 If $(\Psi, \hat{\Psi})$ is a $\geq_D$-maximally joint compatible frame such that $(\Psi, \hat{\Psi}) \geq_D (MOD(FORM(W)), \{MOD(\{q\}) | q \in SUPP(W) \ or \ q = true\})$ then there is a robust set $D' \subseteq D$ such that $\Psi = MOD(FORM(W) \cup CON(D'))$, $\hat{\Psi} = \{MOD(\{q\}) | q \in SUPP(W) \cup CCS(D') \cup \{true\}\}$.

Proof. (outline) Let (d_i) is a sequence of defaults in D such that $(\Psi, \hat{\Psi}) \geq_{(d_i)} (MOD(FORM(W)), \{MOD(\{q\})|q \in SUPP(W) \ \ or \ \ q = true\})$. Let D' be a set of defaults which are in (d_i) . Then it is easy to show that D' is a desired set.

Theorem 5.6 Let $\Delta = (D, W)$ be a well-based ADT, . If E is a CDL extension of (D,W) then there is a $\geq_D$-maximally joint compatible frame $(\Psi, \hat{\Psi})$ such that $(\Psi, \hat{\Psi}) \geq_D (MOD(FORM(W)), \{MOD(\{q\})|q \in SUPP(W) \ \ or \ q = true\})$ and that $\Psi = MOD(FORM(E))$,

$\hat{\Psi} \cup \{MOD(\{\gamma\})|\gamma \in (CON(GD(E, \Delta))\}$

$= \{MOD(\{q\})|q \in SUPP(E) \ \ or \ \ q = true\}$. Vice versa, if $(\Psi, \hat{\Psi})$ is $\geq_D$-maximally joint compatible such that $(\Psi, \hat{\Psi}) \geq_D (MOD(FORM(W)),$ $\{MOD(\{q\})|q \in SUPP(W) \ \ or \ \ q = true\})$ then there is a CDL extension E of (D,W) such that $MOD(FORM(E)) = \Psi$, $\{MOD(\{q\})|q \in SUPP(E)\} = \hat{\Psi} \cup \{MOD(\{\gamma\})|\gamma \in CON(GD\ (E, \Delta))\}$.

Proof. It is obvious by Theorem 5.3 and Lemmas 5.4, 5.5.

It is clear that Schaub's Correctness and Completeness Theorems (a semantics characterization of Brewka's CDL) are an immediate conclusion of Theorem 5.6 here. Finally we can similarly get a semantics characterization of Lukaszewicz' modified extensions [7] in term of the semantics frame. For sake of simplicity we will only state the results since their proofs are easy.

lemma 5.7 Let Φ be any class of first order interpretations and D any set of defaults. A default $\delta \in D$ is Φ-applicable w.r.t. D iff the model $(\Psi, \hat{\Psi})$ of $\{\delta\}$ is distributed compatible and $(\Psi, \hat{\Psi}) \geq_\delta (\Phi, \{MOD(\{\beta\})|\beta \in CCS(D)\})$. A subset D' of D is Φ-applicable iff the model $(\Psi, \hat{\Psi})$ of D' is distributed compatible and $(\Psi, \hat{\Psi}) \geq_D (\Phi, \{\Phi\})$. D' is maximally Φ-applicable w.r.t. D iff the model $(\Psi, \hat{\Psi})$ of D' is maximally distributed compatible and $(\Psi, \hat{\Psi}) \geq_D (\Phi, \{\Phi\})$.

Theorem 5.8 Let (D,W) be a default theory, Ψ the class of all models of W, i.e. $\Psi = MOD(W)$. S is a modified extension of (D,W) iff there is a $\geq_D$-maximally distributed compatible frame $(\Omega, \hat{\Omega})$ such that $(\Omega, \hat{\Omega}) \geq_D (\Psi, \{\Phi\})$ and $\Omega = MOD(S)$.

6. Conclusion

In this paper a uniform semantical framework of various default logics by

extending Etherington's semantics has been presented. We have shown how our framework capture various (potential) application conditions of a default by a preference relation and notions of compatibility of a frame. That apart, we derive semantics characterizations of various default logics, which are corresponded to their syntactic characterizations. By the same method we can deal with Sandewall's accessibility of a BJ-pair [5], Etherington's semantic [6], as well as Giordano and Martelli's commitment to assumption default logic and quasi-default logic, etc.

Compared to Besnard and Schaub's possible worlds semantics, we have not used a modal system to provide a semantical framework. This seems to be natural and simple. Furthermore, it appears to be interesting that the semantics presented in this paper can support the design of a semantical characterization of a particular nonmonotonic reasoning system , which we will discuss in another paper.

References

1 Reiter R. A Logic for Default Reasoning, Artificial Intelligence, Vol.13, No.1-2, 1980.

2 Etherington D.W. Formalizing Non-monotonic Reasoning System, Artificial Intelligence, Vol.31, 1987

3 Etherington D.W. Reasoning from Incomplete Information, Pitman Research Notes in AI, Pitman Publishing limited, London, 1986

4 Etherington D.W. A Semantics for Default Logic, in Proc. of IJCAI-87, Milan, Italy, 1987

5 Sandewall E.J., A Functional Approach to Non-monotonic Logic, Proc. of IJCAI-85, Los Angels, 1985

6 Lukaszewicz W. Two Results on Default Logic, Proc. IJCAI-85,Los Angles, 1985.

7 Lukaszewicz W. Considerations on Default Logic-An Alternative Approach, in Proc. of Nonmonotonic Reasoning Workshop, New Paltz, N.Y, 1984 also Computational Intelligence, 4, 1988.

8 de Guerreiro R.A. and Casanova M.A. An Alternative Semantics for Default Logic, 3-rd International Workshop on Nonmonotonic Reasoning, South Lake Tahoe, CA, 1990

9 Pool D. What the Lottery Paradox Tell Us about Default Reasoning, in Proc. of 1-st ICPKRR, Toronto, 1989

10 Delgrande J.P. and Jackson W.K. Default Logic Revisited , Proc. of 2-nd International Conference on Principles of Knowledge Representation and Reasoning, April, 1991, by Morgan Kaufmann Publishers,Inc.

11 Brewka G., Cumulative Default Logic: in Defense of Nonmonotonic Inference Rules, Artif. Intell. 50, 1991.

12 Schaub T. Assertional Default Theories: A Semantical View, in Proc. of 2-nd International Conference on Principles of Knowledge Representation and Reasoning (KR-2), 1991

13 Brewka G. A Framework for Cumulative Default Logic, TR-92-042, International Computer Science Institute, Berkeley, CA, 1992

14 Zhang Mingyi A Characterization of Extensions of General Default Theories, in Proc. of 9-th Canadian Society for Computational Studies of AI, Vancouver, BC(1992)

15 Zhang Mingyi, On the Existence of Extensions for General Default Theories, Science in China (English Edition), Series A, 10, 1993.

16 Gottlob G. and Zhang Mingyi cumulative Default Logic: Finite Characterization, Algorithms and Complexity, Tech. Report, CP-TR- 93-54, Institute for Informationssystem, Tech. Univ. of Wien, Austria, 1993. also see Artificial Intelligence, vol.96, 1994, pp.829-845.

17 Giordano L. and Materlli A., On Cumulative Default Logic, Artif. Intell., 66, 1994.

18 Besnard P. and Schaub T,, Possible Worlds Semantics for Default Logics, Fundamenta Informaticae, 21,1994.

ADDING EVENTUALLY DIFFERENT REALS

YI ZHANG

Institute of Sorftware, Academca Sinica
Beijing, China
Department of Philosophy, Rutgers University
New Brunswick, New Jersey 08903, USA
cyzhang@math.rutgers.edu

Abstract. In this paper, we will mainly study two cardinal invariants of continuum, $\mathfrak{a}$ and $\mathfrak{a}_{\mathfrak{e}}$, which we will define in the definitions 1.1 and 1.2. Among them, $\mathfrak{a}$ is well-known, $\mathfrak{a}_{\mathfrak{e}}$ looks similar to $\mathfrak{a}$. Futhurmore, $\mathfrak{a}$ and $\mathfrak{a}_{\mathfrak{e}}$ have some important "same-looking" features, for example, $\mathfrak{a}_{\mathfrak{e}}$ stays small in the Cohen forcing model as $\mathfrak{a}$ does. In the second section, we will compare $\mathfrak{a}$, $\mathfrak{a}_{\mathfrak{e}}$ together with some other well-known continuum invariants $\mathfrak{a}_{\mathfrak{s}}$ and $\mathfrak{b}$ in a forcing model of ZFC. Our forcing terminologies are standard, people can find them in either [Kun] or [Jech].

1. Definitions And Some Properties of Cardinals We Study.

Definition 1.1. If $x, y \subset \omega$, x and y are almost disjoint (a.d.) iff $\mid x \cap y \mid < \omega$. An a.d. family is an $A \subset \wp(\omega)$ such that for any $x \in A$, $|x| = \omega$ and any two distinct elements of A are a.d.. Let $\mathfrak{a}$ be the least λ such that there exists a maximal almost disjoint (m.a.d.) family $\mathcal{F} \subseteq \wp(\omega)$ of size λ.

Definition 1.2. Following A. Miller, we say that two functions $f, g \in {}^{\omega}\omega$ are eventually different (e.d.) iff

$$\mid \{n \in \omega \mid f(n) = g(n)\} \mid < \omega.$$

Let $\mathfrak{a}_{\mathfrak{e}}$ be the least λ such that there exists a maximal eventually different (m.e.d.) set of reals of cardinality λ.

The following ressults about $\mathfrak{a}$ are well-known:

(1) Any m.a.d. family $F \subseteq \wp(\omega)$ is uncountable.

The research is partially supported by a grant from "Chinese Climbing Project Foundation for Natural Science" and CNSF–6957303.

1991 Mathematical Subject Classification 03E35, 20A15,20B07,20B35.

(2) $ZFC + MA$ implies that $\mathfrak{a} = 2^\omega$

(3) Let $M \models (ZFC + \neg CH)$, and let κ be a cardinal in M such that $\omega_1 \leq \kappa < 2^\omega$. Then M can be generically extended to an $M[G]$ which preserves 2^ω, and there is, in $M[G]$, a m.a.d. family $A \subseteq \wp(\omega)$ of cardinality κ. Hence

$$Con(ZFC) \to Con(ZFC + \mathfrak{a} < 2^\omega).$$

(4) Let $M \models (ZFC + CH)$. There is a m.a.d. family $\mathcal{F} \subset \wp(\omega)$ of size ω_1 in M such that for any Cohen generic G over M, $\mathcal{F}$ remains to be a m.a.d. family in $M[G]$.

For the proof of (1), (2) and (4), see [Kun]; for the proof of (3), see [Sh:P].

We can prove the corresponding results for $\mathfrak{a}_\mathfrak{e}$ by using the following *c.c.c.* p.o.set (A similar version of the forcing notion was studied by A. Miller in [M]).

Definition 1.3. For any $A \subset {}^\omega\omega$, define a partial order $\mathbb{P}_A$ which consists of all conditions of the form $\langle s, F\rangle$, such that

(1) s is a finite partial function from ω to ω, and

(2) F is a finite subset of A,

where $\langle s_1, F_1\rangle \leq \langle s_2, F_2\rangle$ iff

$$(s_2 \subseteq s_1) \wedge (F_2 \subseteq F_1) \wedge \forall f \in F_2 (f \cap s_1 \subseteq s_2).$$

Theorem 1.4(ZFC). *Any m.e.d. set $A \subset {}^\omega\omega$ is uncountable.*

Theorem 1.5 ($MA(\kappa)$). *Let $A \subset {}^\omega\omega$, where $|A| \leq \kappa$ and $\omega \leq \kappa < 2^\omega$. Then there exists a $g \in {}^\omega\omega$ such that*

$$\forall f \in A(|g \cap f| < \omega).$$

Corollary 1.6 (MA). $\mathfrak{a}_\mathfrak{e} = 2^\omega$.

Theorem 1.7. *Let $M \models (ZFC + \neg CH)$, and let κ be a cardinal in M such that $\omega_1 \leq \kappa < 2^\omega$. Then M can be generically extended to an $M[G]$ which preserves 2^ω, and there is, in $M[G]$, a m.e.d. function set $A^\star \subset {}^\omega\omega$ of cardinality κ.*

Proof. Without lost of generality, let $M \models (MA + \kappa < 2^\omega)$. We can assume that in M, we have a sequence $\langle g_\eta : \eta < \kappa\rangle$, here g_η is a function from ω to ω, and

$$\forall \eta, \zeta < \kappa(\eta \neq \zeta \to | g_\eta \cap g_\zeta | < \omega).$$

Now we proceed with a system of iterated forcing of length ω_1 with finite support as follows. At step α we assume that we have constructed a sequence $\langle g_\eta : \eta < \kappa + \alpha \rangle$ of pairwise e.d. functions in ${}^\omega\omega$. We take

$$A_\alpha = \{g_\eta \mid \eta < \kappa + \alpha\}$$

and we use at this step the forcing notion $\mathbb{P}_{A_\alpha}$. Then we can prove the theorem by standard method.

□

Corollary 1.8. $Con(ZFC) \to Con(ZFC + \mathfrak{a}_\mathfrak{e} < 2^\omega)$.

In the following we prove that $\mathfrak{a}_\mathfrak{e}$ has the similar character as $\mathfrak{a}$ in Cohen forcing, i.e., we force with a *c.c.c.* partially ordered set

$$Fn(I,2) = \{p \mid p \text{ is a finite function, } \mathrm{dom}(p) \subseteq I, \mathrm{rang}(p) \subseteq 2\}.$$

on the ground model M which satisfies $ZFC + GCH$.

We first state a well-known Lemma about Cohen forcing.

Lemma 1.9. *Suppose $I, S \in M$. Let G be $Fn(I,2)$-generic over M, and let $X \subset S$ with $X \in M[G]$. Then $X \in M[G \cap Fn(I_0,2)]$ for some $I_0 \subset I$ such that $I_0 \in M$ and $(|I_0| \leq |S|)^M$.*

Proof. See [Kun].

□

By the above Lemma, it is sufficient to construct some *m.e.d.* family $\mathcal{A} \subseteq {}^\omega\omega$ in extensions via $I_0 = \omega$ which are countable in M.

Theorem 1.9. *Let $M \models (ZFC + CH)$. There is a m.e.d. function set $\mathcal{A}$ of size ω_1 in M such that for any Cohen generic G over M, $\mathcal{A}$ remains to be a m.e.d. function set in $M[G]$.*

Proof. Since $Fn(\omega,2)$ has *c.c.c.* and $M \models CH$, there are at most $\omega^\omega = 2^\omega = \omega_1$ different antichains, there are at most $(2^\omega)^\omega = \omega_1$ nice names for reals.

In M, define a m.e.d. function set $\mathcal{A}$ of size ω_1 as follows:

Let $\langle p_\alpha, \tau_\alpha \rangle$ for $\omega \leq \alpha < \omega_1$ enumerate all pairs $\langle p, \tau \rangle$ such that $p \in Fn(\omega,2)$ and τ is a nice name for a function of ω.

By recursion, pick function $f_\alpha \in {}^\omega\omega$ as follows:

Let $\{f_n \mid n < \omega\}$ be any e.d. function set in ${}^\omega\omega$.

If $\omega \leq \alpha < \omega_1$ and we have functions f_β for $\beta < \alpha$, choose a function f_α so that

(1) $\forall \beta < \alpha(|\{n \mid f_\alpha(n) = f_\beta(n)\}| < \omega)$,

(2) if

$$p_\alpha \Vdash (\tau_\alpha \in {}^\omega\omega), \text{ and}$$

$$\forall\beta < \alpha(p_\alpha \Vdash \tau_\alpha \text{ and } f_\beta \text{ are e.d.})$$

then $p_\alpha \Vdash \tau_\alpha$ and f_α are not e.d..

To see that f_α maybe so chosen, let

$$A_\alpha = \{f_\beta \mid \beta < \alpha\}.$$

Consider the *c.c.c.* p.o.set $\mathbb{P}_{A_\alpha}$, which we defined in Definition 1.3, in M.

We consider the following dense sets:

$$D_f = \{\langle s, F\rangle \in \mathbb{P}_{A_\alpha} \mid f \in F\},$$

$$E_n = \{\langle s, F\rangle \in \mathbb{P}_{A_\alpha} \mid n \in \text{ dom}(s)\},$$

$$C_{n,q} = \{\langle s, F\rangle \in \mathbb{P}_{A_\alpha} \mid \exists x \in \omega \exists m \geq n \exists r \leq q(s(x) = m \wedge r \Vdash \tau_\alpha(\dot{x}) = \dot{m})\},$$

where $q \leq p_\alpha$.

It is easy to see that D_f and E_n are dense in $\mathbb{P}_{A_\alpha}$.

We prove that $C_{n,q}$ is dense in $\mathbb{P}_{A_\alpha}$

Given $\langle s, F\rangle \in \mathbb{P}_{A_\alpha}$, $n \in \omega$ and $q \leq p_\alpha$. By assumption of (2), since $|F| < \omega$ and $q \leq p_\alpha$,

$$q \Vdash ((\exists t \in \omega)(\forall z \geq t)(\tau_\alpha(\dot{z}) \neq f_\beta(\dot{z}), \text{ for all } \beta \in F)).$$

Then there exist $q_0 \leq q$ and $t \in \omega$ with $n \leq t \in \omega$ such that

$$q_0 \Vdash ((\forall z \geq t)(\tau_\alpha(\dot{z}) \neq f_\beta(\dot{z}), \text{ for all } \beta \in F)).$$

Therefore, there exist $r \leq q_0$ and $x \geq t$ and $x \notin \text{dom}(s)$ and $m \geq n$ such that

$$r \Vdash \tau_\alpha(\dot{x}) = \dot{m}.$$

Also, this gives $\tau_\alpha(x) \neq f_\beta(x)$, for all $\beta \in F$. Hence, if $s' = s \cup \{\langle x, m\rangle\}$, then $\langle s', F\rangle \leq \langle s, F\rangle$ and $\langle s', F\rangle \in C_{n,q}$. Thus $C_{n,q}$ is dense in $\mathbb{P}_{A_\alpha}$.

Let $D = \{D_f \mid f \in A_\alpha\} \cup \{E_n \mid n \in \omega\} \cup \{C_{n,q} \mid n \in \omega \wedge q \leq p_\alpha\}$. Then $\mid D \mid \leq \omega$. By $MA(\omega)$, there is a filter $G_\alpha \subseteq \mathbb{P}_{A_\alpha}$ such that for any $d \in D$, $G_\alpha \cap d \neq \emptyset$.

Let $f_\alpha = \bigcup\{s \mid \langle s, F\rangle \in G_\alpha\}$. Then f_α satisfy (1) and (2).

Now, let $\mathcal{A} = \{f_\alpha \mid \alpha < \omega_1^M\}$.

Let G be $\mathbb{P}$–generic over M. Suppose towards contradiction that $\mathcal{A}$ is not maximal in $M[G]$. Then there is a $\langle p_\alpha, \tau_\alpha\rangle$ such that $p_\alpha \in G$,

$$p_\alpha \Vdash (\tau_\alpha \text{ is a function of } \omega) \text{ and } p_\alpha \Vdash \forall f \in \mathcal{A}(\mid \tau_\alpha \cap \dot{f} \mid < \dot{\omega}).$$

Thus the condition of (2) holds at α and $p_\alpha \Vdash (|\ \tau_\alpha \cap \dot{f}_\alpha\ | < \dot{\omega})$, this implies

$$\exists n \in \omega (p_\alpha \Vdash (\tau_\alpha \cap \dot{f}_\alpha) \subset (\dot{n} \times \dot{n})).$$

But this contradicts that

$$p_\alpha \Vdash (\tau_\alpha \text{ and } f_\alpha \text{ are not e.d.}).$$

□

By results in [M], we can easily have the following:

Corollary 1.10. *It is consistant with ZFC that* $\mathfrak{a}_\mathfrak{e} < Cover(meager)$.

2. Adding Eventually Different Reals.

In this section we will prove that it is consistent with ZFC, that $\mathfrak{a} < \mathfrak{a}_\mathfrak{e}$. We will force ω_2 times with finite support by using the following p.o.set on a *c.t.m.* M of $ZFC + GCH$.

Definition 2.1. Let $\mathbb{P} = \{\langle s, F\rangle \mid s \in \omega^{<\omega} \text{ and } F \in [{}^\omega\omega]^{<\omega}\}$, where $\langle s_1, F_1\rangle \leq \langle s_2, F_2\rangle$ iff

$$(s_2 \subseteq s_1) \text{ and } (F_2 \subseteq F_1) \text{ and}$$
$$\forall n \in \omega \forall f \in {}^\omega\omega((n \in \text{dom}(s_1)\backslash\text{dom}(s_2)) \wedge (f \in F_2) \rightarrow (s_1(n) \neq f(n))).$$

Obviously,

Lemma 2.2. $\mathbb{P}$ *has c.c.c..*

Note that $\mathbb{P}$ had been well studied by A. Miller in [M]. Especially the Lemmas 2.3., 2.7.,2.8., 2.10., 2.11., and 2.12 had been stated and proved in [M]. The reason we state them here is just for clarity of our proof of main result in this section.

Lemma 2.3. *Let G be $\mathbb{P}$–generic over M. Then in $M[G]$, there is a $g \in {}^\omega\omega$ such that*

$$\forall f \in {}^\omega\omega \cap M(|\ f \cap g\ | < \omega).$$

Lemma 2.4. *Let $M \models (ZFC + GCH)$. Then M can be generically extended to an $M[G^\star]$ such that $M[G^\star] \models (\mathfrak{a}_\mathfrak{e} = \omega_2 = 2^\omega)$.*

Proof. We proceed with a system of iterated forcing of length ω_2 with finite support as follows:

Let $\mathbb{P}$ be the p.o. set in Definition 2.1. Define $\mathbb{P}_\alpha$ for $\alpha < \omega_2$ as follows:

(1) $\mathbb{P}_0 = \mathbb{P}^M$;
(2) $\mathbb{P}_\alpha = \bigcup_{\beta<\alpha} \mathbb{P}_\beta$, if α is a limit ordinal;
(3) $\mathbb{P}_{\alpha+1} = \mathbb{P}_\alpha * (\dot{\mathbb{P}})^{M^{\mathbb{P}_\alpha}}$, where $(\dot{\mathbb{P}})^{M^{\mathbb{P}_\alpha}}$ is a $\mathbb{P}_\alpha$ name such that $\Vdash_{\mathbb{P}_\alpha} (\dot{\mathbb{P}}^{M^{\mathbb{P}_\alpha}} = \mathbb{P})$.

It is easy to see that $M[G_{\omega_2}] \models 2^\omega = \omega_2$.

We know that $ZFC \vdash$ (there is no m.a.d. family in ${}^\omega\omega$ has cardinality ω).

We claim that there is no m.a.d. $\mathcal{A}^\star \subset {}^\omega\omega$ in $M[G_{\omega_2}]$ such that $\mathcal{A}^\star$ has cardinality ω_1.

Assume that there exists one m.a.d. $\mathcal{A}^\star \subset {}^\omega\omega$ in $M[G_{\omega_2}]$ which has cardinality ω_1. Consider the nice name of $g \in \mathcal{A}^\star$. Since for each $n \in \omega$, there is a maximal antichain B_n^g which decides $\dot{g}(n)$, where $\dot{g}$ is a name of g. $\mathbb{P}_{\omega_2}$ is *c.c.c.*, then $\mid B_n^g \mid = \omega$. Let

$$B^\star = \bigcup_{g \in \mathcal{A}^\star} (\bigcup_{n \in \omega} B_n^g).$$

Obviously $\mid B^\star \mid \leq \omega_1$. For each $p \in B^\star$, supt(p) is a finite subset of ω_2. Hence there is an $\alpha < \omega_2$ such that

$$\forall p \in B^\star(\text{supt}(p) \subset \alpha).$$

If G_α is the component of G_{ω_2} in the iterated forcing up to (not including) α, then we have $\mathcal{A}^\star \in M[G_\alpha]$. But this implies that $\mathcal{A}^\star$ is not maximal by our construction of forcing.

Therefore we proved that $M[G_{\omega_2}] \models \mathfrak{a}_\mathfrak{e} = \omega_2 = 2^\omega$.

□

The idea of proving the following Lemma 2.5. was given by A. Miller in [M], the proof we give here is just a clearer version of the proof in [M].

Lemma 2.5. *Suppose* $\Vdash_\mathbb{P} \tau \in \omega$, $s \in \omega^{<\omega}$, *and* $n < \omega$. *Then there exists a* $H \in [\omega]^{<\omega}$ *such that for any* $F \in [{}^\omega\omega]^{<\omega}$, *if* $\mid F \mid \leq n$, *then* $\exists p \leq \langle s, F\rangle (p \Vdash_\mathbb{P} \tau \in H)$.

Proof. For any $H \in [\omega]^{<\omega}$, let

$$G_H = \{\langle g_0, g_1, ..., g_{n-1}\rangle \in ({}^\omega\omega)^n \mid \exists p \leq \langle s, \{g_0, ..., g_{n-1}\}\rangle (p \Vdash \tau \in H)\}.$$

$\{G_H \mid H \in [\omega]^{<\omega}\}$ is a cover of $({}^\omega\omega)^n$.

Give ω the cofinite topology and give ${}^\omega\omega$ and $({}^\omega\omega)^n$ the product topology. Then by Tychonoff's theorem, $({}^\omega\omega)^n$ is compact.

Now we will prove that G_H is open. For any $\langle g_0, ..., g_{n-1}\rangle \in G_H$, if $p = \langle t, F\rangle \leq \langle s, \{g_0, ..., g_{n-1}\}\rangle$ and $p \Vdash \tau \in H$, then let

$$U_H = \{\langle f_0, ..., f_{n-1}\rangle \mid \forall m \in \omega\ (m \in \text{dom}(t) \setminus \text{dom}(s) \to \forall k < n(t(m) \neq f_k(m))\}.$$

By Definition 2.1, we know that $\langle g_0, ..., g_{n-1}\rangle \in U_H$. Also, $U_H \subseteq G_H$, since for any $\langle f_0, ..., f_{n-1}\rangle \in U_H$,

$$p' = \langle t, F \cup \{f_0, ..., f_{n-1}\}\rangle \leq \langle s, \{g_0, ..., g_{n-1}\}\rangle,$$

by $p \Vdash \tau \in H$, and $p' \leq p$, we know that $p' \Vdash \tau \in H$. And U_H is open in $({}^\omega\omega)^n$. Hence G_H is open.

Since $G_H \cup G_K \subseteq G_{H\cup K}$, therefore, there exists $H \in [\omega]^{<\omega}$, such that

$$G_H = ({}^\omega\omega)^n.$$

We get the conclusion.

□

We do not actually need the following Lemma 2.6 to prove the main results in this section. For clarity, this is just a warm-up exercise for Lemma 2.9.

Lemma 2.6. *Let $M \models (ZFC+GCH)$. Let G be $\mathbb{P}$–generic over M. Then, in M, we can construct a m.a.d. family $\mathcal{F} \subseteq \wp(\omega)$ of size ω_1, and $\mathcal{F}$ remains to be maximal in $M[G]$.*

Proof. Within M, we shall define a m.a.d. family $\mathcal{F}$ of size ω_1 such that $\mathcal{F}$ remains to be maximal in $M[G]$ when G is $\mathbb{P}$–generic over M. From now on, all forcing terminology refers to the p.o. set $\mathbb{P}$.

Within M, do the following. By CH, let

$$E = \{\langle p_\xi, \tau_\xi\rangle \mid \omega \leq \xi < \omega_1\}$$

enumerate all pairs $\langle p, \tau\rangle$ such that $p \in \mathbb{P}$ and τ is a nice name for subset of $\dot{\omega}$. By recursion, pick infinite $F_\xi \subset \omega$ as follows.

Let $\{F_n \mid n < \omega\}$ be any almost disjoint sets in $\wp(\omega)$. If $\omega \leq \xi < \omega_1$, and we have F_η for $\eta < \xi$, choose F_ξ so that

(1) $\forall \eta < \xi(|F_\eta \cap F_\xi| < \omega)$, and
(2) if

$$p_\xi \Vdash (|\tau_\xi| = \dot{\omega}) \text{ and } \forall \eta < \xi(p_\xi \Vdash (|\tau_\xi \cap F_\eta| < \dot{\omega})),$$

then

$$\forall n \in \omega \forall q \leq p_\xi \exists r \leq q \exists m \geq n((m \in F_\xi) \wedge (r \Vdash \dot{m} \in \tau_\xi)).$$

To see that F_ξ maybe so chosen, assume that the condition of (2) holds, since if it fails then only (1) need to be considered, and we can simply apply the fact that there is no m.a.d. family of size ω (see [Kun]).

Let E_n, for $n \in \omega$, re-enumerate $\{F_\eta \mid \eta < \xi\}$.

We shall build an F_ξ such that $\forall n \in \omega(|F_\xi \cap E_n| < \omega)$ and $p_\xi \Vdash (|\tau_\xi \cap F_\xi| = \dot{\omega})$.

We shall build F_ξ in stage as follows:

$$F_\xi = \bigcup_{n<\omega} K_n,$$

where $K_n \in \wp(\omega)$ and K_n is finite. When we choose each K_i, for $i \in \omega$, we shall make sure that $\forall m \geq n(K_m \cap E_n = \emptyset)$, therefore for any n, $|F_\xi \cap E_n| < \omega$. Also, we shall arrange that

$$\forall n \in \omega \forall p \leq p_\xi \exists q \leq p \exists m \in F_\xi (m \geq n \text{ and } q \Vdash m \in \tau_\xi),$$

i.e., $p_\xi \Vdash (|F_\xi \cap \tau_\xi| = \dot{\omega})$.

We list all $\langle s_n, l_n, m_n \rangle \in \omega^{<\omega} \times \omega \times \omega$ such that $s_{p_\xi} \subseteq s_n$. To set K_n we consider $\langle s_n, l_n, m_n \rangle$ in our list.

Using Lemma 2.5, we can find a finite $H_n \in [\omega]^{<\omega}$ such that for any $F \in [({}^\omega\omega)]^{<\omega}$, if $|F| \leq l_n$, then

$$\exists r_0 \leq \langle s_n, F \rangle (r_0 \Vdash \tau^n \in H_n).$$

here, $\tau^n = \min\{\tau_\xi \setminus \max((\tau_\xi \cap \bigcup_{i<n} E_i) \cup (\dot{m}_n + 1) \cup (\bigcup_{i<n} K_i))\}$. We can define that

$$K_n = H_n.$$

And we set

$$F_\xi = \bigcup_{n<\omega} K_n.$$

By construction, for any $n < \omega$, $F_\xi \cap E_n \subseteq \bigcup_{i<n} K_i$, i.e., $| F_\xi \cap E_n | < \omega$.

We show that $\forall p \leq p_\xi (p \Vdash F_\xi \cap \tau_\xi$ is infinite). Suppose otherwise, there exists some $p \leq p_\xi$ and some $m < \omega$ such that there is no $r \leq p$ such that for any $t \in \omega$, if $m < t \in F_\xi$,

$$r \Vdash t \in \tau_\xi,$$

i.e., $p \Vdash F_\xi \cap \tau_\xi \subseteq m + 1$. Let $p = \langle s_n, F \rangle$, where $| F | \leq l_n$ and $m_n = m$. By Lemma 2.5, there exists some $r \leq p$ such that there exists $t \in K_n$, and $t > m_n$, $r \Vdash (t \in \tau_\xi)$. We get contradiction.

Now let $\mathcal{F} = \{F_\xi \mid \xi < \omega_1^M\}$.

We claim that $\mathcal{F}$ is a m.a.d. family in $M[G]$.

Let G be $\mathbb{P}$–generic over M. If $\mathcal{F}$ failed to be maximal in $M[G]$, there would be a $\langle p_\xi, \tau_\xi\rangle$ such that $p_\xi \in G$,

$$p_\xi \Vdash |\tau_\xi| = \dot{\omega} \;\text{ and }\; p_\xi \Vdash \forall x \in \mathcal{F}(|\tau_\xi \cap x| < \dot{\omega}).$$

Thus, the condition of (2) holds at ξ, but also $p_\xi \Vdash (|\tau_\xi \cap F_\xi| < \dot{\omega})$. So there is a $q \leq p_\xi$, and there is a $m \in \omega$ such that

$$q \Vdash (\tau_\xi \cap F_\xi \subset \dot{m} + 1).$$

This implies that

$$\neg \exists r \leq q \forall t > m(t \in F_\xi \rightarrow r \Vdash \dot{t} \in \tau_\xi).$$

But this is a contradiction. Therefore $\mathcal{F}$ is a m.a.d. family in $M[G]$.

□

Lemma 2.7. *$\forall p \in \mathbb{P}_{\omega_2} \exists q \leq p \forall \alpha \in supt(q) \exists s_\alpha \in \omega^{<\omega}$ injective $\exists n_\alpha < \omega$*

$$\Vdash_{\mathbb{P}_\alpha} (q(\alpha) = \langle s_\alpha,\ G\rangle \textit{ for some } G \in [{}^\omega\omega]^{n_\alpha}).$$

Call such q canonical, from now on, we assume that, for any $p \in \mathbb{P}_{\omega_2}$, p is canonical, since the canonical ones are dense.

Lemma 2.8. *Suppose $\Vdash \tau \in \omega$, and given $F \in [\omega_2]^{<\omega}$ and $n_\alpha < \omega$ and $s_\alpha \in \omega^{<\omega}$ injective for every $\alpha \in F$, then there exists a $H \in [\omega]^{<\omega}$ such that*

$$\forall q \in \mathbb{P}_{\omega_2}(supt(q) = F \wedge \forall \alpha \in F(n^q_\alpha = n_\alpha \wedge s^q_\alpha = s_\alpha) \rightarrow \exists p \leq q(p \Vdash \tau \in H)).$$

Using Lemmas 2.7. and 2.8., we can prove our crucial Lemma:

Lemma 2.9. *Let $M \models (ZFC + GCH)$. There is a m.a.d. family $\mathcal{F}$ in M such that $\forall \alpha < \omega_1$ if G_α is $\mathbb{P}_\alpha$–generic over M, $\mathcal{F}$ remains to be maximal in $M[G_\alpha]$.*

Proof. Within M, we shall define a m.a.d. family.

By CH, let

$$E = \{\langle p_\xi, \tau_\xi, \xi\rangle \mid \omega \leq \xi < \omega_1\}$$

enumerate all triples $\langle p, \tau, \alpha\rangle$ such that $p \in \mathbb{P}_{\omega_1}$ and τ is a $\mathbb{P}_{\omega_1}$ –nice name for subset of $\dot{\omega}$. Without lost of generality, we may assume that in our list, for each $\xi < \omega_1$, $\sup(\text{supt}(p_\xi) \leq \xi)$ and τ_ξ is a $\mathbb{P}_\xi$-name. By recursion, pick infinite $F_\xi \subset \omega$ as follows:

Let $\{F_n \mid n < \omega\}$ be any almost disjoint set in $\wp(\omega)$.

If $\omega \leq \xi < \omega_1$, and we have F_η for $\eta < \xi$, choose F_ξ so that

(1) $\forall \eta < \xi(| F_\eta \cap F_\xi |< \omega)$, and

(2) if

$$p_\xi \Vdash_{\mathbb{P}_\xi} (|\ \tau_\xi\ | = \dot{\omega}) \text{ and } \forall\eta < \xi(p_\xi \Vdash_{\mathbb{P}_\xi} |\ \tau_\xi \cap F_\eta\ | < \dot{\omega}),$$

then for any $n \in \omega$ and for any $q \leq p_\xi$ with $q \in \mathbb{P}_\xi$ there exists some $r \leq q$ with $r \in \mathbb{P}_\xi$ and there exists some $m \geq n$ such that

$$(m \in F_\xi) \wedge (r \Vdash_{\mathbb{P}_\xi} \dot{m} \in \tau_\xi).$$

To see that F_ξ maybe so chosen, assume that the condition of (2) holds, since if it fails then only (1) need to be considered, and we can simply apply the fact that there is no m.a.d. family of size ω.

Let E_n, for $n < \omega$, re-enumerate $\{F_\eta \mid \eta < \xi\}$.

We shall build an F_ξ such that $\forall n \in \omega(|F_\xi \cap E_n| < \omega)$ and $p_\xi \Vdash_{\mathbb{P}_\xi} (|\tau_\xi \cap F_\xi| = \dot{\omega})$.

We shall build F_ξ in stage as follows:

$$F_\xi = \bigcup_{n<\omega} K_n$$

where $K_n \in \wp(\omega)$ and K_n is finite. When we choose each K_i, for $i \in \omega$, we shall make sure that, for any $p \in \mathbb{P}_\xi$ with $p \leq p_\xi$, there exists a $q \in \mathbb{P}_\xi$ with $q \leq p$, such that

$$\exists m \in F_\xi(m \geq n \text{ and } q \Vdash_{\mathbb{P}_\xi} m \in \tau_\xi),$$

i.e., $p_\xi \Vdash_{\mathbb{P}_\xi} (|\ F_\xi \cap \tau_\xi\ | = \dot{\omega})$.

We may assume $\text{supt}(p_\xi) = \{\alpha_i^\xi \in \omega_1 \mid 1 \leq i < m_\xi, \text{ for some } m_\xi\}$.

Let $\bar{s} = \langle s^{\alpha_1}, ..., s^{\alpha_k}\rangle$ and $\bar{l} = \langle l^{\alpha_1}, ..., l^{\alpha_k}\rangle$ with

$$\{\alpha_1^\xi, ..., \alpha_{m_\xi}^\xi\} \subseteq \{\alpha_1, ..., \alpha_k\},$$

here all $\alpha_i \leq \xi$.

We list all

$$\langle \bar{\alpha}_n, \bar{s}_n, \bar{l}_n, m_n\rangle \in \xi^{<\omega} \times (\omega^{<\omega})^k \times \omega^k \times \omega,$$

with infinite many repetitions, such that

$$\{\alpha_1^\xi, ..., \alpha_{m_\xi}^\xi\} \subseteq \bar{\alpha}_n, \text{ and}$$

$$\forall i \leq k(\alpha_i \leq \xi),$$

$$|\text{supt}(p_\xi)| \leq k < \omega,$$

$$\forall n \in \omega \forall \alpha_i^\xi \in \operatorname{supt}(p_\xi)(s_{p_\xi(\alpha_i^\xi)} \subseteq s_n^{\alpha_j}, \text{ if } \alpha_i^\xi = \alpha_j \text{ for some } 1 \leq j \leq k).$$

Let $\tau^n = \min\{\tau_\xi \setminus \max\{\tau_\xi \cap \bigcup_{i<n} E_i, \dot{m}_n, \bigcup_{i<n} K_i\}\}$. To set K_n, we consider $\langle s_n^{\alpha_1}, ..., s_n^{\alpha_k}, l_n^{\alpha_1}, ..., l_n^{\alpha_k}, m_n\rangle$. Pick $q \leq p_\xi$ with $\operatorname{supt}(q) = \{\alpha_1, ..., \alpha_k\}$ such that

$$\Vdash (q(\alpha_i) = \langle s_n^{\alpha_i}, F\rangle \text{ for some } F \in [{}^\omega\omega]^{l_n^{\alpha_i}} \text{ where } 1 \leq i \leq k).$$

Then by Lemma 2.8, there exists a $H_n \in [\omega]^{<\omega}$, such that

$$\exists p \leq q(p \in \mathbb{P}_\xi \wedge p \Vdash_{\mathbb{P}_\xi} \tau^n \in H_n).$$

We can define that

$$K_n = H_n.$$

And we set

$$F_\xi = \bigcup_{n<\omega} K_n.$$

By construction, for any $n < \omega$, $F_\xi \cap E_n \subseteq \bigcup_{i<n} K_i$, i.e., $|F_\xi \cap E_n| < \omega$.

We show that $\forall p \leq p_\xi (p \Vdash_{\mathbb{P}_\xi} F_\xi \cap \tau_\xi$ is infinite). Suppose otherwise, there exist some $p \leq p_\xi$ and some $m < \omega$ such that there is no $r \leq p$ such that for any t, if $m < t \in F_\xi$, then

$$r \Vdash_{\mathbb{P}_\xi} t \in \tau_\xi,$$

i.e., $p \Vdash_{\mathbb{P}_\xi} F_\xi \cap \tau_\xi \subseteq m+1$.

We may assume that

$$\operatorname{supt}(p) = \{\alpha_i \mid 1 \leq i \leq k\}, \quad \text{and}$$
$$\Vdash (p(\alpha_i) = \langle s_n^{\alpha_i}, F\rangle \text{ for some } F \in [{}^\omega\omega]^{l_n^{\alpha_i}}), \quad \text{and}$$
$$m = m_n.$$

Then there exists some $r \leq p$ such that there exists $t > m_n$ and $t \in K_n$ and $r \Vdash_{\mathbb{P}_\xi} (t \in \tau_\xi)$. We get contradiction.

Now let $\mathcal{F} = \{F_\xi \mid \xi < \omega_1^M\}$.

We claim that $\mathcal{F}$ is an m.a.d. family in $M[G_\alpha]$ for any $\alpha < \omega_1$.

We assume that there exists some $\alpha < \omega_1$ such that G_α is $\mathbb{P}_\alpha$-generic over M, and $\mathcal{F}$ failed to be maximal in $M[G_\alpha]$. Then there would be a $\langle p_\xi, \tau_\xi, \xi\rangle$ such that

$$p_\xi \Vdash_{\mathbb{P}_\alpha} |\tau_\xi| = \dot{\omega}, \text{ and}$$
$$\forall \eta < \xi (p_\xi \Vdash_{\mathbb{P}_\alpha} |\tau_\xi \cap F_\eta| < \dot{\omega}).$$

Thus the condition of (2) holds at ξ, but also

$$p_\xi \Vdash_{\mathbb{P}_\alpha} |\tau_\xi \cap F_\xi| < \dot{\omega}.$$

This is a contradition. Therefore, $\mathcal{F}$ is a m.a.d. family in $M[G_\alpha]$.

□

Lemma 2.10. *Suppose $M \subseteq N$ are transitive models of ZFC. Then if G is $\mathbb{P}^N$–generic over N then $G \cap \mathbb{P}^M$ is $\mathbb{P}^M$–generic over M.*

Proof. We claim that for any $A \subseteq \mathbb{P}^M$, if A is a maximal antichain in $\mathbb{P}^M$, then A is an maximal antichain in $\mathbb{P}^N$. This is because of that being maximal antichain of a *c.c.c.* p.o. set is a Π^1_1 property, therefore, it is absolute.

If G is $\mathbb{P}^N$–generic over N, then for any maximal antichain A in $\mathbb{P}^M$ in M, A is a maximal antichain in $\mathbb{P}^N$ in N, therefore $G \cap A \neq \emptyset$ in N, we have $(G \cap \mathbb{P}^M) \cap A \neq \emptyset$ in M. Hence $G \cap \mathbb{P}^M$ is $\mathbb{P}^M$–generic over M.

□

Call a p.o. $\mathbb{P}$ absolute just in case it is definable (possibly with parameters in M) and given any $N \supseteq M$, a transitive model of ZFC if G is $\mathbb{P}^N$–generic over N, then $G \cap \mathbb{P}^M$ is $\mathbb{P}^M$–generic over M. Suppose $\mathbb{P}_\alpha$ for $\alpha \leq \beta$ is a finite support iteration of sbsolute partial order over M, i.e. $\mathbb{P}_{\alpha+1} = \mathbb{P}_\alpha * \mathbb{P}^\alpha$ where $\mathbb{P}^\alpha$ is some name for an absolute partial order in $M[G_\alpha]$. Given $X \subseteq \beta (X \in M)$ define the iteration $P^*_\alpha(X)$ for $\alpha \leq \beta$ as follows:

For $\alpha \notin X$, let $\mathbb{P}^*_{\alpha+1} = \mathbb{P}^*_\alpha * 1$ (where 1 is the one element order).

For $\alpha \in X$, let $\mathbb{P}^*_{\alpha+1} = \mathbb{P}^*_\alpha * (\mathbb{P}^\alpha)^{M[H^*_\alpha]}$ where H^*_α is $\mathbb{P}^*_\alpha$–generic over M.

For G_α $\mathbb{P}_\alpha$–generic over M define $G^*_\alpha = G_\alpha \cap \mathbb{P}^*_\alpha$.

Lemma 2.11. *For any $\alpha \leq \beta$, if G_α is $\mathbb{P}_\alpha$–generic over M, then G^*_α is $\mathbb{P}^*_\alpha$–generic over M.*

Lemma 2.12. *If $\Vdash \tau \subseteq \omega$, then there exists a $X \subseteq \omega_2$ countable, $X \in M$ and for any G $\mathbb{P}_{\omega_2}$–generic over M, $\tau_G \in M[G^*]$.*

Thus we conclude that

$$M[G_{\omega_2}] \cap \wp(\omega) = \bigcup \{M[H_\alpha] \cap \wp(\omega) \mid \alpha < \omega_1, H_\alpha \in M[G_{\omega_2}], \text{ and } H_\alpha \text{ is } \mathbb{P}_\alpha\text{–generic over } M\}.$$

Therefore we proved that

Theorem 2.13. *It is consistent with ZFC that $\mathfrak{a}_\mathfrak{e} = \omega_2 = 2^\omega$ and $\mathfrak{a} = \omega_1$.*

Now we try to compare $\mathfrak{a}$ and $\mathfrak{a}_\mathfrak{e}$ with some other wellknown continuum invariants in our forcing model.

Definition 2.14. Let f, g be functions in ${}^\omega\omega$, $f \leq^* g$ iff $f(n) \leq g(n)$ for all but finitely many $n \in \omega$. A subset of ${}^\omega\omega$ is called *unbounded* iff it is unbounded in $\langle {}^\omega\omega, \leq^* \rangle$. We define $\mathfrak{b}$ to be the least λ such that there exists an unbounded subset of ${}^\omega\omega$ which has size λ.

It is well-known that $\mathfrak{b} \leq \mathfrak{a}$. Therefore, in our forcing model $\mathfrak{b} = \omega_1$.

It is easy to prove the following:

Remark (ZFC)**.** $\mathfrak{b} \leq \mathfrak{a}_{\mathfrak{e}}$

Proof. Assume that $\mathfrak{a}_{\mathfrak{e}} < \mathfrak{b}$. Then there exists a m.e.d. set $\mathcal{A} \subseteq {}^{\omega}\omega$ such that $\mid \mathcal{A} \mid = \kappa < \mathbf{b}$. Let

$$\mathcal{A} = \{f_\alpha \in {}^{\omega}\omega \mid \alpha < \kappa\}.$$

Then $\mathcal{A}$ is domited, i.e., there is a function $g \in {}^{\omega}\omega$, for any $f \in \mathcal{A}$, $f <^* g$. Therefore,

$$\forall f \in \mathcal{A} \exists N_f \in \omega \forall n > N_f (f(n) \neq g(n)).$$

Thus, $\mathcal{A}$ is not a m.e.d. set of functions. We get contradiction.

□

Definition 2.15. Let $\mathfrak{a}_{\mathfrak{s}}$ be the least λ such that there exists a maiximal family of almost disjoint subsets of $\omega \times \omega$ that are graphs of partial functions from ω to ω.

It is well-known that $\mathfrak{a} \leq \mathfrak{a}_{\mathfrak{s}}$. Shelah provided a model in which $\mathfrak{a} < \mathfrak{a}_{\mathfrak{s}}$ (see [S:207]). We can easily prove that in our forcing model, $\mathfrak{a}_{\mathfrak{s}} = \omega_2$.

Theorem 2.16. $M[G_{\omega_2}] \models \mathfrak{a}_{\mathfrak{s}} = \omega_2$.

Proof. Assume that in $M[G_{\omega_2}]$, there exists a maximal almost disjoint family $\mathcal{A}_s$ of partial functions from ω to ω, $\mid \mathcal{A}_s \mid = \omega_1$. For any $f_\alpha \in \mathcal{A}_s$, here $\alpha < \omega_1$, we can extend f_α to a function $g_\alpha \in {}^{\omega}\omega$, such that $g_\alpha(n) = f_\alpha(n)$, if $n \in \mathrm{dom}(f_\alpha)$; $g_\alpha(n) = 0$, otherwise. Then we get

$$\bar{\mathcal{A}}_s = \{g_\alpha \in {}^{\omega}\omega \mid \alpha < \omega_1\}.$$

Let $g_\alpha \in \bar{\mathcal{A}}$ in $M[G_{\omega_2}]$. We know that there exists an $\alpha < \omega_2$, such that $\bar{\mathcal{A}} \in M[G_\alpha]$. Then for any β such that $\alpha < \beta < \omega_2$, the $\mathbb{P}_\beta$–generic function g_β is e.d. from any function in $\bar{\mathcal{A}}$. Hence, in $M[G_{\omega_2}]$, we can find some fuction g_β which is e.d. from every fuction in $\mathcal{A}$. This is contradicting that $\mathcal{A}_s$ is maximal in $M[G_{\omega_2}]$. Therefore, we proved the theorem.

□

Thus we conclude that it is consistent with ZFC that $\mathfrak{a} = \mathfrak{b} = \omega_1$ and $\mathfrak{a}_{\mathfrak{e}} = \mathfrak{a}_{\mathfrak{s}} = \omega_2 = 2^\omega$.

3. Questions.

We define $\mathfrak{a}_{\mathfrak{p}}$ as the following:

Definition 3.1. Two permutations $f, g \in Sym(\omega)$ are a.d. iff

$$|\{n \in \omega \mid f(n) = g(n)\}| < \omega.$$

Let $\mathfrak{a}_\mathfrak{p}$ be the least λ such that there exists a m.a.d. set of permutations of cardinality λ.

We can prove that MA implies that $\mathfrak{a}_\mathfrak{p} = 2^\omega$. It is also not hard to prove that

$$Con(ZFC) \rightarrow Con(ZFC + \mathfrak{a}_\mathfrak{p} < 2^\omega)$$

and in Cohen foring $\mathfrak{a}_\mathfrak{p}$ stays small. Using a similar (but much harder, since the space $Sym(\omega)$ is not compact, the method of proving Lemma 2.5 does not work) argument, we can prove that

Theorem 3.2. *It is consistent with ZFC that $\mathfrak{a} < \mathfrak{a}_\mathfrak{p}$.*

Proof. Force with the following *c.c.c.* p.o.set in Definition 3.3. For a detailed proof, see [Z].

□

Definition 3.3. Let

$$\mathbb{P} = \{\langle s, F\rangle \mid s \in \omega^{<\omega} \text{ is injective and } F \in [\mathrm{Sym}(\omega)]^{<\omega}\},$$

where $\langle s_1, F_1\rangle \leq \langle s_2, F_2\rangle$ iff

$$(s_2 \subseteq s_1) \text{ and } (F_2 \subseteq F_1) \text{ and}$$

$$\forall n \in \omega \forall f \in \mathrm{Sym}(\omega)((n \in \mathrm{dom}(s_1) \setminus \mathrm{dom}(s_2)) \wedge (f \in F_2) \rightarrow (s_1(n) \neq f(n))).$$

Interesting questions would be what the relationships are among $\mathfrak{a}_\mathfrak{e}$ and $\mathfrak{a}_\mathfrak{p}$ and $\mathfrak{a}_\mathfrak{s}$. It is also interesting to know that if any of $\mathfrak{a}_\mathfrak{e}$ and $\mathfrak{a}_\mathfrak{p}$ and $\mathfrak{a}_\mathfrak{s}$ has an upper-bound, which is a cardinal invariants of continuum.

Acknowledgments.

I would like to thank:

Professor Boban Veličković, for initially suggesting me to investigate $\mathfrak{a}_\mathfrak{e}$, also for his hospitality during the spring and summer, 1995 when I was visiting Paris University 7, Paris, France;

Professor Renling Jin, for various discussions I had with him about this paper;

my friends Luca Bonatti, Martine Gueguen, Aitze Peng and Ming Xu, for all kinds of helps they offered me;

last but most, Professor Simon Thomas, for his consistant and patient helps.

REFERENCES

[vD] E. van Douwen, *The integers and topology*, in Handbook of Set Theoretic Topology (ed. K.Kunen and J. Vaughan), 1984, pp. 111–167, North-Holland, Amsterdam.

[Jech] T. Jech, *Set Theory*, Academic Press.

[Kun] K. Kunen, *Set Theory. An Introduction to Independence Proofs*, North Holland, Amsterdam.

[M] A. Miller, *Some properties of measure and category*, Transactions of the American Mathematical Society **266, Number, 1** (July 1981), 93–114.

[Sh:P] S. Shelah, *Proper Forcing, Lecture Notes in Mathematics 940*, Springer Verlag.

[Sh:207] S. Shelah, *On cardinal invariants of the continuum.*, in Axiomatic Set Theory (ed. J. E. Baumgartner, D. A. Martin, and S. Shelah), Contemporary Mathematics 31., 1984, pp. 183–207, American Mathematical Society, Providence.

[V] J. E. Vaughan, *Small uncountable cardinals and topology*, Open Problems in Topology (ed. J. van Mill and G. M. Reed), 1990, pp. 197–218, North-Holland, Amsterdam.

[Z] Y.Zhang, *Cofinitary Groups and Almost Disjoint Families.*

Asian Logic Conference:ALC

Mariko Yasugi *

The Mathematical Society of Japan (MSJ) is playing a more prominent role in international activities in Asia. I thus feel that it is a good time to report on The Asian Logic Conference (ALC) , whose aims and objectives are along the same line as those of the MSJ. Chong (Singapore), Tamthai (Thailand), Yang (China), Motohashi (Tsukuba:Japan) have provided the author with valuable materials in preparing for this article. These will also be kept for future reference.

1 The Beginning

In the year 1980, Crossley (Australia), Chong, Müller (Germany) and Motohashi were participating in Logic Colloquium 80 in Greece. They were concerned with the progress of logic in Asia, and thought that an international conference held in that region would be useful. Later that year Chong was spending his sabbatical leave at MIT, and had discussions with Anil Nerode who was also visiting MIT then. Nerode suggested, on the occasion of a visit by Sacks to Singapore scheduled for the following year, that a workshop be organized in Singapore. Chong approached various organizations for possible support. Several responded, offering cooperation and financial aid. The Department of Mathematics at the National University of Singapore also agreed to host the conference. The plan then developed into an international meeting. Less than one year had lapsed before the vision of the four logicians who met in Greece became a reality. The communication within Japan was handled by Motohashi.

In the beginning, the name of the conference was a temporary one, and it was not certain if there would be a second meeting. In fact it was succeeded by a conference in Thailand three years later, and one in China three years after that.

The ALC has since evolved into the major meeting ground for specialists in logic and the foundations of mathematics in Asia. This was not anticipated by those such as Chong and others who organized the first meeting in 1981.

* Translated from the original Japanese article in "Sugaku," edited by the Mathematical Society of Japan, vol 46, No.1(1994), 57-62.

2 The Outline of the Conference Series

ALC aims at promoting activities of mathematical logic in the Asia-Pacific, as mentioned above, so that logicians both from within Aisia and elsewhere would get togeother and exchange information and ideas. Its purpose is therefore to establish a system of mutual acquaintance and collaboration among Asian logicians. There is no formal superorganization; the detailed plans of activities are discussed each time during one of the meetings, or through e-mail. We are still groping. The tacit assumption is that ALC be a conference on mathematical logic 'by the Asians and for the Asians,' as Chong put it at the Beijing meeting. The host country would be in Asia, and the organizing commitee members would be mostly from the host country. Active logicians from the region are invited to give talks, and special encouragement, in the form of financial assistance if necessary, is given to young logicians in the region. These are unwritten agreements. As Crossley emphasized at the beginning of his lecture in Beijing (1987), it is no less important for the participants to be heard than to listen to lectures. The contributed papers are accepted without filtering for this reason. Papers are refereed only during the publication process of the Proceedings. ALC accepts papers from all areas of mathematical logic. While mathematical logic originally grew from the study of the foundations of mathematics, an essentially intellectual pursuit, it has, however, developed and expanded into other areas and now bears close relationship with computer science (through computability theory), and so there are many conferences on the border areas of the two subjects. ALC has that tendancy too, but there are different opinions on how far we should be involved in computer science.

What follows is an up-to-date report on the ALC. I will include the following items.

1 Name 2 Time and Place 3 Organization 4 Cooperating bodies 5 Finantial Aid 6 Invited Speakers 7 Number of Contributed Papers 8 Number of Participants 9 Publication 10 Prominent features 11 Miscellaneous

3 Record of Conference: from The First through The Third

The First Meeting 1 The First Southeast Asian Conference in Mathematical Logic

2 November 9-13, 1981 National University of Singapore (NUS), Singapore

3 Organizing Committee C. T. Chong, H. H. Teh, M. J. Wicks (NUS), J. N. Crossley (Monash)

4/5 Singapore Mathematical Society, National University of Singapore, Department of Mathematics (NUS), Association for Symbolic Logic (ASL), International Union of History and Philosophy of Science, Division for Logic, Methodology and Philosophy of Science (IUHPS)

6 C. Ash, J. N. Crossley (Monash), M. Lerman (U. Conn), A. McIntyre (Yale), A. Nerode (Cornell), G. E. Sacks (Harvard-MIT), M. Tamthai (Chulalongkorn)

7 9 (30 minutes each)

8 35 (From Japan: Susumu Hayashi, Toru Kawai, Nobuyoshi Motohashi, Koji Nakatogawa, Kanji Namba, Makoto Takahashi)

9 Proceedings, Southeast Asian Conference on Logic, Studies in Logic and the Foundations of Mathematics, vol. 111, editied by C. T. Chong and M. J. Wicks, 1983, North-Holland

10 The themes of the tutorial sessions were recursion theory and model theory. Many of the papers were in these subjects.

11 In 1981, formal letters from Chong were sent to Motohashi, Crossley, Müller and B.F.Nebres (the Phillipines). The letter said: 'I consulted people at MIT and NUS acording to our conversation in Greece. I have got feeling that we had better start with a small scale meeting. I will set the the dates in accordance with the visit of G.Sacks.' This was the start of ALC. Chong later recalls: 'It grew bigger than I had thought.' Even then, I hear, it was a more relaxed gathering than recent conferences. One of the participants, Hayashi, who had just got his degree, tells me a story which proves it. One of the speakers switched from English to Chinese in the course of his talk, and Chong interpreted the lecture into English for him. At the end there was a big applause. We would like to keep up such an atmosphere.

<u>The Second Meeting</u> 1 The Second Southeast Asian Logic Conference

2 October 29 - November 2, 1984 Bangkok, Thailand

3 Chair of the Organizing Committee Mark Tamthai (Chulalongkorn)

4 ASL

5 IUHPS, IBM (Thailand), Southeast Asian Mathematical Society, Mathematical Association of Thailand, Chulalongkorn University

6 C. T. Chong, J. N. Crossley, J. L. Loyd, A. Nerode, R. Soare, G.Takeuti

7 13

9 Annals of Pure and Applied Logic (APAL): Special Issue, vol. 31, 1986, Guest Editor: M. Tamthai, North-Holland. The proceedings of the conference series has continued to be in this journal up till 1993. This is a journal of high quality in mathematical logic, and the standard of review is strict. Reverse mathematics, hierarchy of sets, recursion theory, logic programming, extensions of Peano arithmetic are covered. The authors are from Singapore, Japan, USA, Britain, Canada, Australia and New Zealand.

10 The theme of the conference was logic as well as its applications to mathematics and computer science. The themes of the tutorial session consisted of extensions of ZFC set theory and the programming language of the Fifth Generation Computer.

11 According to the letter by Tamthai, until then the conference had been organized in Southeast Asia, and they were not sure if anybody in Asia outside this district would be interested in organizing such a meeting. So the conference was named after Southeast Asia.

The Third Meeting 1 The Third Asian Conference in Mathematical Logic

2 October 26 - 30, 1987 Beijing, China

3 Sponsor Institute of Software, Academia Sinica (Director: K. Xu)

Organizing Committee K.Xu, D. Yang (Secretary) (Beijing), C. T. Chong, J. N. Crossley

4 ASL

5 IUHPS, International Centre for Theoretical Physics, Nature Science Foundations Academia Sinica, Institute of Software Academia Sinica, National Science Foundation of China

6 R. I. Soare, R. Solovay, A. Nerode, S. G. Simpson, M. B. Pour-El, (all USA), C. T. Chong, G. Müller, M. Yasugi (Japan)

7 72 Due to a large number of participants, there were many contributed papers and the meeting was a lively one. Although the majority of the participants was from the host country, there were a fair number of Japanese and Euro-Americans.

9 APAL, vol.44,1989. Guest Editor:D.Yang

The contents cover most subjects of mathematical logic. I was a bit concerned that the authors were from outside Asia except by Chong and two from Japan.

10 Thanks to the large number of participants, talks in various areas were presented.

11 I attended the meeting for the first time. The conference was much larger than the preceding two. This fact made us realize the size of China. I heard that Yang had worked for preparation almost all by himself. During the conference, his students helped him a lot. The site of the conference as well as the lodging of the overseas participants was a hotel near Beijing University. It was clear that a lot of hard work had been put into organizing the conference.

4 Towards the Session in Japan

Just prior to the Beijing Conference, a new committee called Committee on Logic in Japan and East Asia was formed by ASL. ASL was one of the co-operating organizations of ALC. Tugue (Nagoya University) chaired the committee, and Yasugi and Chong served as committee members. I guess that the possibility of holding a Logic Conference in Japan after Beijing had been tapped through Tugue. Chong related that Tugue had raised the possibility of hosting the fourth ALC during a conversation they had in Kyoto Hotel in 1987, just before the conference on Mathematical Logic and Its Applications held at the Research Institute of Mathematical Sciences in Kyoto that year. Considering the economic power of Japan (to the overseas eyes), it was a natural course of events. Japan was prosperous, maybe, but the community of Japanese mathematical logicians was rather indigent. Some of us, therefore, expressed anxiety over this. In Beijing, Tugue invited several of us, Yang, Chong, Yasugi and others to a meeting. There we discussed the possibility of the next conference. It was then that Japan accepted the responsibility. All the details were decided to be left to the host country. Chong emphasized the need of financial support. I was silently worried about the whole thing.

At the meeting table in Beijing, I expressed my concern on the absence of invited speakers from the host country this time. According to Chong, it had been the same with the preceding two conferences. "It's perhaps due to oriental modesty, but you can break that rule in Japan," he said.

Having returned from Beijing, we had to discuss a concrete measure. I happened to be on a committee in ASL, and at the same time a council of the Mathematical Society of Japan (MSJ). For this reason I was placed to start the matter. In December of 1987, I called Tugue as well as Shirai (Shizuoka University), then another council of MSJ, and Motohashi, a council of MSJ of the next term, to get together. We thus held the first consultation. There was no prospect of financial support at that time. At any rate, it was decided to put the base on the Section of the Foundations of Mathematics in MSJ. We circulated the news to the members thereof, and held the meeting involving persons interested on the occasion of the 1988 Spring Meeting of the MSJ at Rikkyo University. There Simauti (Rikkyo University) was elected the chair of the organizing committee. The conference site was chosen to be Tokyo or the surrounding area, in view of manpower, transportaion and visa application considerations, concerning which Tokyo seemed to have the greatest advantage..

In July 1988, the organizing committee was formed, chaired by Simauti. The committee members were selected according to the following principle. The major force be from Asia, especially from Japan with emphasis on Tokyo. There would also be one from Oceania and one from Euro-America. We included computer scientists in the committee. Hirose together with Kakehi and Kikyo, of Waseda University, volunteered to take the responsibility for administration. The date of the conference was set to be the beginning of Septemer, taking into account the need to travel to Tokyo from the site of ICM'90 in Kyoto, which was to be held in late August.

The restriction of "South East Asia" area had been lifted as of the preceeding conference, and the management of the meetings had more or less stabilized. Considering these, a permanent name for the conference was proposed: *Asian Logic Conference*, Ajia Ronri Kaigi. This name is convenient. It has a nice rythm in English, and can be shortened to ALC. The practice of holding the conference every three years became definite from then on.

We did not set any central subject, while we took into consideration that the invited speakers come from a variety of regions. As for the invited speakers, we asked all the program committee members for suggestions, and then contacted the candidates. All in all, things went quite smoothly. Each speaker kindly came on his/her own grant except one. We had reached a general agreement to start a precedent that henceforth the organizers do not suffer from too much financial load. That is, the most critical issues be taken care of, and things not of direct bearing to the conference be left aside. This would help to reduce the total costs. In our case, therefore, the party was of buffet type, and there was no organized sightseeing. Nevertheless, it seemed, the participants toured Tokyo and the surrounding areas by themselves.

The conference site, especially the accomodation, was a big problem. It has to be a place which minimizes the financial burden of the participants, and yet has to have the capacity of accomodating approximately 150 people. Eventually we succeded in securing the use of the *CSK Educational Center* in Tama-Shi. We owe it to the effort and the kindness of Hirose. Throughout, we enjoyed tremendous service by *Director* Kazuo Hirono of CSK. During the period of the conference, he even supplied us with manpower service. Thanks to this, we were able to operate the conference efficiently without much expence. We received heartfelt admiration from the overseas participants with words like "It has been a wonderful conference." Such a service was due to the policy of the CSK management as well as to Hirono's principle: "We need not be remembered by the participants; we only hope that they feel the visit to Japan has been rewarding upon returning home."

The registration fee was payable at the conference site in Japanese Yen in cash. Those who were affiliated with educational or research institutions would pay 15,000yen, while those who were affiliated with companies would pay 30,000yen. We set such a distinction not to discriminate against companies, but to respond to the wishes of some companies that they would like to contribute in this way. The speakers, and only speakers, were exempt of meal and hotel expence. This policy was intended to encourage the participants to read papers. Modest amount of honororium was offered to invited speakers.

The first announcement was sent out in September 1988, the second in September 1989 and the last one in June 1990. Tugue and Simpson respectively asked ASL and AMS so that they would announce the event.

It was also the time when e-mail usages had spread, and so the efficiency of communications was greatly upgraded. It was of help that we were able to manage the trouble of visa in real time thanks to e-mail.

The committee members residing in Japan forwarded preparation,meeting regularly in Simauti's office at Rikkyo University. Simauti's health had been deteriorating, and, in December 1989, we experienced the misfortune of his sudden death. Inspite of that, we preserved the organization with Hirose as the acting chairman. We bore in mind to realize Simauti's long cherished wishes to 'esteem Asian mathematicians and observe proprieties.' We thus proceeded to the Fourth Asian Logic Conference.

5 Record of Conference:The Fourth

The Fourth 1 The Fourth Asian Logic Conference (ALC)

2 September 3 - 6, 1990 CSK Educational Center, Tama-Shi, Tokyo

3 General Chair Tugue, Tosiyuki (Aichi Inst. of Tech.)

Organizing Committee Simauti, Takakazu (Rikkyo U.,Chair), Hirose, Ken (Waseda U.,Secretary, later Acting Chair), Namba, Kanji (U.of Tokyo, Program Committee Chair), Motohashi, Nobuyoshi (U. of Tsukuba,Financial Affairs), Yasugi, Mariko (Kyoto Sangyo U.,External Affairs), Hirono, Kazuo (CSK), Chong, Chi Tat (NUS), Crossley, John N. (Monash U.), Simpson, Stephen G. (Penn.State U.), Tamthai, Mark (Chulalongkorn), Xu, K. (Academia Sinica), Yang, Dongping (Academia Sinica)

Program Committee All the members of the organizing committee (Chair Namba), Doi, Norihisa (Keio U.), Kakehi, Katsuhiko (Waseda U.)

4 The Mathematical Society of Japan, Japan Society for Software Science and Technology, Japan Association for Philosophy of Science, ASL

5 The Commemorative Association for the Japan World Exposition, CSK Corporation, International Information Science Foundation, Kawai Institute for Culture and Education, Calbee Corporation

6 S. R. Buss (U. of California, USA), R. G. Downey (Victoria U. of Wellington, New Zealand), J-Y.Girard (CNRS, U. Paris VII, France), M. Gitik (Tel Aviv U., Israel), S. Koppelberg (Freie U.,Germany), S.G.Simpson, A. Tsuboi (U. of Tsukuba, Japan), D. Yang (absent)

7 69

8 140 (100 from Japan) Overseas participants are from (then)Soviet Union, Autsralia, France, New Zealand, Singapore, Israel, Germany, China, Yugoslavia, Taiwan, Holland, Canada, Thailand, United Kingdom. It was regrettable that the organizers of the previous conferences, Chong, Tamthai, and Yang were absent for various reasons. From within the country, researchers of wide range areas, including analysis and computer science, made the conference a lively one.

9 APAL, vol. 59, 1993, Guest Editors: Ken Hirose and Kanji Namba

10 Wide range area was covered.

11 It was our wish that all the participants, foreign or not, would feel it enjoyable. One of the functions of the ALC is that the participants make new acquaintances so that future communications on research would be easier. We paid attention so that there would be nobody who came alone left without meeting and interacting with others. Since we all stayed at the same place, we were able to enjoy Karaoke-party and to converse over beer. It was a friendly meeting. We decided to reduce the number of invited talk in order to emphasize the importance of the contributed talks. For this reason we scheduled just one invited talk at each session.

We sent invitation letters to our seniors. Unfortunately, each had already some engagement. Professor Iyanaga, however, kindly attended the conference party. His reunion there with an old acquaintance Kurepa was a moving scene.

We had asked Maehara (University of the Air) to give a party speach, offering an interpreter. He gave a talk on a history-let of the Foundations of Mathematics in post-war Japan. He told us that most of the Japanese logicians were disciples or grand disciples of Iyanaga, who was present then. We had the honor to have Iyanaga's speech as well. There were addresses of thanks by Girard and unscheduled persons. Otherwise, the party continued with pleasant chats.

Maehara attended the conference everyday in good humor. In March 1992, we were all astounded by the news of his death. Even now, I feel as if I could see his smiling face. Hirose continued to work to form a new department. Just after his efforts had borne fruit, he suddenly passed away from our world in August this year, having faught against his illness only for two months. I recall the occasion last year when we discussed about the 5th ALC. We have lost three persons who contributed to ALC Tokyo.

6 Towards the Fifth Conference

Even during the preparation for the Tokyo Conference, we were anxious about the viability of holding the next one. We anticipated difficulties, judging from the statistics of participations from Asian countries. To host a conference, manpower, economic resources, policital stability etc. are necessary. During the Tokyo Conference, we held a preparatory meeting involving individuals closely linked to organizing the ALC series, as well as a business meeting addressed to all the participants. At the business meeting, the invitatin by Chong saying "Welcome to Singapore once again!" was communicated. His invitation was accepted unanimously.

The organizing committee of the 5th ALC consisted of several members of NUS. They set a policy that one hour invited talks were all from outside Asia, and forty minutes invited talks were from Asian countries. As for the latter, we on the advisory committee were asked for recommendations. I recommended a few candidates, and our opinions were well taken care of. At first, I felt that the distinction as above was rather strange. As a matter of fact, this is a kind of oriental consideration. Being a person of too much rationality, I felt urged to reconsider. I was struck with admiration afresh for the human capacity of Chong. He is naturally capable of well balancing the traditional oriental spirit and maintaining of academic standard.

7 Record of Conference: The Fifth

1 The Fifth Asian Logic Conference

2 June 14th - 17th, 1993 National University of Singapore, Singapore

3 Organizing Commitee C. T. Chong (Chair), Q. Feng, C.H. L. Ong, Y. N. Sun, Y. C. Tay, M. J. Wicks (NUS)

Advisory Board J.N.Crossley, N.Motohashi, M.Tamthai, T.Tugue, D.P.Yang, M.Yasugi,

4 National UniverSity of Singapore

5 Lee Kong Chian Centre for Mathematical Research, Lee Foundation, Singapore Mathematical Society, Professor Tosiyuki Tugue

6 One hour H.P.Barendregt (Nijmegen), L.Blum (ICSI), H.J.Keisler (Madison), T.A.Slaman (Chicago), W.J.Mitchell (Florida), van den Dries (Urbana), H.Woodin (Berkeley)

Forty minutes D.Ding (Nanjing), S.Hayashi (Ohtsu), J.Shinoda (Nagoya), K.Tanaka (Tohoku), S.P.Tung (Taiwan), J.Wang (Beijing)

7 50

8 99 About half the members of Department of Mathematics of NUS attended. There were 52 overseas participants, among them 11 from Japan and USA each and 8 from China. Besides the countries listed in the previous conference, there were participants from Turkey, Kazakstan, the Philippines, Vietnam, Switzerland, Italy, and Poland.

9 APAL Volume 84, No.1, 1997

10 Although the subjects covered a wide range, many of the methodologies were in modern styles. This made us feel the new era in logic had arrived.

11 We met in a new Conference Hall. Everything was administered with efficiency and in good order. The food supplied during the morning and afternoon breaks was fine. All in all, I felt the economic and academic power of this young nation. Some of the old acquaintances had returned from their studies in Europe and USA, and some had turned from being students to faculty members. I was pleased with such changes. It was regrettable that there were less participants from Japan compared with those in Beijing. It could not be helped, as June is a particularly busy season in Japanese universities. It is a big problem to secure 6 days off,

including travelling time. Another problem is that we cannot force others to pay themselves the travel expence. In my case, I was able to get subsidized by our university. I wish we could use at least part of the "General C" category of the Science Grants of Japan for overseas travel expence. [1]

8 Towards the Sixth Conference

At a dinner hosted by Chong, we discussed the ALC of 3 years later. The members present at the dinner were young Chinese from various countries except Crossley, Tugue and myself. Even though those Chinese belonged to different countries now, they exhibited an atmosphere of a family gathering. They respect traditions, while going through modernization.

A message from Yang saying "We would like to invite you to Beijing next time" was communicated. Chong whispered to me "How about Japan?" I replied that we had just hosted the last conference. I broke into cold sweat. In fact we had no perspective of hosting another ALC. It might be that in China they may get subsidized by the state through the Academia Sinica. Even in that case, I felt I had caught a glimse of the power of China. In our case, we had totally depended on private funds. I do not know which was better. At any rate, it seems to be mistaking the means for the end to sacrifice research for collecting funds. It was agreed that China would be hosting the Sixth ALC in 1996, with the venue to be either in Beijing or Nanjing, to be decided at a later date by the hosts.

In 1994, we heard that it had been decided that the Institute of Software, Academia Sinica, would again organize the Sixth ALC in Beijing. There was concern expressed on the issue of funding. Chong wrote to the organizers that he would contact various sources for some additional funding.

9 Record of the Sixth Conference

1 The Sixth Asian Logic Conference

2 May 20th–24th, 1996 Academia Sinica, Beijing, China

3 Organizing Committee

Yang, Dongping (Academia Sinica, Chair), Ding Decheng (Nanjing),Gao Hengshan (Academia Sinica), Huang Qieyuan (Academia Sinica), Li Wei (Beijing University of Aeronautics and Astronautics), Li Xiang (Guizhou), Lu Yizhong (Nanjing Aeronautics and Astronautics University), Moh Shaokui (Nanjing), Shen Engshao (Shanghai Jiaotong), Shen Fuxing (Beijing Normal), Wang Shiqiang (Beijing Normal)

[1] Remark added at translation: The new system of the Science Grants is more flexible in its use.

Advisory Committee

C.T.Chong (NUS), J.N.Crossley (Monash), T.Tugue (Konan Women's Junior College), M.Yasugi(Kyoto Sangyo)

4 Institute of Software, Academia Sinica

5 Institute of Software, Lee Foundation of Singapore

6 K.Ambos-Spies(Heidelberg), S.Buss(UC San Diego), S.B.Cooper(Leeds), D.Ding(Nanjing), Q.Feng(NUS), K.Hauser(UC Berkeley), A.Kechris(Caltech), S.Lempp(Wisconsin-Madison), W.Li(Beijing Univ of Aeronautics and Astronautics), K.Ono(JAIST), G.Sacks(Harvard), M.Yasugi(Kyoto Sangyo)

7 73

8 150 There were many participants from China. Japan and Singapore were also represented. There was also one participant from Taiwan.

9 World Scientific Publishing Company

10 The most significant event was that this was the first time that a representative from Taiwan attended a logic meeting in Beiing.

Towards the end of the conference, a meeting was held and it was agreed that Taiwan would host the 7th ALC in Taipei in 1999. We see this as an indication of hope that there will be close scientific cooperation between experts across the Straits of Taiwan, and a symbol that the next century will bring further progress to logic in Asia.

A few years from now, the political situation may change, and trips from China to Taiwan may become common. More specialists may grow in other countries. There may be more candidates for the host country. There will occur change in generation. I think that we need various kinds of effort in order to make ALC a real common property of Asia. The most important is to nurture mathematical logicians in each country. It is also desirable if we can establish a cooperative system as a result of our conferences.

It is unfortunate that there has been no participant from Korea. I sent the announcement of the Fourth to several institutions in Korea. In a later day, I was introduced to the President of Korean Mathematical Society by Professor Hong of Nihon University. He was visiting Japan then. He promised me to distribute the announcement of the Fifth ALC within Korea. In fact, there were several entries from Korea for the Fifth, but nobody showed up. According to Chong's guess, the subjects of ALC are too science-techno-oriented, which does not appeal to the Korean logicians, who are more oriented towards natural languages and philosopy. The content of a conference shoud be, however, made up by the participants. We have to promote a better understanding of that.

Finally, I would like to express my hope that ALC will continue to be not too stiff, while keeping up quality, and that any interested individual can attend the conference at ease, feeling it worthwhile to have been there.

(First drafted September 14, 1993; revised and updated by the editors

August 15, 1997)
(Yasugi,Mariko Kyoto Sangyo University, Faculty of Science)